Inorganic Materials Series

Series editors:
Duncan W. Bruce, *University of York, UK*
Elizabeth A. Gibson, *Newcastle University, UK*
Dermot O'Hare, *University of Oxford, UK*
Richard I. Walton, *University of Warwick, UK*

Titles in the Series:

1: Pre-combustion Carbon Dioxide Capture Materials
2: Post-combustion Carbon Dioxide Capture Materials
3: Solar Energy Capture Materials
4: Bioinspired Inorganic Materials: Structure and Function
5: Energy Storage and Conversion Materials
6: Two-dimensional Inorganic Nanomaterials for Conductive Polymer Nanocomposites
7: The Chemistry of Inorganic Biomaterials
8: Computer Simulation of Porous Materials: Current Approaches and Future Opportunities
9: Inorganic Thermoelectric Materials: From Fundamental Concepts to Materials Design
10: Biomedical Applications of Inorganic Materials
11: Photocatalysis Using 2D Nanomaterials
12: Mechanical Behaviour of Metal–Organic Framework Materials
13: Flexible Metal–Organic Frameworks: Structural Design, Synthesis and Properties
14: Unconventional Green Synthesis of Inorganic Nanomaterials
15: Elemental Carbon

How to obtain future titles on publication:
A standing order plan is available for this series. A standing order will bring delivery of each new volume immediately on publication.

For further information please contact:
Book Sales Department, Royal Society of Chemistry, Thomas Graham House, Science Park, Milton Road, Cambridge, CB4 0WF, UK
Telephone: +44 (0)1223 420066, Fax: +44 (0)1223 420247
Email: booksales@rsc.org
Visit our website at books.rsc.org

Elemental Carbon

Edited by

Peter J. F. Harris
Bristol University, UK
Email: p.j.f.harris@bristol.ac.uk

ROYAL SOCIETY OF **CHEMISTRY**

Inorganic Materials Series No. 15

Print ISBN: 978-1-83916-451-4
PDF ISBN: 978-1-83916-998-4
EPUB ISBN: 978-1-83916-999-1
Print ISSN: 2472-3819
Electronic ISSN: 2472-3827

A catalogue record for this book is available from the British Library

The Royal Society of Chemistry is a charity, registered in England and Wales, Number 207890, and a company incorporated in England by Royal Charter (Registered No. RC000524), registered office: Burlington House, Piccadilly, London W1J 0BA, UK, Telephone: +44 (0) 20 7437 8656.

For further information see our website at www.rsc.org

For general enquiries, please contact books@rsc.org

For EU product safety enquiries, please email books@rsc.org or contact Royal Society of Chemistry Worldwide (Germany) GmbH, Römischer Hof, Unter den Linden 10, 10117 Berlin.

Printed in the United Kingdom by CPI Group (UK) Ltd, Croydon, CR0 4YY, UK

Series Preface

Materials chemistry has grown hugely in the past few decades and continues to evolve rapidly. This is especially the case since the subject now finds itself central to many of the major challenges that face global society, such as those concerning energy and the environment; but applications of materials are also found in many everyday situations, with ongoing developments in technology around electronics and the miniaturisation of devices. Although the study of materials is a highly interdisciplinary topic, the role of innovative, fundamental chemistry continues to lie at the heart of the discovery of new materials, the understanding of their atomic-scale structure and the relationship of structure to a useful property.

With the *Inorganic Materials Series*, our aim is to provide a series of themed volumes, each reflecting the diversity of the topic as it continues to develop over the coming years, with many chosen to relate closely to emerging practical applications. The use of the title *Inorganic Materials* is by no means limiting, but instead emphasises the study of the chemistry of all elements in the Periodic Table and how they impact the development of new materials. The chapters in each book will have a pedagogical flavour with a target audience from first-year postgraduate student upwards, but with wide subject coverage ranging from continuous inorganic solids, to molecular and soft matter systems. We will use expert, active researchers as editors and contributors to provide an up-to-date perspective of their field. At the same time our aim is to provide an international perspective, so to reflect the diversity and interdisciplinarity of the now very broad area of the subject.

Duncan W. Bruce
Elizabeth A. Gibson
Dermot O'Hare
Richard I. Walton

Inorganic Materials Series No. 15
Elemental Carbon
Edited by Peter J. F. Harris

Published by the Royal Society of Chemistry, www.rsc.org

Preface

Carbon is the element of life and the basis of the huge disciplines of organic chemistry and biomolecular sciences. However, until relatively recently, the science of elemental carbon was something of a backwater. It was widely believed that research on graphite, diamond, soot and other forms of solid carbon would be unlikely to yield any great surprises, let alone Nobel Prizes. In recent decades the situation has been transformed, not only by the Nobel Prizes for buckminsterfullerene (1996) and graphene (2010), but by major developments in the synthesis and applications of diamond, and by the development of intriguing new materials such as carbon dots. Today the study of solid carbon in all its varieties is one of the most exciting areas of science.

This book brings together experts from five fields of research on solid carbon to provide a wide-ranging and up-to-date overview of the subject. The first chapter covers graphite and carbon fibres. As well as outlining the chemical and physical properties of graphite, two of its most important current applications are reviewed, namely its use in nuclear power and in lithium-ion batteries. Carbon fibres, which have essentially graphitic structures, are then discussed. The structure of carbon fibres is described and the preparation, properties and applications of carbon fibre composites reviewed.

Chapter 2 covers perhaps the most exciting new form of carbon, graphene. Since its isolation in 2004, graphene has captured the imagination of scientists the world over, and many of its extraordinary properties are beginning to lead to commercial applications. In this chapter, the chemical modification of graphene, and its structural variants are discussed in detail. Graphane, graphyne and graphone are among the materials covered.

Chapter 3 provides a comprehensive review of carbon black. This soot-like carbon is produced on an enormous scale by the pyrolysis of carbonaceous precursors, mainly for use as a reinforcement in automobile tyres and as a pigment. Manufacturing processes, structure, properties and applications of

Inorganic Materials Series No. 15
Elemental Carbon
Edited by Peter J. F. Harris

Published by the Royal Society of Chemistry, www.rsc.org

this important material are described, while efforts to improve the environmental sustainability of carbon black production are emphasised, including the use of recycled materials as feedstocks, as well as making the production processes themselves more energy efficient.

Like graphene, carbon dots are a new form of carbon with a wide range of possible applications, as discussed in Chapter 4. Consisting of tiny carbon particles (<10 nm), their tuneable optical properties and good biocompatibility make them potentially useful in a range of biomedical applications including bioimaging, cancer therapy and drug delivery. They may also have potential in electrocatalysis, optoelectronic devices, energy storage and other areas.

While the first four chapters are concerned primarily with sp^2, graphite-type, carbon, the final chapter discusses sp^3 carbon: diamond. Specifically, the chapter describes the low-pressure synthesis of diamond by chemical vapour deposition (CVD). Great strides have been made in this field in the past 20 years, to the point where CVD diamond is now an industrially manufactured product. The chapter covers recent developments in the CVD synthesis of diamond, notably the 'static flow' method. The production of semiconducting and functionalised diamond is described and some applications of the diamond films are outlined.

Today, when used in phrases such as 'carbon footprint' and 'carbon emissions', the word 'carbon' has acquired negative connotations. This is of course unfair, since the cause of global warming is not carbon itself, but carbon dioxide. In fact, many of the carbon-based technologies described in this book can contribute to the alleviation of the climate crisis. As mentioned already, graphite is an essential component of many nuclear reactors and of lithium-ion batteries, both of which can be considered sources of clean energy, while advances in carbon black synthesis are helping to improve the environmental sustainability of the production process, as also noted above. Carbon dots have great potential in electrocatalytic applications including fuel cells and water electrolysis, and CVD diamond has a number of clean energy applications, including thermionic energy converters and beta voltaic batteries. In all of these ways, and in many others, carbon itself is helping to reduce humankind's 'carbon footprint'.

Peter Harris

Contents

Inorganic Materials Series No. 15
Elemental Carbon
Edited by Peter J. F. Harris

Published by the Royal Society of Chemistry, www.rsc.org

CHAPTER 1

Graphite and Carbon Fibres

DONG LIU[a], MING JIANG[a], KEVIN D. POTTER[b] AND PETER J. F. HARRIS[c*]

[a]Department of Engineering Science, University of Oxford, Parks Road, Oxford, OX1 3PJ, UK; [b]Bristol Composites Institute ACCIS, University of Bristol, Bristol BS8 1TR, UK; [c]Department of Physics, University of Bristol, Tyndall Ave, Bristol BS8 1TL, UK
*E-mail: p.j.f.harris@bristol.ac.uk

1.1 Introduction

Graphite's unique properties have made it a valuable technological material for hundreds of years. In the Elizabethan era, graphite from the Borrowdale mine in Cumbria was used as a refractory material to line moulds for cannon balls, apparently producing smoother, rounder shot with a longer range. The British Crown quickly recognised that graphite was an important strategic material, and strictly controlled its mining and use. Perhaps graphite's most familiar application is an even older one, namely as a component of pencils. It seems that the first pencils were produced in 1565 by the Swiss scientist and philosopher Conrad Gessner, again using Borrowdale graphite. Pencils remain as popular as ever, and most pencil manufactures still employ natural graphite; currently pencils account for some 4% of the world's natural graphite use. Two of graphite's most important applications today are both relevant to combatting the climate crisis, namely in nuclear power plants and in lithium-ion batteries. As moderators in nuclear reactors, graphite helps to produce low-carbon electricity, while lithium-ion batteries with graphite anodes power electric cars, as well as a plethora of electronic devices. Carbon

Inorganic Materials Series No. 15
Elemental Carbon
Edited by Peter J. F. Harris

Published by the Royal Society of Chemistry, www.rsc.org

fibres, which are graphitic in structure, also help to mitigate the climate crisis: the increasing use of carbon fibre composites has enabled the production of lighter, more fuel-efficient aircraft, and they are also widely used in wind turbine blades.

In this chapter, we begin by describing naturally occurring graphite, and some of its early applications. We then outline the different methods which can be used to synthesise graphite. The structure of graphite is then described and the different kinds of defects that occur in graphite crystals are reviewed. We then cover the chemistry of graphite: its oxidation by gases, the formation of graphite oxide and the synthesis and properties of graphite intercalation compounds. The physical and electronic properties of graphite are then summarised. The two applications of graphite mentioned above are reviewed, namely its use in nuclear power and in lithium-ion batteries. The sections on nuclear graphite cover its manufacture, microstructure and thermomechanical properties. The impact of irradiation on the properties of nuclear graphite is discussed in detail. The role of graphite in lithium-ion batteries is then briefly described, and some of the problems with graphite anodes highlighted. In the final sections, we discuss carbon fibres. We describe their synthesis, their structure and the preparation, properties and applications of carbon fibre composites.

1.2 Natural Graphite

1.2.1 Early Applications of Natural Graphite

The earliest application of graphite may have been in the decoration of pottery in south-eastern Europe in the 4th millennium BCE.[1] However, it was the discovery of a large deposit of natural graphite in the English Lake District in the 16th century which led to its widespread use. Legend has it that a severe storm uprooted an oak tree in the Borrowdale valley, exposing a mysterious black/grey substance underneath.[2] At first it was thought to be a form of lead, and was given the name plumbago, after the Latin for lead; hence the term 'pencil lead'. In fact it was in pencils that graphite found its first significant use. As already noted, the Swiss scientist Conrad Gessner obtained some graphite from the Borrowdale mine in 1565 and showed that encasing rods of the material in wooden tubes produced a convenient and useful writing implement. Graphite's ability to leave marks on paper is a result of its layered structure and the relatively weak bonds between the layers. As the pencil moves across the paper, flakes become detached and adhere to the surface. In today's pencils a mixture of graphite and clay is employed rather than pure graphite, with the amount of clay determining the hardness or softness of the pencil.

Another early application of natural graphite was in the manufacture of cannon balls, as also mentioned above. Due to its strategic importance, graphite's price increased rapidly and a black market developed (it has been suggested that this was actually the origin of the phrase 'black-market', although this seems questionable). This eventually led, in 1752, to the passage of an Act of Parliament making unauthorised entry into a graphite mine, a crime. Other applications followed. Graphite was first used to make laboratory crucibles in Bavaria in the 16th century, while graphite's use as a dry lubricant dates back to the 1800s.[3] Today, refractory materials represent approximately 16% of graphite consumption,[4,5] and graphite lubricants are still widely used, as of course are pencils.

1.2.2 Classes of Natural Graphite

There are generally considered to be three types of natural graphite: vein graphite, flake graphite, and amorphous graphite.[6,7] Vein graphite, sometimes called lump graphite, is the highest quality and highest purity form. It occurs in fissures or veins in granulites (high-grade metamorphic rocks). These veins are generally around 10–20 cm thick.[8] The formation of vein graphite is rather poorly understood, but is thought to involve precipitation from carbon-rich fluids at fractures in the rock.[9–11] Today vein graphite is only produced in commercial quantities in Sri Lanka.

Flake graphite is the most commonly occurring class of natural graphite and the most widely used. It occurs most frequently in metamorphic rocks and is formed when an organic deposit is subjected to high pressure and temperature inside the rock. A micrograph showing graphite flakes in a metapelite (a metamorphosed sedimentary rock) from northern Norway is shown in Figure 1.1.[12] Commercial flake graphite deposits occur in the USA, Canada, India, China, Brazil and elsewhere.

Amorphous graphite is not truly amorphous but is less crystalline than the other classes of natural graphite. It is formed by the metamorphism of anthracite coal at temperatures of around 250 °C and relatively low pressures. It consists of extremely fine flakes and usually has higher impurity

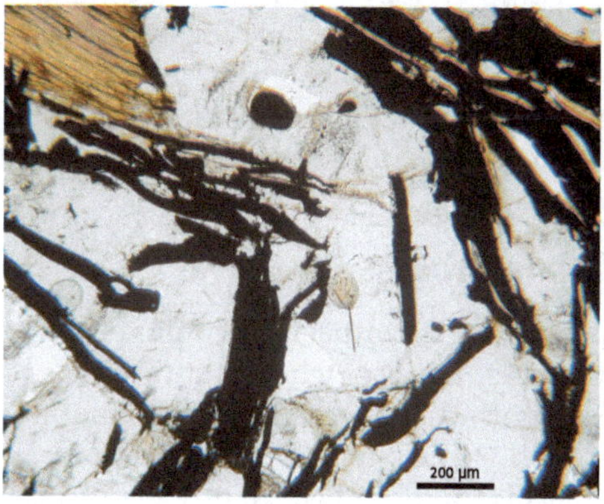

Figure 1.1 Polarised light image of graphite-bearing metapelite from a metamorphic region in northern Norway. Reproduced from ref. 12, http://doi.org/10.17850/njg96-1-03, under the terms of the CC BY 4.0 license, https://creativecommons.org/licenses/by/4.0/deed.en.

levels than other types of natural graphite. China is the largest producer of amorphous graphite.[13]

1.2.3 Processing of Natural Graphite

Natural graphite occurs with a range of impurities, mainly clay minerals, which are generally physically attached to the graphite crystals rather than chemically bonded to them. Extracting the graphite from the ore, therefore, mainly involves physical rather than chemical processing methods. The particular techniques used to process, or beneficiate, natural graphite ore depend on the class of graphite and the ultimate application,[14] but there are typically three stages. In the first stage, the rock is ground or milled to reduce the particle size, a process known as comminution. Next, to separate the graphite from the unwanted 'gangue', the most common method is froth flotation. This involves suspending the powder in water or oil and bubbling air through the suspension. A frothing agent is usually added to the water to promote bubble formation. The hydrophobic graphite particles stick to the air bubbles, which rise to the surface, forming a froth. The froth is then skimmed from the surface. Finally, the product is refined by treatment with acid or by heating.

In addition to the papers already cited, reviews of the purification and applications of natural graphite have been given by Jara *et al.*[15] and by Duan *et al.*[16] An online lecture by Sharma can also be recommended.[17]

1.3 Synthesis of Graphite

In the following sections, we describe some of the methods which are used to synthesise graphite, beginning with a description of the Acheson process, which is the most commercially important method. We then outline the mechanism of graphite formation in the Acheson process, emphasising the role of the carbonaceous mesophase. Brief descriptions of the production and properties of Kish graphite and pyrolytic graphite are then given. The preparation of graphite for the nuclear industry is discussed in Section 1.9.1.

1.3.1 The Acheson Process

A commercial process for the synthesis of graphite was developed by the US scientist and industrialist Edward Acheson at the end of the 19th century.[18] Earlier workers had synthesised graphite in the laboratory by passing electrical currents through various forms of carbon (*e.g.* ref. 19), but Acheson was the first to scale this up for industrial production. Like other workers in the field, Acheson's original aim had been to create artificial diamond, but instead he found a method for producing silicon carbide, which in turn led to the synthesis of graphite. His method involved passing a current through a mixture of powdered carbon and fused clay in an iron crucible. This produced

SiC crystallites which proved to have excellent abrasive properties. Acheson commercialised the new material as 'carborundum'. He also found that pure graphite was frequently formed as a by-product of the carborundum process, and this led him to think about ways of intentionally synthesising graphite. He found that if the SiC was heated to a sufficiently high temperature, the Si vaporised leaving pure graphite. Acheson patented this process 1896.[20] He believed that a carbide was an essential intermediate in the formation of graphite but later found that this was not required, and that graphite could be synthesised directly from pure carbon starting materials.

As already mentioned, the Acheson process remains the most important method for producing artificial graphite. The main raw material for the process is petroleum coke, obtained by carbonising heavy hydrocarbon fractions obtained in petroleum processing. Before being used for graphite production, the coke must be purified by calcining in a rotary kiln at approximately 1300 °C. The calcined coke particles are then usually mixed with coal tar pitch at high temperature and formed into the desired shapes, often rods. The pitch helps to bind the particles together. The rods are then loaded into the Acheson furnace, which is typically 12–15 m long and 3 m wide. By passing a current through the carbon charge, temperatures in the range 2800–3000 °C are obtained. Approximately 3 days of heat treatment are needed to produce graphite and a further 8–10 days are required for cooling the furnace.

1.3.2 Mechanism of Graphite Formation from Coke

It has been known since the early days of graphitisation research that not all carbons can be graphitised, although the subject of graphitisability was only put on a scientific footing in the early 1950s, with the work of Rosalind Franklin.[21,22] Franklin's experiments involved carrying out a series of heat treatments on a range of organic precursors at temperatures up to 3000 °C and investigating their structures using X-ray diffraction. She found that while some of the carbons, the cokes, could be graphitised by heat treatments above about 2200 °C, others, the chars, could not be transformed into crystalline graphite, even at 3000 °C. Instead, they formed a porous, isotropic material which only contained tiny domains of graphite-like structure. She christened the two kinds of carbon graphitising and non-graphitising. Among the graphitising carbons were petroleum coke and carbons made by pyrolysing polyvinyl chloride (PVC) while non-graphitising carbons included those derived from polyvinylidene chloride (PVDC). For discussions of current thinking on the nature of graphitising and non-graphitising carbons see ref. 23–25.

An important difference between graphitising and non-graphitising carbons is that the former pass through a fluid phase when heated, while the latter do not. This intermediate material is known as the mesophase, a term first employed in this context by Brooks and Taylor in the mid-1960s.[26] These workers had been studying coal from a seam in New South Wales, Australia. Parts of the seam had been heated to high temperatures by ancient magma to form a coke, while other areas had been heated to lower temperatures.

Thus, a geothermal gradient existed within the seam, providing a record of the carbonisation process. When Taylor and Brooks studied the carbon that had experienced the highest temperatures in a polarised light microscope, they observed optically anisotropic spheroids embedded in an isotropic matrix. A typical image of carbonaceous mesophase is shown in Figure 1.2(a) and the anisotropy, which results from the structure of the spheres, is shown in Figure 1.2(b). The spheres contain parallel arrays of polynuclear aromatic hydrocarbons (PAHs) forming a nematic liquid crystal (Figure 1.2(c)). As the heat treatment proceeds, higher molecular weight hydrocarbons are formed by condensation and incorporated into the mesophase. Eventually the mesophase spheres coalesce, becoming the dominant phase, and finally solidifying into a coke. A number of reviews of the carbonaceous mesophase and its role in graphitisation are available.[26–28]

The process whereby coke is transformed into graphite has been widely discussed, notably by Marsh and colleagues[29,30] and Oberlin.[31] Figure 1.3(a)

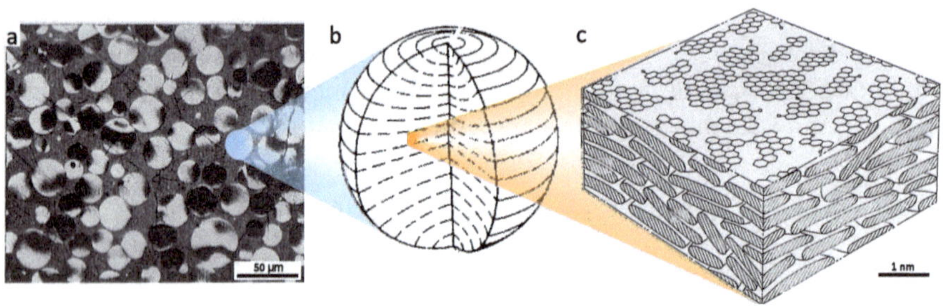

Figure 1.2 (a) Optical image of a polished sample of carbonaceous mesophase, (b) equatorial, or bipolar, structure of mesophase spheres, (c) layered nematic molecular structure in the interior of the spheres. Reproduced from ref. 27 with permission from AIP.

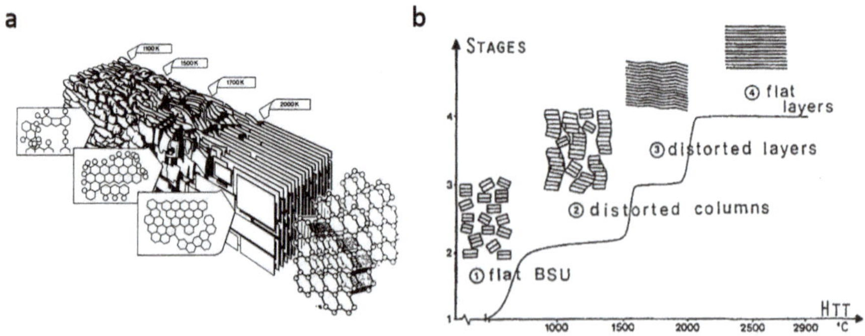

Figure 1.3 Diagrams illustrating models of the graphitisation process, from the work of (a) Marsh and Griffith[29,30] and (b) Oberlin.[31] (a) Reproduced from ref. 30 with permission from Elsevier, Copyright 1984. (b) Reproduced from ref. 31 with permission from Elsevier, Copyright 1984.

shows the Marsh–Griffith model of the graphitisation process.[29] The left-hand side of the diagram shows the stacking of the PAHs in the mesophase. These have buckled structures with significant defects, vacancies and so on. With increasing heat treatment, the planarity increases and the defect density decreases. At the same time, cross-linking can occur across the layers resulting in further buckling, until at about 2000 K, the planarity of the layers begins to become established, leading to the formation of crystalline graphite. The Oberlin model[31] (Figure 1.3(b)) envisages a four-stage process, beginning with roughly flat 'basic structural units', which coalesce into buckled layers before finally forming graphite. More recent work has emphasised the role of screw dislocations in graphite formation.[32]

1.3.3 Kish Graphite

Kish graphite is a high-quality form of synthetic graphite that is a by-product of iron- or steel-making. The graphite forms during the cooling of molten iron as it is tapped from a blast furnace. At this point the iron is supersaturated with carbon, and this comes out of solution as graphite flakes which float to the surface together with lime-rich slag and other waste products. Kish graphite samples often contain large single crystallites (up to 5 mm in size); XRD and electron microscope studies show that the crystallites have a highly perfect structure, superior to that of highly ordered pyrolytic graphite and similar to that of natural graphite.[33] The high quality of kish graphite crystals made them well suited for experimental studies of graphite intercalation[34] and, more recently, for the preparation of graphene.[35] However, the relatively high costs of preparing purified kish graphite have so far prevented any significant commercial applications.

1.3.4 Pyrolytic Carbon and Graphite

Pyrolytic carbon (PyC) is carbon formed by the decomposition of hydrocarbon gases on a substrate. Its main application is in the form of coatings, although free-standing structures can also be made. The structure of PyC varies from rather disordered to highly graphitic; the most perfect form is known as highly oriented pyrolytic graphite (HOPG). Here we summarise the preparation and applications of pyrolytic carbon.

The production of carbon by decomposing hydrocarbon gases has a history which extends back to the 19th century and the work of Thomas Edison and others. In the 1880s, Edison experimented with processes for coating light bulb filaments with pyrolytic carbon in order to improve their performance. However, there was little further interest in PyC until the 1950s and the development of rocket technology. This led to a demand for materials with exceptional mechanical properties at high temperatures. Research in the defence and aerospace industries led to techniques for producing massive PyC structures up to several cm thick and several m^2 in extent,[36] for use in missile nose cones and rocket motors. The production method involved depositing the pyrolytic carbon onto a graphite mandrel and then removing the mandrel by machining.[37]

Pyrolytic carbon has also been successfully used in biomedical devices. This application began as a spinoff of research on PyC coatings for nuclear applications by Bokros *et al.*[38] In the 1960s, these workers showed that PyC displayed excellent blood and tissue biocompatibility;[39] this led to the development of a PyC cardiac valve in 1969. Prosthetic heart valves are still the most important medical applications of PyC, although it is also used for some orthopaedic implants.[40]

Pyrolytic carbon is made up of small domains of randomly oriented turbostratic graphite. The sizes of the domains and their degree of alignment both increase with temperature.[41] In the highest quality conventional PyC, the average crystallite size, L_c is typically ~10 nm and the interlayer spacing ~0.343 nm.[42,43] The degree of alignment of the crystallites, known as the mosaic spread, generally ranges from 2° to 5°.

A more highly crystalline form of graphite can be produced by applying tensile stress to a PyC specimen in the direction of the *c*-axis while heating to about 3000 °C and allowing the specimen to expand in the basal plane direction. The production of highly oriented pyrolytic graphite in this way was pioneered by Ubbelohde *et al.* in the early 1960s.[44] The best HOPG samples can have mosaic spreads as small as 0.2°, with interplanar spacings approaching the ideal value for graphite of 0.3354 nm. Highly oriented pyrolytic graphite has important applications in X-ray optics,[45] in electrochemistry[46] and in scanning probe microscopy, for magnification calibration and as a substrate.[47] Scientifically, its most important scientific application was arguably its use by Geim, Novoselov *et al.* to prepare graphene,[48] as discussed in Chapter 2.

1.4 Structure of Graphite

The structure of graphite was determined by J. D. Bernal and W. L. Bragg in the early days of X-ray crystallography.[49] Their structure, now known as the Bernal structure, is shown in Figure 1.4. Here the stacking sequence of the graphene layers is ABAB, with an interlayer {0002} spacing of 0.334 nm; the space group is $P6_3/mmc$ (D_{6h}) and the unit cell contains four atoms. In 1942 Lipson and Stokes reported a new form of graphite with the stacking sequence is ABCABC.[51] This form of graphite has rhombohedral symmetry and its structure can be thought of as an extended stacking fault in hexagonal graphite. Work by Bacon in 1950 showed that shear deformation could increase the proportion of the rhombohedral form.[52] The mechanism of the transformation from Bernal to rhombohedral graphite was analysed theoretically by Nery *et al.* in 2020.[53] It has also been established that rhombohedral graphite transforms to the Bernal form on heating above 1600 °C.[54]

The discovery of superconductivity in twisted bilayer graphene,[55] as discussed in Chapter 2, prompted renewed interest in rhombohedral graphite. If superconductivity could be produced by simply reorientating a layer of

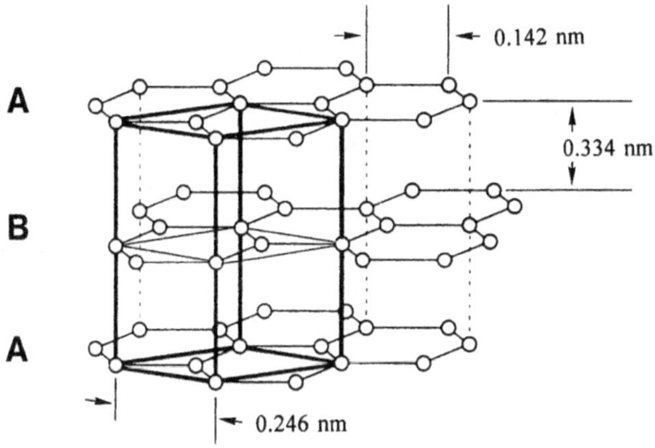

Figure 1.4 The structure of Bernal graphite, showing the unit cell. Reproduced from ref. 50 with permission from Cambridge University Press.

graphene with respect to another layer, might interesting electronic effects occur in rhombohedral graphite, where the stacking differs from the Bernal form? With this in mind, Mishchenko *et al.* used van der Waals heterostructure technology in 2020 to produce high-quality rhombohedral graphite films up to 50 layers thick.[56] Using these samples they observed behaviour that they attributed to the emergence of strongly correlated electronic surface states, although definite signatures of superconductivity were not seen. A US–Japanese group did report superconductivity in trilayer rhombohedral graphene in 2021.[57]

In addition to the well-established ABAB and ABCABC forms of graphite, a metastable form with AA stacking has been discussed theoretically (*e.g.* ref. 58 and 59) and is sometimes seen experimentally during lithium intercalation (*e.g.* ref. 60). Small domains of AA graphite have been grown on the (111) surface of diamond[61] and AA stacking is quite commonly seen in bilayer graphene.[62]

A form of carbon with a structure close to graphite, without being fully crystalline, is known as turbostratic carbon.[63] This consists of rather disordered layer planes, stacked approximately parallel to each other with random translation about the *a*-axis and rotation about the *c*-axis. The interplanar spacing in turbostratic carbon is significantly larger than for single crystal graphite (typically ~0.344 nm). Graphitising carbon that is not completely converted to graphite has this structure.

1.5 Defects in Graphite

In this section we discuss the point defects and dislocations that occur in graphite crystals. The basic kinds of point defects found in graphite are vacancies, defects with non-hexagonal rings and interstitials.

Vacancies occur when one or more carbon atom is missing from the lattice. The simplest vacancy is where a single atom is missing, as shown in Figure 1.5(a) but many other vacancy configurations occur such as the divacancy shown in Figure 1.5(b). Vacancies can arise in graphite at low concentrations during defective growth[66] and can also be produced (together with interstitials) by irradiation, as discussed in Section 1.9.2.1. The simplest example of a defect containing non-hexagonal rings is the 5:7 defect illustrated in Figure 1.5(c). The structure of vacancies in graphene has been studied in detail in recent years; these and other possible configurations of non-hexagonal rings in the graphene lattice are discussed in Chapter 2. Interstitials are individual carbon atoms lying between the layer planes.

Dislocations in graphite fall into four categories,[67] as described in the following sections.

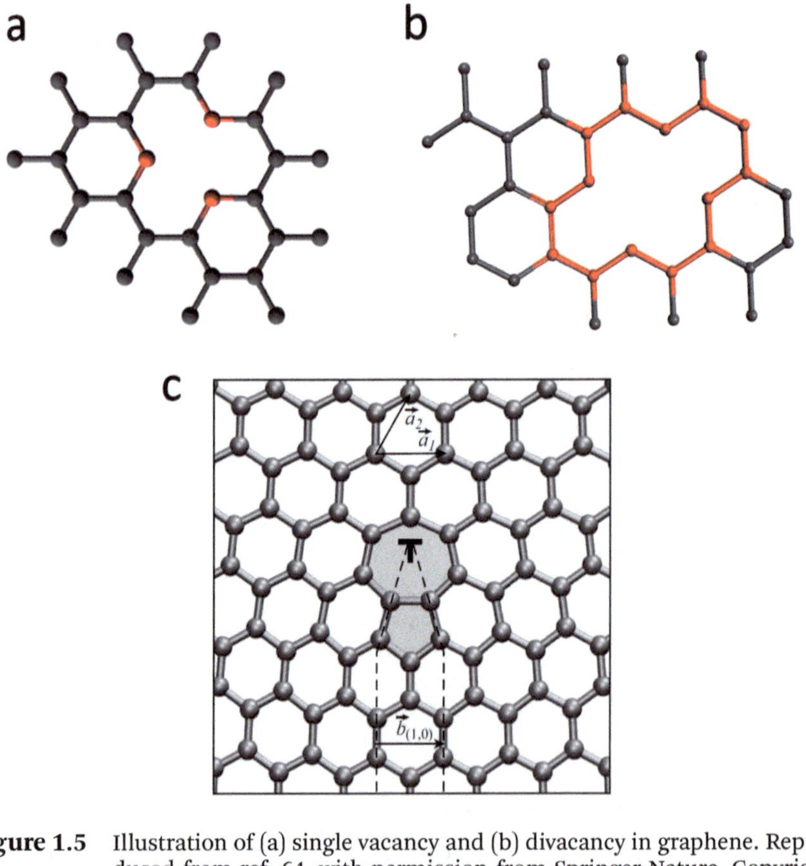

Figure 1.5 Illustration of (a) single vacancy and (b) divacancy in graphene. Reproduced from ref. 64, with permission from Springer Nature, Copyright 2019. (c) 5:7 defect in graphene, also known as a (1, 0) dislocation. Reproduced from ref. 65, with permission from AIP.

1.5.1 Basal Dislocations

These are dislocations in which the Burger's vector and dislocation line are both in the basal plane. They occur abundantly in graphite. Early TEM studies of these dislocations showed that they often occurred as pairs, approximately 100 nm apart, enclosing ribbons.[68–71] The ribbons could be randomly distributed, or form networks. Regular arrays of parallel ribbons were sometimes seen, as in Figure 1.6.

A model for basal dislocations in graphite put forward by Amelinckx *et al.*[70] is shown in Figure 1.7, with (a) showing the view along the graphene planes and (b) showing a plan view of the layers above and below the glide plane of the dislocation (indicated by red arrows in (a)). There are two partial dislocations here, whose positions are indicated by the dotted lines in (b). The dislocation towards the right results from an extra row of hexagons at the circled position. The incorporation of extra carbon atoms results in hexagons, in the vicinity of the partial dislocation, becoming distorted: those in the layer containing the extra atoms are compressed, while those in the layer below are stretched. This results in a second partial dislocation, shown by the dotted line on the left in Figure 1.7(b).

Between the two partials is a fault in which the stacking is rhombohedral. In this region the two upper A layers become C layers, as can be seen in Figure 1.7(a). Note that the model shown in Figure 1.7 is schematic: experimentally the width of the stacking faults is much greater than indicated in the model, typically around 100 nm.

Figure 1.6 Transmission electron micrograph of ribbons of dislocations in graphite. Reproduced from ref. 71, with permission from Taylor & Francis.

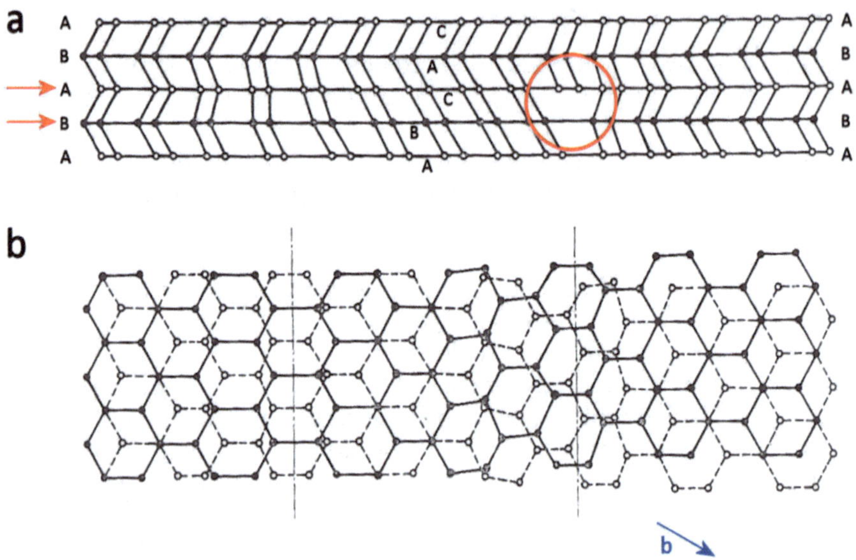

Figure 1.7 Schematic diagrams showing model of 60° basal dislocation in graphite proposed by Amelinckx *et al.* (a) View down the graphene planes, (b) plan view of the two planes indicated by red arrows in (a). The two vertical dotted lines indicate the core regions of the two partial dislocations. The arrow at bottom left indicates the total Burger's vector of the dislocation. Reproduced from ref. 70, with permission from AIP.

Telling and Heggie have analysed the motion of basal dislocations in graphite.[72] They showed that the barrier for dislocation motion is extremely small and suggested that this may have implications for understanding how graphite acts as a lubricant. In their view, lubrication might involve a step-by-step sliding of layers through dislocation motion, rather than through the movement of entire sheets at once.

1.5.2 Non-basal Edge Dislocations

In non-basal dislocations the dislocation line is perpendicular to the basal planes and the Burger's vector lies in the basal plane. These kinds of defects can result from the presence of a stack of defects, such as the 5 : 7 defect illustrated in Figure 1.5(c). Ewels *et al.* analysed the motion of non-basal dislocations and found that the barrier to motion for these dislocations is high (7.64 eV), since it requires bond reconstruction.[73] However, this barrier is considerably reduced if an adatom is present at the core.

1.5.3 Prismatic Screw Dislocations

In prismatic screw dislocations, the Burger's vector and dislocation line are both perpendicular to the basal plane. The structure is illustrated schematically in Figure 1.8. Screw dislocations are difficult to image by TEM, and have

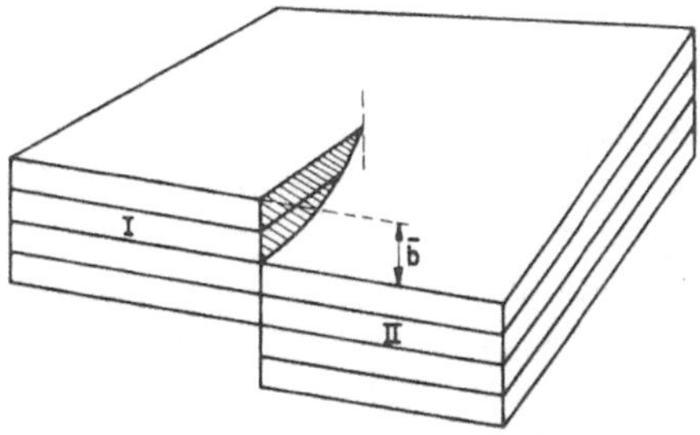

Figure 1.8 Screw dislocation in graphite. Here a slip of two-layer planes is indicated. This is necessary with ABAB stacking if the layers are to knit together perfectly. Reproduced from ref. 71, with permission from Taylor & Francis.

generally been studied with optical microscopy or SEM. In the early 1960s, Hennig developed a method of revealing screw dislocations which involved cleaving the graphite crystal, etching in a gas to expand the defect and then decorating the surfaces with gold nanoparticles.[74–76] This technique revealed few, if any, screw dislocations in crystals of natural Ticonderoga graphite, whereas pyrolytic graphites annealed above 3000 °C contained many such defects: the approximate concentration was 10^6 to 10^8 cm^{-2}. The screws consisted of monatomic steps, so had Burger's vectors of 0.335 nm.

Important studies of screw dislocations in graphite were also carried out by Thomas *et al.*[77–79] Like Hennig, Thomas *et al.* also used partial oxidation to reveal the surface defects, but unlike Hennig, they found that Ticonderoga graphite contained up to 10^6 screw dislocations per cm^2. In contrast to the screws observed by Hennig, these dislocations had large Burger's vectors, ranging from 40 nm to 350 nm; such defects are sometimes termed 'superscrews'.[80]

Oxidation of graphite by gases is discussed further in Section 1.6.1.

1.5.4 Prismatic Edge Dislocations

In prismatic edge dislocations, the Burger's vector is perpendicular to the basal plane and the dislocation line parallel to the basal plane. These dislocations occur when one or more layers of atoms are missing from a graphite crystal or when extra layers of atoms are incorporated. Such defects can be produced by the accumulation of vacancies and interstitial atoms as a result of irradiation, as discussed further in Section 1.9.2.1.

1.6 Chemistry of Graphite

1.6.1 Oxidation of Graphite by Gases

Oxidation by O_2 and other gases can be a problem when graphite is used in crucibles, in electrodes, in the nuclear industry and in many other applications. For this reason, as well as out of scientific curiosity, the reaction of graphite with oxygen has been studied intensively over many decades and remains an active subject of research. Here, we focus mainly on fundamental studies of graphite oxidation using ideal (single-crystal) specimens. As noted above, important early work in this field was conducted by Hennig *et al.*[74–76] and Thomas *et al.*[77–79] Both groups showed that etching with O_2 and other gases could reveal non-basal screw and edge dislocations, thus demonstrating that oxidation occurs preferentially at surface defects. These workers also demonstrated that oxygen could react with the edges of vacancies one atomic plane deep and enlarge them to create circular or hexagonal etch pits.[76,78] Further studies of the phenomenon were reported by Yang and Wong in 1981.[81] An optical microscope image of some hexagonal etch pits is shown in Figure 1.9(a). Transmission electron microscope images of circular monolayer etch pits from work by Hennig[76] are shown in Figure 1.9(b).

Thomas *et al.* obtained information about the rate of oxidation of natural graphite by measuring the rate of growth of etch pits and the rate of motion of surface steps using optical microscopy.[77–79,82] A small difference was found between the oxidation rates in the armchair and zigzag directions, with the former having a higher activation energy. Measuring oxidation rates normal to the basal plane is more difficult. However, by determining the size of etch pits produced on graphite crystals by O_2 and NO, Thomas *et al.* were able to show that the ratio of oxidation rates along the *ab* and *c* directions was 26 : 1 (although note that this refers solely to the rate of oxidation within etch pits).[82] In 1973, Acharya and Olander measured the same ratio using pristine pyrolytic graphite, and found that it ranged from 2.5 : 1 to 29 : 1.[83]

Several scanning tunneling microscopy (STM) studies of the reaction of graphite with O_2 and other gases have been reported (*e.g.* ref. 84–88). In 1992, Chu and Schmidt described STM studies of the oxidation of HOPG by various gases.[84] At low temperatures, the reactions originated at point defects on the basal planes, forming hexagonal or circular pits one monolayer deep whose diameters varied linearly with time. Similar observations had been made in early work by Thomas and Hughes[77] and by Hennig.[74] In another STM study, Hahn investigated the oxidative etching of graphite parallel and perpendicular to the basal plane at temperatures up to 950 °C.[88] He found that oxidation was initiated from point defects at temperatures below 875 °C while at higher temperatures it could also be initiated at basal plane atoms. It was found that oxidation produced monolayer-depth pits with round shapes, as in the early work by Hennig. Examples are shown in Figure 1.10.

Oxidation of the edge (prismatic) surface of graphite has been less thoroughly studied than oxidation of basal planes. However as already noted, it is evident that edge surfaces are much more reactive than basal surfaces.[83,89,90] This is to be expected, since atoms at the edges of graphene planes contain dangling bonds, defects, and functional groups, due to the lattice termination.

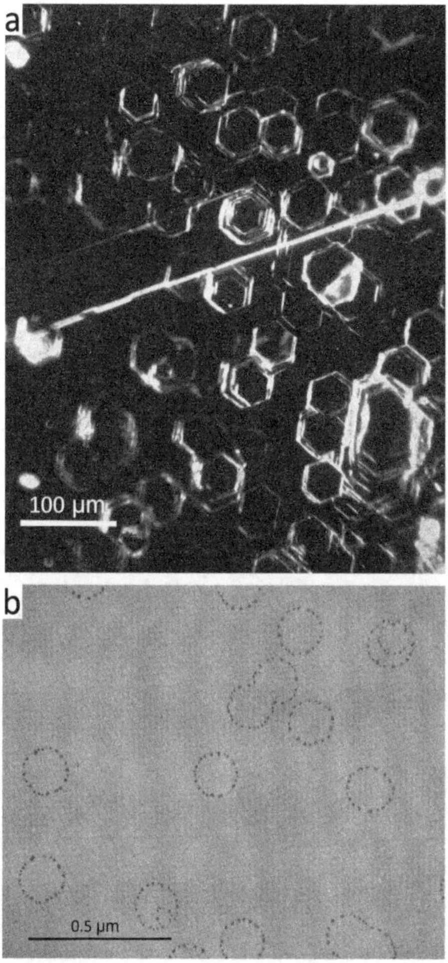

Figure 1.9 (a) Optical microscope images of hexagonal etch pits in the basal plane of single crystal graphite following oxidation by O_2 at 800 °C. The bright diagonal striation is a twin band. Reproduced from ref. 79, with permission from Taylor & Francis. (b) Transmission electron microscope images of circular monolayer etch pits. Reproduced from ref. 76, with permission from Taylor & Francis.

The mechanism of the reaction between graphite and oxygen has been a subject of debate for decades.[91–94] It is established that the reaction produces both CO_2 and CO, and can be written as:

$$C+\left(1-\frac{y}{2}\right)O_2 \rightarrow (1-y)CO_2 + yCO$$

where the parameter y can vary between 0 and 1.

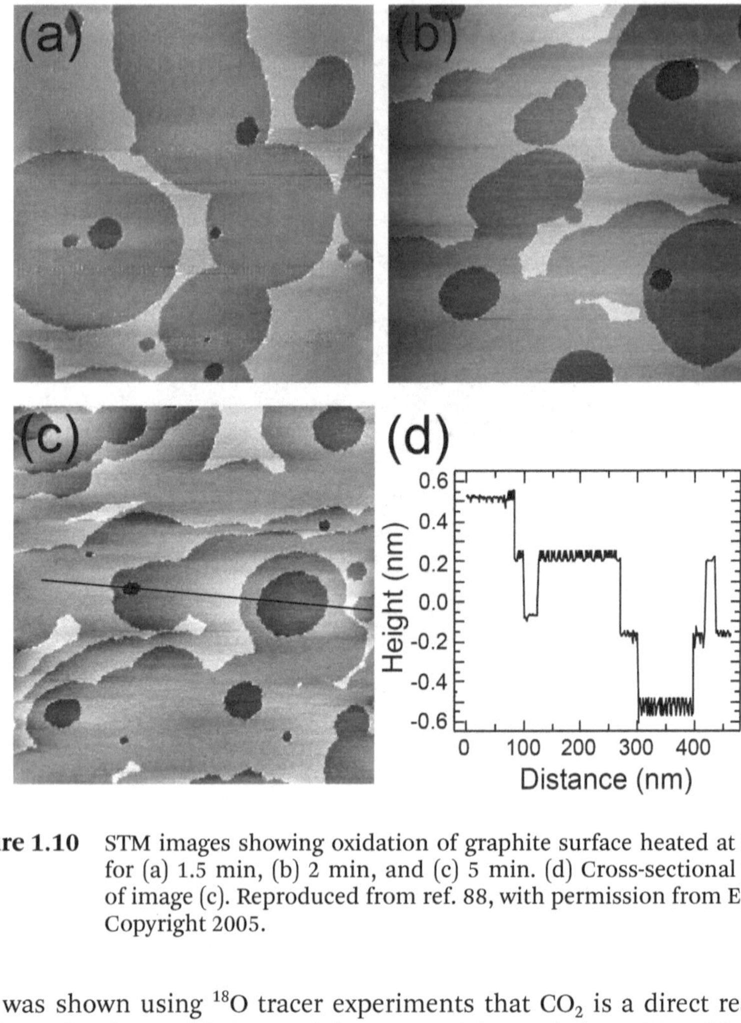

Figure 1.10 STM images showing oxidation of graphite surface heated at 900 °C for (a) 1.5 min, (b) 2 min, and (c) 5 min. (d) Cross-sectional profile of image (c). Reproduced from ref. 88, with permission from Elsevier, Copyright 2005.

It was shown using ^{18}O tracer experiments that CO_2 is a direct reaction product of carbon oxidation and does not solely result from CO oxidation.[95] It was also demonstrated that while irreversibly bound oxygen is present on the graphene basal plane,[96] O_2 chemisorption only occurs at graphene edges.[97] In recent years, Radovic *et al.* have attempted to rationalise these observations by proposing the reaction scheme set out below, supported by density-functional theory calculations.[98–102] These authors have considered the roles played by carbon atoms in the three types of graphene sites, namely zigzag and armchair edges, labelled C_{zz} and C_{ac} respectively and within the basal plane, C_b. In their scheme, C_e denotes atoms in edge sites that can be either zigzag or armchair, while $C_b(O)$ represents temporarily stable surface complexes, notably epoxide groups, and $C_e(O)$ represents reactive intermediates. The first stage is the adsorption of O_2 on zigzag and armchair edge sites:

$$\text{adsorption(non-dissociative)}: C_{zz} + O_2 \rightarrow C_e(O_2)$$

$$\text{adsorption(dissociative):} \quad 2C_{zz} + O_2 \rightarrow 2C_e(O)$$

$$\text{adsorption(dissociative):} \quad 2C_{ac} + O_2 \rightarrow 2C_e(O)$$

$$\text{surface diffusion(hopping):} \quad C_e(O) \leftrightarrow C_b(O)$$

$$\text{desorption(direct):} \quad C_e(O_2) \rightarrow CO_2(+2C_e)$$

$$\text{desorption(direct):} \quad C_e(O) \rightarrow CO(+2C_e)$$

$$\text{desorption(indirect):} \quad C_e(O) + C_b(O) \rightarrow CO_2 + C_e$$

$$\text{nascent site deactivation:} \quad 2C_e \rightarrow 2C_b$$

The reaction labelled 'surface diffusion (hopping)' is an important stage in the process and involves the formation of epoxide groups on the basal planes as a consequence of migration, or 'spillover' of atomic oxygen from graphene edges.[100,101] This is followed by the desorption of CO_2 and CO from the basal planes. The final stage, nascent site deactivation, involves the stabilisation of free radicals formed as a consequence of pyrolytic bond cleavage (*e.g.* as a result of the evolution of CO_2 and CO).[102] The reaction shown here, $2C_e \rightarrow 2C_b$, is the conversion of an edge to a basal-plane site, a process which involves the formation of a pentagonal ring.

1.6.2 Graphite Oxide

Graphite oxide was first prepared by Brodie in the mid-19th century. In a paper published in 1859, he described treating natural graphite with a mixture of $KClO_3$ and fuming HNO_3 at 60 °C for 3–4 days.[103] After washing and drying, this produced a flaky material with an approximate composition of $C_{11}H_4O_5$. The product was slightly acidic and so Brodie gave it the name 'graphitic acid'. Today this material is known as graphite oxide (GO), although the description is not strictly accurate, as many different functional groups are typically present (see below). Currently the most widely used methods for synthesising graphite oxide are based on a technique described by Hummers and Offeman in 1958 which involves treating graphite with sodium nitrate, sulfuric acid and potassium permanganate.[104]

Graphite oxide has a rather disordered structure and its chemical composition varies according to the synthesis method. For this reason, there is no single agreed structural model for GO. One model which has become popular was put forward by Lerf, Klinowski *et al.* in 1998.[105,106] In the model, there are 'islands' of unoxidised carbon rings and rings functionalised with hydroxyl groups, epoxide groups. The edges are terminated with hydroxyl and carboxyl groups. This and other models are discussed in Chapter 2. One aspect of the structure of graphite oxide which has been well established is that the interlayer spacing increases considerably as a result of oxidation. X-ray diffraction showed that the 0.34 nm graphite interlayer spacing increased to approximately 0.68 nm following a Hummers'-type oxidation treatment.[107]

Graphite oxide has a host of actual and potential applications, including catalysis, water purification and in sensors.[107] The discovery of graphene has stimulated renewed interest in graphite oxide, since it can be used as a source of graphene oxide, or of graphene itself. This has led to the development of a number of 'modified Hummers' methods' for generating graphite oxide. These are also discussed in Chapter 2.

1.6.3 Graphite Intercalation Compounds

Graphite's layered structure gives it the ability to accommodate atoms, ions or molecules between the layers, producing graphite intercalation compounds (GICs). The bonding in these compounds involves charge-transfer between the intercalate and the graphite and this can radically change their properties. Charge transfer can occur either from graphite layers to intercalated species or the other way round. Here we summarise the synthesis, properties and possible applications of GICs, beginning with a brief summary of the historical background.

The German scientist, C. Schafhäutl, reported in 1841 that immersing graphite in a mixture of H_2SO_4 and HNO_3 produced a dramatic swelling of the material.[108] The reason for this swelling was not clear until X-ray diffraction was applied to the problem by Hofmann and Rüdorff in the 1930s.[109] These workers established that the structure consisted of graphite layers intercalated by HSO_4^- ions and H_2SO_4 molecules. Further work on graphite intercalation was carried out in the 1950s. Intercalation with bromine,[110] halogen compounds[111,112] and alkali metals[113] was demonstrated. The structure of potassium-intercalated graphite was determined by Rüdorff and Schulze in 1954.[114] They showed that in the 'saturated' structure, with the formula C_8K, the graphite sheets are stacked directly above each other with the K atoms forming ordered arrays which have a p(2 × 2) structure, as shown in Figure 1.11(a). In this structure the spacing between the layers is expanded to 0.54 nm. The intercalated K atoms occupy sites over the hollows of every fourth carbon hexagon, but each layer of atoms is shifted slightly from the one below it, with the 4th layer lying above the 1st. This structure is an example of a 1st-stage intercalation compound, where intercalation has reached a maximum. In order to reach this final structure, the material passes through several preliminary stages as illustrated in Figure 1.11(b).

In the 1960s and 1970s graphite intercalation became a fashionable field of research. This was partly a result of the availability of highly oriented pyrolytic graphite, an almost perfect form of graphite (see Section 1.3.4), which was an ideal host for intercalation experiments. Much of the research in this period focused on the potential of GICs as 'synthetic metals'. Leaders in the field were the groups of Dresselhaus[34] and Fischer.[116] These groups demonstrated increases in basal plane conductivity of up to 13 times that of pure graphite as a result of intercalation.[116] Other groups reported evidence of superconductivity in GICs, with T_cs of 1.4 K[117] and 2.7 K.[118] At the same time,

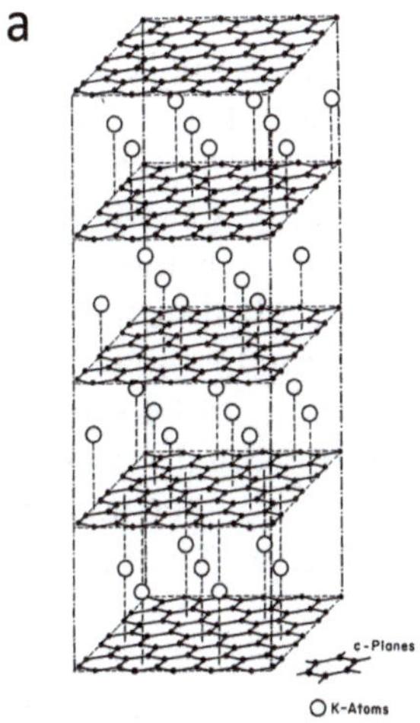

Figure 1.11 (a) Structure of potassium-intercalated graphite, C_8K, according to Rüdorff and Schulze. Reproduced from ref. 114, with permission from John Wiley & Sons, Copyright 1954. (b) Schematic illustration of stages of graphite intercalation. Dark dots and lines represent graphene sheets while red circles represent intercalated species. Reproduced from ref. 115, https://arxiv.org/abs/1701.06493, under the terms of the CC0 1.0 Deed, https://creativecommons.org/publicdomain/zero/1.0/.

interest in the intercalation of lithium into graphite, in connection with lithium-ion batteries, continued to grow, as discussed in Section 1.10.

Only a very brief discussion of the methods that have been used to intercalate graphite is possible here; several excellent reviews are available.[34,116,119] Vapour transport methods are commonly used, these can either be isothermal or 'two-zone', in which the host and intercalant are at different temperatures. These methods are effective for intercalating alkali metals and halogen compounds, into graphite. The use of electrochemistry to intercalate graphite, as occurs in lithium-ion batteries, is also popular, while non-electrochemical solution methods, as employed in the original work of Schafhäutl, are also still used.

Today, there is great interest in graphite intercalation as a method of exfoliating graphite to produce graphene. This is discussed in Chapter 2.

1.7 Physical Properties of Graphite

The electronic properties of graphite are discussed in the next section. Here we briefly summarise the thermal and mechanical properties of crystalline graphite. The properties of nuclear graphite, and the effect of irradiation on these properties, are discussed in Section 1.9.

Considering graphite's thermal conductivity first, many experimental studies have been carried out, all giving values which are highly anisotropic.

This is to be expected, since phonons will propagate far more quickly along the planes than between planes. Thus, room-temperature thermal conductivity parallel to the basal planes is generally found to be >1000 W m^{-1} K, while the value perpendicular to the basal planes is typically <10 W m^{-1} K.[120] The thermal conductivity shows a maximum at a temperature of ~100 K. At higher temperatures, the thermal conductivity decreases with increasing temperature due to phonon scattering. Turning to the specific heat capacity of graphite, measurements of this extend back to 1915.[121] Studies up to 1973 were reviewed by Butland and Maddison.[122] These showed that the specific heat varied significantly with temperature and the authors fitted a polynomial to the data, which was valid for the temperature range 250–3000 K. More recent work gives the room temperature specific heat capacity of graphite as 706.9 J K^{-1} kg^{-1}.[123]

Graphite has relatively low thermal expansivity. Measurements on single crystal specimens show that the thermal expansion coefficient (TEC) parallel to the *c*-axis is positive, and that in the basal plane is small, being in fact negative at low temperatures but becoming positive at about 400 °C. This was established in early work by Nelson and Riley who used powder XRD to measure the lattice parameters of natural graphite parallel and perpendicular to the basal plane in the temperature range 15–800 °C.[124] Subsequent studies confirmed these findings.[125] A 2011 study of graphene showed that this has a negative thermal expansion coefficient in the temperature range 200–400 K.[126] The negative TEC apparently results from out-of-plane phonon modes effectively reducing the in-plane carbon–carbon distances. The effect of irradiation on graphite's thermal properties is discussed in Section 1.9.2.4.

Considering now graphite's elastic properties, five independent constants are required to define the stress–strain relationships for a hexagonal crystal-like graphite.[127] A study by Blakslee *et al.* in 1970 produced figures for the constants which have generally been confirmed in later work.[128] Measurements were made on compression-annealed pyrolytic graphite specimens using ultrasonic, sonic-resonance and static-test methods. The values they found were as follows:

C_{11}	1060 ± 20 GPa
C_{12}	180 ± 20 GPa
C_{13}	15 ± 5 GPa
C_{33}	36.5 ± 1 GPa
C_{44}	4.0 – 4.5 GPa

These elastic constants are components of the stiffness tensor, which relates the stress tensor and the strain tensor *via* Hooke's law. C_{11}, C_{13} and C_{44} are diagonal components of the tensor and can be interpreted as follows: C_{11} is the elastic constant in the direction of the layer planes and C_{33} is the constant along the *c*-axis. The meaning of the off-diagonal components C_{12}, C_{13} is more difficult to state simply but they are related to the Poisson's ratio of the material. The value that Blaksee *et al.* found for the shear modulus, C_{44}, was 0.2 GPa. Later experimental work gave the much higher value of 4–5 GPa.[129,130] Theoretical work by Cousins and Heggie produced similar results.[131] The important point here is that the figures illustrate the extreme anisotropy of the elastic properties of graphite: at 1060 GPa the in-plane modulus is higher than that of any known material, while the modulus perpendicular to the layer planes is a mere 36.5 GPa. The shear modulus of 4–5 GPa indicates the weakness of the interplanar bonds.

The exceptionally high in-plane modulus indicates the great potential of graphite-related structures as high-stiffness materials. This is illustrated in the exceptional properties of carbon fibres (see Section 1.11), of carbon nanotubes[50] and of graphene (Chapter 2).

Relatively few studies of the tensile strength of single-crystal graphite have been carried out. Measuring the in-plane strength of graphite crystals would be experimentally very challenging, although measurements have been made on carbon nanotubes and graphene, as discussed below. One parameter which has been measured using graphite crystals is the interlayer shear strength. Work by the Soule and Nezbeda produced values in the range 0.25–0.75 MPa for natural graphite[132] while Blakslee *et al.* found values of 0.9–2.5 MPa.[128] A much more recent study using a micromanipulator found the far higher figure of 0.14 GPa.[133] The sample here was HOPG, and the authors claim that their value is more accurate than earlier ones because the specimens they employed were small, highly perfect crystals.

A notable study of the tensile strength of multiwalled carbon nanotubes (MWNTs) was reported by Ruoff *et al.* in 2000.[134] This involved using a micromanipulator inside a scanning electron microscope to attach individual MWNTs to the opposing tips of atomic force microscope cantilever probes. One of the cantilevers was then moved away from the other in order to apply tensile load. With sufficient stress, the outer layers fractured, leaving the inner layers intact; tensile strengths as high as 63 GPa were found for the outer layers. Still higher values have been reported for the tensile strength of single-layer graphene. In 2008, Hone *et al.* described a study in which the breaking strength of graphene membranes was measured by nanoindentation in an atomic force microscope.[135] They reported a value of 130 GPa.

1.8 Electronic Properties of Graphite

1.8.1 Electronic Band Structure of Graphite

Before discussing the band structure of three-dimensional graphite we briefly consider the band structure of graphene. This has of course become a topic of great interest since the isolation of graphene by Geim, Novoselov and their colleagues in 2004,[48] and is discussed in more detail in Chapter 2 of this book. The first band structure calculations for graphene, were carried out as early as 1947 by Wallace using the tight binding approximation.[136] He showed that graphene is a semi-metal with zero density of states at the Fermi level. The energy–momentum dispersion relations are linear at the points where the valence and conduction bands touch the Dirac points, as shown in

Figure 1.12. In three dimensions, the conduction and valence bands are conical in shape and are known as Dirac cones. It is the linearity of the dispersion relations near the Dirac points that gives rise to graphene's remarkable electronic properties.[137,138]

The first band structure calculations for three-dimensional graphite were carried out by Slonczewski, Weiss and McClure in the 1950s;[140–143] these will now be summarised. In three-dimensional Bernal graphite, there are four atoms per unit cell, each carbon atom having four valence electrons. There are thus 16 energy bands, 12 σ and 4 π. The Slonczewski–Weiss–McClure (SWMcC) band structure is shown in Figure 1.13.

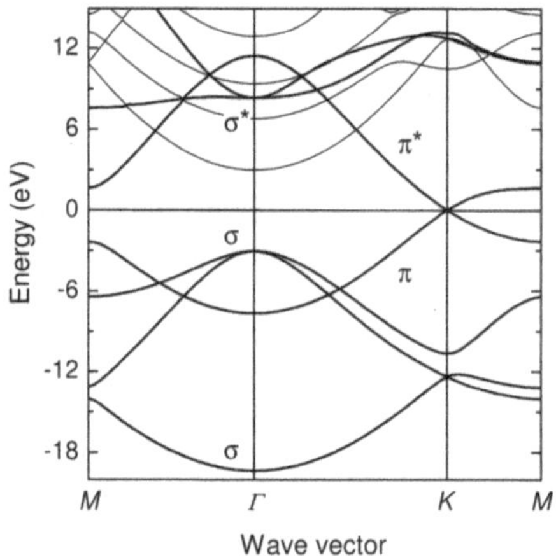

Figure 1.12 Electronic band structure of graphene. Reproduced from ref. 139, with permission from John Wiley & Sons, Copyright 2004.

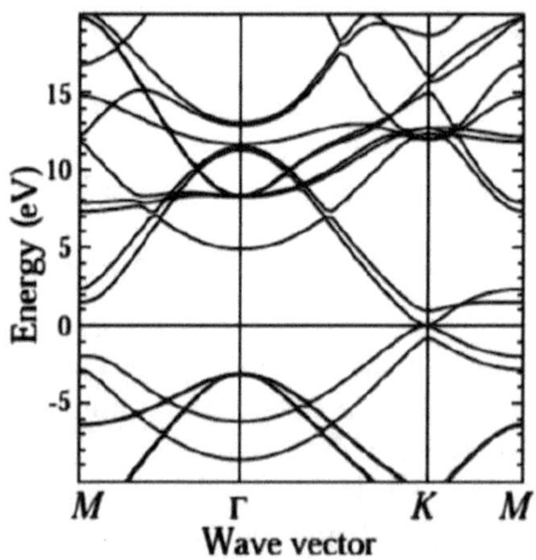

Figure 1.13 Band structure of graphite. Reproduced from ref. 144, with the permission of the University of Toronto.

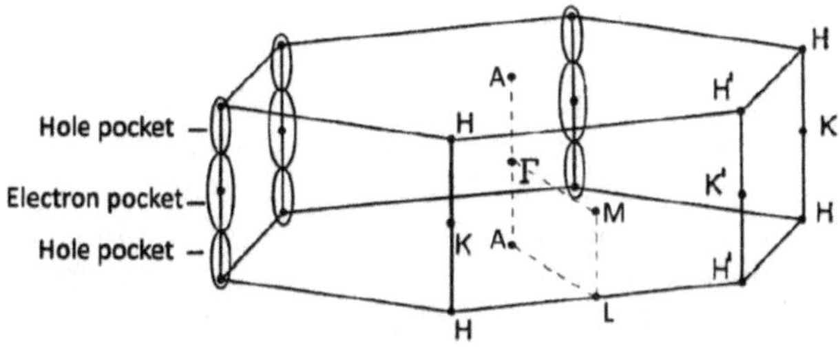

Figure 1.14 The Slonczewski–Weiss–McClure model of the first Brillouin zone for graphite, showing the Fermi surfaces for holes and electrons. Adapted from ref. 142, with permission from IBM.

As can be seen, the π-bands lie between the groups of σ-bands and the Fermi level lies between the 4 π-bands. The Brillouin zone according to this model is a hexagonal prism, as shown in Figure 1.14.

The Fermi surfaces are situated at the six edges of the zone. Each surface comprises a pocket of electrons centred at point K and two half pockets of holes centred at point H.

The isolation of graphene in 2004 has focused attention once again on the electronic properties of graphite, and prompted further studies. In 2006 Louie *et al.* reported the first direct observation of Dirac fermions in graphite.[145] Angle-resolved photoemission spectroscopy (ARPES), a technique which probes the allowed energies and momenta of electrons in a solid,[146] was applied to single-crystal graphite specimens. A linear dispersion relation was observed near the H point of the BZ, while a parabolic dispersion was seen at the K point, as predicted in the theoretical calculations illustrated in Figure 1.13. In 2008, Orlita *et al.* also detected the signature of Dirac fermions at the H point, using far-infrared magneto-absorption.[147]

As well as Bernal graphite, the band structures of AA-stacked and rhombohedral graphite have been calculated.[58,59,148–151] These will not be discussed in detail here, but we note that they differ considerably from the band structure of Bernal graphite. The fact that graphite variants which differ only in their stacking can have very different electronic properties is to be expected, given recent discoveries concerning the electronic properties of bilayer graphene with different twist angles.[55]

1.8.2 Electronic Transport Properties of Graphite

Given its layered structure, it is unsurprising that the electronic transport properties of graphite are highly anisotropic. Within the planes, electron mobility is high, due to overlap between the π orbitals on adjacent atoms. However, perpendicular to the planes the mobility is low. Calculating the electronic transport properties of graphite using the SWMcC model, is difficult, and a close agreement with experiment is not always obtained.[152] The calculations show that graphite has a carrier density of the order of 10^{18} cm^{-3}, *i.e.* about one carrier per 10^4 atoms. Thus the conductivity is low compared with a metal like copper, which has one free carrier per atom. The low carrier density is partly offset by high carrier mobilities in the basal plane.

Measuring the electronic transport properties of graphite experimentally is not easy, mainly due to the challenge of preparing defect-free single-crystal graphite specimens. Making good electrical contacts with specimens can also be difficult. In most cases, resistivity measurements have been made parallel to the layer planes.[152–156] Despite the difficulties, these resistivity measurements have produced quite consistent results, with values generally close to 0.4 μΩ m. Resistivity perpendicular to the layer planes is far higher: the anisotropy ratio ρ_c/ρ_{ab} varies from 10^2 to 10^5.[157] Unsurprisingly, polycrystalline graphite samples have higher resistivities than single crystal graphite: typical values are 9–17 μΩ m.[120,158–160]

In 2013 Esquinazi and co-workers reported detecting superconductivity in HOPG.[161–163] The superconductivity was observed when the specimen contained interfaces between the HOPG domains which were twisted with respect to each other. Superconductivity could arise in the same way as for twisted bilayer graphene[55] (see also Chapter 2). The report of possible superconductivity in trilayer rhombohedral graphene at low temperatures was mentioned previously.[57]

1.9 Nuclear Applications of Graphite

Currently, nuclear power generates about 10% of the world's electricity, totalling 2679 TW h.[164,165] In the UK, 15% of total electricity generation is supplied by the nine operating nuclear reactors of which eight are graphite-moderated thermal fission reactors.[166] The graphite materials used in these reactors are synthetic composites with complex structures which will be introduced briefly in this section.[167–170] The idea of designing and constructing a self-sustained thermal fission reactor using graphite as a fast-neutron moderator was first explored by Enrico Fermi's team in the 1940s.[171] Since the first criticality of the Chicago Pile 1, later Oak Ridge X-10 Pile (part of Manhattan Project) and the Hanford Production Reactors, which used AGX and AGOT graphite,[171–174] high-purity synthetic polycrystalline graphite has been used as a fast-neutron moderating and reactor-core structural material.

Graphite-moderated thermal fission reactors have evolved into several different variants around the world, mainly in the Soviet Union, United Kingdom, France, United States, Japan, Germany, China and South Africa. Typical examples include the Soviet Union's (and later Russia's) RBMK (Reaktor Bolshoy Moshchnosty Kanalny (High-Power Channel Reactor)) water-cooled graphite-moderated reactors in which GR-280 graphite was used,[176–182] the UK's first generation gas-cooled graphite-moderated Magnox reactors at Calder Hall using highly anisotropic medium-grained Pile Grade A (PGA)

Figure 1.15 Typical design of the Magnox reactor core. Consists of octagon-square PGA graphite bricks that are radially keyed together.[175]

graphite,[183–186] and the French UNGG (Uranium Naturel Graphite Gaz) reactor that was designed in parallel to the Magnox reactor in the 1950s.[187,188] The typical design of the UK's Magnox reactor core is shown in Figure 1.15. Understanding graphite irradiation behaviour and damage mechanism is of paramount importance and this is best demonstrated by the Windscale Pile fire accident in the UK in 1957. The deliberate increase in core temperature served to initiate the controlled release of the stored energy and anneal the irradiation damage in graphite thereby achieving a partial annealing of the graphite core. It was during such routine annealing procedure that the metal fuel cartridges were damaged and burnt due to the uncontrolled sharp release of Wigner energy from the graphite core.[168–170,189]

In the UK's second-generation Advanced Gas-cooled graphite-moderated reactors (AGR), medium-fine grained near-isotropic Gilsocarbon graphite is currently used in significant quantities as moderator bricks.[190–192] Fast neutrons released from the fission reaction are slowed down by scattering with carbon atoms during moderation, to maximise the ${}^{235}_{92}U - {}^{1}_{0}n$ cross section for neutron economy and fission sustainability.[167,198] Fuel sleeves and reflector bricks and various interlock keys for structural integrity are all made of nuclear graphites (SM2-24 and Gilsocarbon IM1-24 and GCMB grades).[168,170,172,192,199] A typical AGR structure/design is shown in Figure 1.16.

Nuclear graphite is also one of the key materials in some of the advanced fission reactor systems, where higher thermal efficiency, safety and thermochemical hydrogen production can be achieved.[200–203] This is exemplified by the next generation high temperature gas-cooled reactors (HTGR) that are the successors of the historical HTGRs, which were first proposed and developed in the 1960s. The first of these were the Dragon test reactors in the

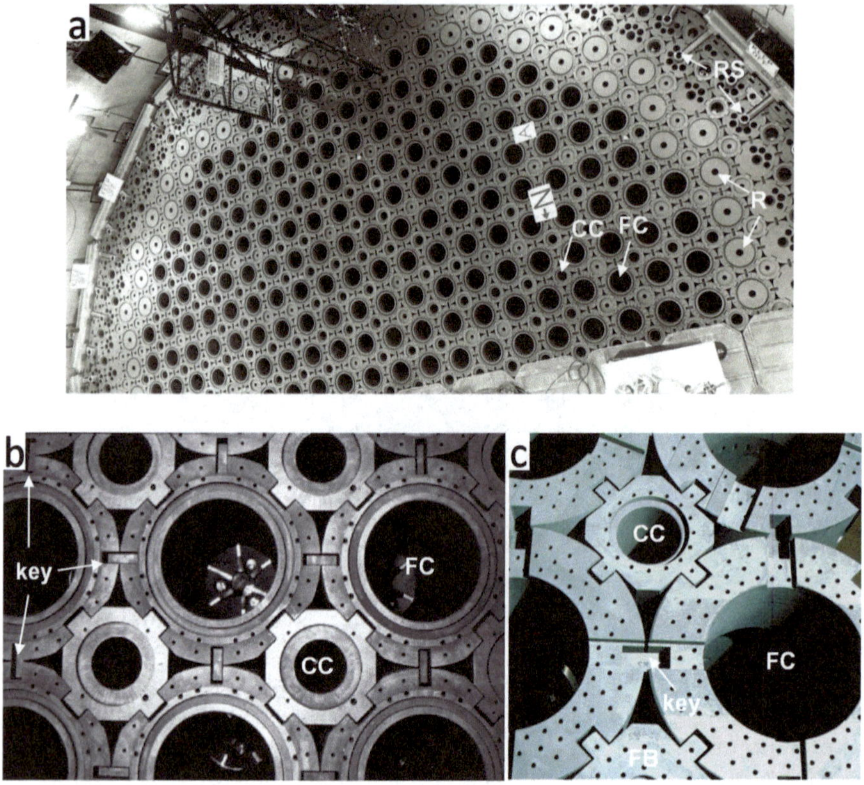

Figure 1.16 Typical design of AGR reactor core. (a) Graphite core lattice made of Gilsocarbon graphite bricks that are radially keyed together. Reflector bricks (R) and core restraining system (RS) are also shown.[193] Typical active graphite core measures 9 m in diameter and 8 m in height.[194,195] (b) Graphite brick arrangement (Heysham2-Torness type).[196,197] (c) Another graphite brick arrangement (Heysham1-Hartlepool type).[196,197] FC: fuel channel, FB: filler brick, CC: control channel, small circular holes on each brick are methane holes. Reproduced from the references cited, with permission.

UK and the German AVR (ARS/AMT graphite) reactor. These were succeeded by the pebble bed thorium high-temperature reactor, THTR-300, and the US Peach Bottom HTGR, which was in turn succeeded by the Fort St Vrain reactor (H-327 and H-451 graphite).[203–211] Relatively modern designs, such as the HTTR in Japan (shown in Figure 1.17), the HTR-10 and HTR-PM in China, are using fine-grained IG-110 graphite (PGX graphite is also used by HTTR).[212–216] Other grades, such as medium grained NBG-10, NBG-18 and ATR-2E, medium-fine grained NBG-17, PCEA and PGX, fine-grained IG-430, SNG742, SNG623 and ETU-10 and many others, are also potential candidates for next generation HTGRs.[172,217–226]

Further, two designs of the Generation IV reactor systems, namely, the molten-salt reactor (MSR) and the very high temperature reactor (VHTR) will

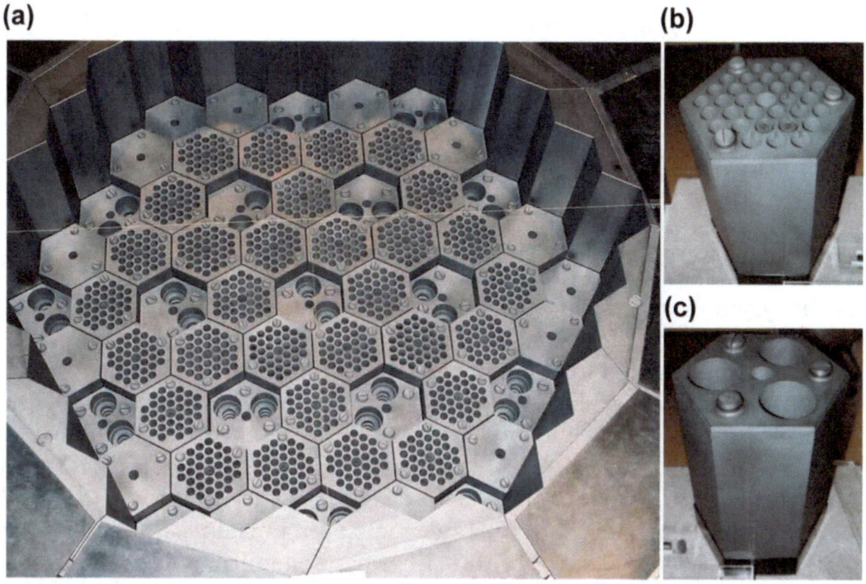

Figure 1.17 (a) Image of Japan's HTTR prismatic graphite core with diameter and height of about 3 m.[236] (b) A prismatic graphite block for fuel elements with holes being coolant and fuel element channels. Each block measures about 360 mm across prismatic planes and 580 mm in height.[236] (c) A prismatic graphite block for providing channels for control rods.[194] These blocks are further surrounded by graphite reflector blocks that are in contact with the reactor wall. Reproduced from the references cited, with permission.

also employ fine-grained nuclear graphite, such as IG-430, NBG-25, SNG623 and T220. In these applications, nuclear graphite materials serve various purposes including moderating and reflecting fast neutrons, physically accommodating fuel (assemblies) and control rods, providing primary circuit coolant channels for heat exchange, as well as being a load-bearing structural support comprising the majority of the core volume.[167,203,223,227–231] In addition to being directly used as structural material, POCO graphite grades are also considered to be a surface sealant for nuclear graphite components in the MSR owing to its superior resistance to molten salt infiltration and gas permeation due to its very fine porosity.[232–235]

In most applications, these graphite materials are non-replaceable, hence limiting the lifetime of the reactors. Decommissioning of graphite-moderated reactors could result in a large quantity of nuclear graphite waste, of which more than 300 000 tonnes have already been accumulated worldwide.[237] It is therefore important that the lifetime of the graphite components is extended to the maximum possible degree. However, the extreme service conditions of nuclear reactors pose severe challenges to graphite materials by virtue of the potentially high temperatures, up to 1100 °C, fast-neutron irradiation and accident conditions.[200,238] The importance of fine-grained graphite to the structural integrity of reactor cores cannot be overstated, both from an industrial and a scientific research perspective, especially for such lifetime safety-critical applications.

1.9.1 Manufacture and Microstructure of Nuclear Graphite

1.9.1.1 Manufacturing Route

The microstructure and properties of nuclear graphites are highly affected by the way they are produced.[183,185,190,239,240] Hence it is necessary to briefly introduce the raw material and manufacturing route used for nuclear graphite production in order to better appreciate their physical and/or thermomechanical properties. Nuclear graphite at industrial scale is produced by the Acheson process (see Section 1.3.1 above) in which the shaped mixture of filler and binder materials are graphitised in a high temperature Acheson furnace to form product billets before final machining into the desired geometry and shapes.[169,189,241,242] Under most circumstances, the raw filler particle materials are sourced from either petrochemical and coal industry (*e.g.* coal-tar pitch coke) or petroleum coke derived from naturally occurring aromatics-containing feedstock associated with petroleum reservoirs. Petroleum coke is considered superior due to its high content of aromatic carbon (>70 wt%) and high crystallinity achieved after heat treatment to about 2800–3000 °C.[169,243,244] An example of a source of petroleum coke is the hydrocarbon gilsonite bitumen extracted from the Uinta Basin in Utah, USA,[245] which is the direct feedstock for producing Gilsocarbon nuclear graphite for UK's AGR fleet.[192,199,246,247] Petroleum coke (acicular or needle-shaped laths after delayed coking and grinding) is also used for the historic PGA graphite employed in the Magnox reactors in the UK,[183,248,249] while coal-tar pitch coke is the source of the NBG-17 graphite that is proposed to be the structural and moderator material for HTGRs.[172,250,251] More examples are given in Table 1.1.

A binder material provides the connection between ground filler cokes (filler particles). Binder materials are often pitch materials that are by-products of petrochemical process (*i.e.*, residual products from high temperature coking process).[169,243] Both coal-tar and petroleum pitches are often used for mixing with filler cokes. Gilsocarbon graphite and PGA graphite both use coal-tar pitch binder. Apart from these two phases, the binder matrix often includes finely milled filler flour.[172,249,252] Unfortunately, the exact binder materials used for the vast majority of nuclear graphite grades listed in Table 1.1 are not disclosed in the open literature, as details of the raw materials used in manufacturing nuclear graphite vary across different manufacturers are often regarded as proprietary.

The manufacturing and processing techniques used can produce variation in the properties of the final nuclear graphite products.[37,172,183,194,244,249] Regarding the manufacturing process shown in Figure 1.18, more comprehensive reviews are included in ref. 168–170, 189, 241, 242, 244 and 252 and hence only brief descriptions of the conventional polygranular nuclear

Table 1.1 Summary of physical and mechanical properties of manufactured nuclear graphite extracted from various open literature and manufacturers' datasheets.

	POCO ZXF-5Q	IG-430	IG-110	SNG623	SNG742	T220
Apparent (bulk) density/g cm^{-3}	1.78	1.82	1.78	1.81	1.83	1.87
Young's modulus/GPa	14.5	10.8	9.8	11	11	11
Bulk CTE/10^{-6} K^{-1}	8.1	4.8	4.5 (25–500 °C)	4.6 (20–600 °C)	4.8 (20–600 °C)	4.4 (20–500 °C)
Thermal conductivity/W m^{-1} K^{-1}	70	140	116–130	—	—	83.4
Tensile strength/MPa	79	37.2	24.5–28	—	28.5	33.2
Compressive strength/MPa	175	90	75–78.5	93	77	120
Flexural strength/MPa	112	54	34–39.2	50	45	41
Mode I fracture toughness/MPa m$^{0.5}$		1.04	0.9–1.07	1.6–1.9	1.12–1.54	0.8–1.4, 1.4–1.6
Bulk porosity/vol%	20 (total), 16 (open), 22.8	14–21, 14.2	17.3, 19–21.9 (total), 13.9 (open),	17	12	5.5 (2D)
Average grain size/μm	1	10	20 (40–50 max)	20	20	5
Classification (ASTM D7219-19)	Microfine	Ultrafine	Superfine	Superfine	Superfine	Ultrafine
d_{002} spacing/Å	3.3833,[280] 3.389 from XRD	3.3681,[280] 3.3736 from XRD	3.3646,[281] 3.3634[282]	3.3767	—	3.3758
Graphitisation temperature/°C	2500	≥2800	3000 max.	≥2800	≥2800	—
Filler type	Petroleum coke	Coal-tar pitch coke	Petroleum coke	Petroleum coke[283]	Petroleum coke	N/A
Forming method	Isostatic pressing	Isostatic moulding	Isostatic moulding	Isostatic pressing	Isostatic pressing	Isostatic moulding
Isotropic ratio based on CTE ratio	<1.03	1.09	1.09	—	—	—

(*continued*)

Table 1.1 (*Continued*)

	POCO ZXF-5Q	IG-430	IG-110	SNG623	SNG742	T220
Manufacturer	POCO – Entegris, USA	Toyo Tanso, Japan	Toyo Tanso, Japan	Sinosteel, China	Sinosteel, China	Sinosteel, China
Main application/usage	MSR, proton beamline target	VHTR, MSR, GTHTR-300, proton beam-line target	HTTR, HTR-10, HTR-PM, Gen IV reactors	HTGR, MSR	HTGR, MSR	MSR
References	224, 232 and 284–287	172, 220, 223, 231 and 287–290	172, 189, 225, 226, 254 and 290–292	283 and 293–295	283, 294 and 296–299	296 and 300–302

	SGL R7650	NBG-18	NBG-17	NBG-25	G347A	ETU-10
Apparent (bulk) density/g cm^{-3}	1.84	1.85	1.89	1.82	1.85	1.75
Young's modulus/GPa	12.5	11.5	11.9	11	10.8	9.8
Bulk CTE/10^{-6} K^{-1}	3.9–4.1(20–200 °C)	4.6 (WG), 4.5 (AG), (20–200 °C)	4.8 (WG), 4.6 (AG), (20–200 °C)	3.9–4.3 (20–200 °C)	5.5 (25–500 °C)	3.8 (50–400 °C)
Thermal conductivity/W m^{-1} K^{-1}	90–95	140	142	140	128	104
Tensile strength/MPa	46	20–22.9	22	—	31.4	34.3
Compressive strength/MPa	150	80	90	95	98	98
Flexural strength/MPa	65	30	32	46–48	49	58.8
Mode I fracture toughness/MPa m$^{0.5}$	—	1.17–1.27, 1.4	1.4	0.9–0.99	—	—
Bulk porosity/vol%	10 (open), 11.1	12.3, 17.8 (total)	14	19.1 (total), 12.3 (open)	12	15

Average grain size/μm	7	1600 max.	800 max.	50 (60 max.)	50	15
Classification (ASTM D7219-19)	Ultrafine	Medium	Medium-fine	Superfine (fine)	Superfine	Superfine
d_{002} spacing/Å	3.3794 [280]					
3.3801 from XRD	3.3647 [281]	3.3651 [281]	3.3684 [303]	—	—	
Graphitisation temperature/°C	—	—	2900		—	—
Filler type	—	Coal-tar pitch coke	Coal-tar pitch coke	Petroleum coke	Coal-tar pitch coke	Coal-tar pitch coke
Forming method	Isostatic pressing	Vibrational moulding	(Unidirectional) vibrational moulding	Isostatic moulding	Isostatic pressing	Isostatic moulding
Isotropic ratio based on CTE ratio	—	1.04	1.04	1.10	1.06	1.08
Manufacturer	SGL Carbon, Germany	SGL Carbon, Germany	SGL Carbon, Germany	SGL Carbon, Germany	Tokai Carbon, Japan	Ibiden, Japan
Main application/usage	Proton beam-line baffle, battery/electrode [304,305]	HTGR-PBMR, VHTR	HTGR-PBMR, VHTR	VHTR, MSR, Gen IV reactors	Gen IV reactors	VHTR, HTGR, MSR
References	234 and 290	172, 189, 250, 254, 281, 292, 306 and 307	174, 215, 250 and 290	172, 222, 225, 250, 308 and 309	172, 189, 263, 310 and 311	174, 214, 225, 226 and 312

			PCEA			PGA graphite		ATR-2E	
	Mersen 2020	Mersen 2114	WG	AG	Gilsocarbon graphite	WG	AG	WG	AG
Apparent (bulk) density/g cm^{-3}	1.77	1.81	1.83		1.80	1.70		1.80	
Young's modulus/GPa	9.2	10	11.3	9.9	10.6–11	11.7	5.4	9.6	8.4
Bulk CTE/10^{-6} K^{-1}	4.3–4.5	5.3	3.5	3.7	4.7 (WG), 4.9 (AG)	0.9	3.0	3.6	4.1

(continued)

Table 1.1 (*Continued*)

	Mersen 2020	Mersen 2114	PCEA		Gilsocarbon graphite	PGA graphite		ATR-2E	
			WG	AG		WG	AG	WG	AG
Thermal conductivity/W m^{-1} K^{-1}	75–85	104	162	159	138	200	109	179	163
Tensile strength/MPa	31.5	29	22	17	20–28	17	11	12.6	12.4
Compressive strength/MPa	98	103	61	68	70	25–30	25–30	55.9	57.8
Flexural strength/MPa	45	52	32	23	26.7 (WG), 26.9 (AG)	12–20	9–14	23	18.9
Mode I fracture toughness/MPa $m^{0.5}$	—	1.15	1.33	1.48	1–1.5	—	—	—	—
Bulk porosity/vol%	9, 12.1	10, 19.7(total), 12.9 (open)	13.3 (total), 9.3 (open)		20	20			
Average grain size/μm	15	13	800		500	800		1000	
Classification (ASTM D7219-19)	Superfine	Superfine	Medium-fine		Medium-fine	Medium-fine		Medium	
d_{002} spacing/Å	—3.3807 from XRD	—	3.3664,[281] 3.3613–3.3626[282]		3.3754[191]	3.3575, 3.3574–3.3600[282]		—	
Graphitisation temperature/°C	—	—	>3000		~2700–3000	~2800		—	
Filler type	Petroleum coke	Non-petroleum coke	Petroleum		Gilsonite coke	Shell H100 Petroleum coke		Pitch coke	
Forming method	Isostatic moulding	Isostatic moulding	Extrusion		Vibrational moulding	Extrusion		Extrusion	
Isotropic ratio based on CTE ratio			1.06		1.04	3.33		1.14	
Manufacturer	Mersen, France	Mersen, France	GrafTech International, USA		Anglo Great Lakes, British Acheson Electrodes UK; GrafTech International, USA	British Acheson Electrodes; Anglo Great Lakes, UK		SGL Carbon (previously SIGRI GmbH)	
Main application/usage	HTGR, proton beamline target	HTGR, Gen IV reactors	HTGR, VHTR, Gen IV reactors		AGR	Magnox reactors		HTGR-PB	
References	214, 224, 252–255 and 313	172, 214, 292 and 314–316	174, 189, 218, 254, 260, 281, 282 and 292		194, 190–192, 243, 246, 275 and 317	183, 194, 243, 248 and 282		173, 194 and 318	

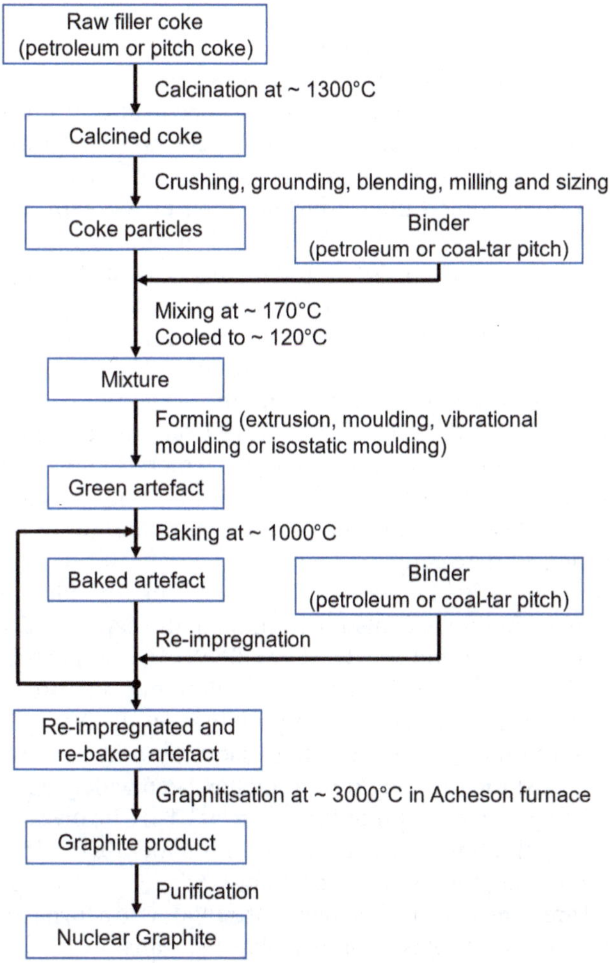

Figure 1.18 Typical manufacturing process of nuclear graphite. Adapted from ref. 244, with permission from Springer Nature, Copyright 1983.

graphite manufacture route are outlined here. To start with, filler cokes, either derived from the product of delayed coking processes (for petroleum coke) or from coal-tar pitch coke, are calcined (heat-treated) at ~1300 °C in a calciner to remove excess moisture, volatile hydrocarbon and hydrosulfide contents (H_2, CH_4, C_2H_6, H_2S and CH_3SH) which are burned and released as volatile gases. As a consequence, cokes lose weight (due to loss of water, evolution of volatile matter and dusts), densify (shrink in volume) and form a porous structure during this process. Note that it is extremely important to calcine the raw coke before mixing with pitch and entering the baking stage, as the shrinkage of raw coke volume will likely result in a weak structure with cracks if the raw coke-binder artefact were directly baked without calcination. In addition, the pores formed during calcination allow better penetration of pitch binder through the coke in the coke-binder mixing stage forming a stronger bond between the two.

The calcined cokes are then crushed, ground and sorted into a range of different sizes according to different needs and specifications. It is this milling and sizing step that predominantly determines the nominal grain sizes of the nuclear graphite products. These crushed and blended coke filler particles (together with finely milled filler flour) are then mixed with liquified pitch binder at ~170 °C to ensure the compatibility of liquified pitch binder viscosity with filler particles to allow for full mixing. The mixture is then cooled to ~120 °C, slightly above the softening temperature of the pitch binder, and passed to the forming stage. The forming operation is conventionally conducted either by extrusion *via* pressing through a hollow cylinder (a large die) at ~120 °C (formerly the most common method); or alternatively, by uniaxially compressing the enclosed mixture from one or both ends using movable platens within an enclosed chamber, which is known as moulding/pressing. The forming process fosters strong bonding between the pitch-coated filler particles, resulting in a substantial reduction in the mixture's porosity and hence increasing its density and compactness, and producing the artefact in a comparable size and shape to the final product to minimise cost. Vibrational moulding is also a commercially viable method that has been used in the production of NBG-17 and NBG-18 and near-isotropic Gilsocarbon graphite.[192,250] In this method, graphite hot mixture paste undergoes compaction through vertical mould vibration against a heavy metal covering.[253] Additionally, to achieve high isotropy and density in the finished artefact, isostatic pressing/moulding has been widely used in producing modern fine-grained graphite.[172,189,224,254,255] This involves pressing the graphite paste in all directions through a rubber membrane immersed in a liquid-filled chamber (cold isostatic pressing).[37,168]

The formed green artefact is then baked at ~1000 °C in a furnace. During the baking stage, the thermoplastic binder phase is transformed into an infusible solid that is carbonised *via* a pyrolysis process, releasing large amounts of hydrogen and volatile gases that are vented by the formation of so-called gas-run bubbles that are mostly open porosity connected to the graphite surface. Baked artefacts are often re-impregnated, with liquified petroleum pitch or coal-tar pitch, with normally no more than six rounds of impregnation to increase the density. Each round may be followed by re-baking for pyrolyzing the impregnant pitch. The effectiveness of impregnation on reducing the porosity reduces with more implementation rounds.[169,189,242,244]

The final stage of the manufacturing process prior to machining into the desired geometry is the high temperature graphitisation stage. An Acheson electric furnace is often used for heating the baked graphite artefact up to ~3000 °C to grow graphite crystals to their thermodynamically stable phase. For nuclear graphite, the graphitisation process usually takes about 15 days from loading to cooling down and unloading.[169,242] During this process lenticular cracks at graphite crystal lattice level (known as Mrozowski cracks at the nanoscale) are formed due to the anisotropic crystal lattice thermal shrinkage during cooling from the graphitisation temperature to room temperature. For nuclear-grade graphite, an additional purification stage at the end is required to remove impurity elements, such as boron, aluminium, chromium, iron, sulfur and so on.[168,169,243] The final product can then be machined and tailored into various component shapes for industrial applications.

1.9.1.2 Microstructure of Nuclear Graphite

As explained above, synthetic nuclear graphite materials are polycrystalline aggregates comprising filler particles and solidified binder matrix, incorporating various forms of porosity.[228,240,241,249,256,257] Distinct filler particle shapes observed across different nuclear graphite grades predominantly arise from different degrees of crystallisation in the raw coke materials and thereafter how they fracture during calcination and crushing.[243] A typical example is the globular Gilsonite cokes in Gilsocarbon graphite that have conchoidal fractographic surfaces and therefore 'onion-like' layered calcination cracks aligned along the periphery of these filler particles.[191,228,245,258] Acicular flakes or needle-shaped filler particles, *e.g.* filler particles in PGA graphite, often result from the breaking of petroleum cokes with high levels of crystal alignment, which is believed to make them more susceptible to microcracking along basal planes *via* cleavage fracture.[170] On the other hand, less ordered raw cokes tend to be equiaxial in shape.[243,252] Fine-grained graphite products that use starting filler grain sizes less than 100 μm (ASTM7919-19), in general do not have clearly distinguishable boundaries between filler particles and binder matrix. Instead, finely-milled filler particles tend to either cluster together to form relatively dense regions that appear to be free of pores,[172,224] or form small mosaic domains that strongly reflect optical light.[172,242] Some exceptions exist, such as IG-110 and NBG-18, where needle-shaped filler (and fragments) and globular filler have been found.[172,259,260] The binder matrix in nuclear graphite is intricate in nature with constituent phases containing domains or mosaics of crystallites, finely ground filler flours dispersed into the binder matrix and porosity.[172,199,261,262] Sub-structures, such as quinoline insoluble (QI) particles, and subsequently rosette structures that are formed by concentric graphitic ribbons with no orientation preference have been identified.[172,230,259,261–265] Examples of nuclear graphite microstructures are given in Figure 1.19.

Early nuclear-grade graphites manufactured in the UK (PGA graphite), USA (*e.g.* TSX and H-451 graphite) and Germany (ARS and ATR-2E) were anisotropic not only because of the needle-shaped coke fillers used but also because the subsequent extrusion forming method preferentially aligned the longer axes of the needle grains with the extrusion direction.[169,173,191,211,266] Since the longer dimensions of these acicular grains are in general parallel to their crystallographic *a*-axes direction, the extrusion direction is then parallel to the crystallographic *a*-axis and this direction is defined as the with-grain (WG) direction in the graphite billet.

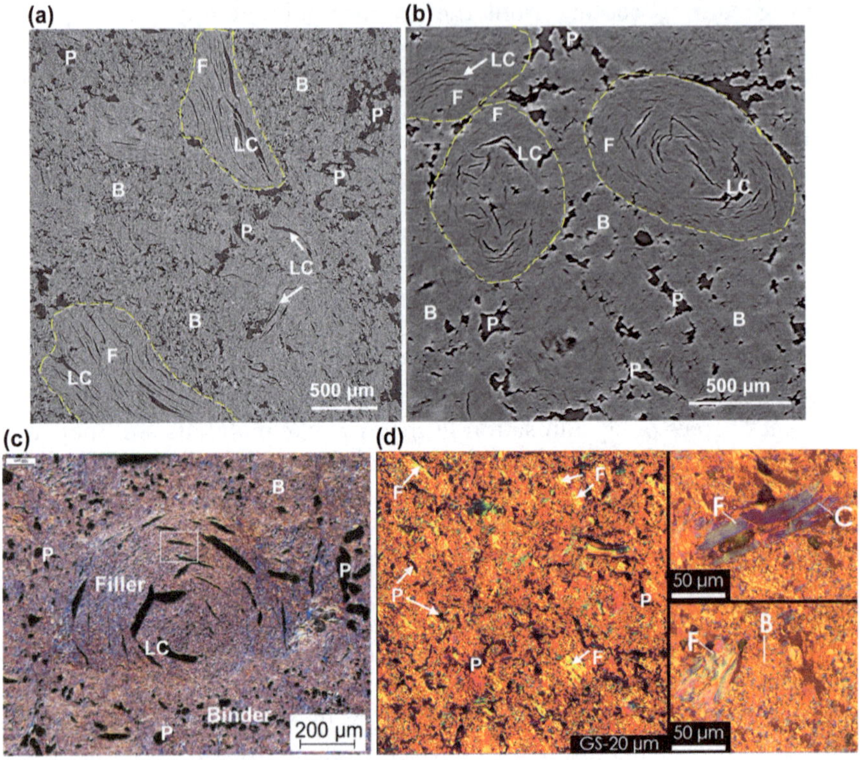

Figure 1.19 Nuclear graphite microstructure. (a) X-ray tomography image of PGA graphite showing the needle-like filler particles (and fragments), (b) X-ray tomography image of Gilsocarbon graphite showing its spherical filler particles and fragments with lenticular calcination cracks aligned along periphery direction; image courtesy of Dr Dong Liu, University of Bristol. (c) Polarised light micrograph of NBG-18 graphite showing its spherical filler particles also with layered 'onion-like' appearance; figure adapted from ref. 259 with permission from Elsevier, Copyright 2022. (d) Polarised light micrograph of IG-110 graphite showing its acicular filler particles with mosaic domains of fragmented filler grains; figure adapted from ref. 172 with permission from Elsevier, Copyright 2022. Average grain size (GS) in (d) is 20 μm. F: filler particle; B: binder matrix; P: porosity; LC: lenticular crack.

In contrast, the against-grain (AG) direction is orthogonal to WG, and hence, orthogonal to the extrusion direction. The resulting thermomechanical properties, such as Young's modulus, thermal conductivity, coefficient of thermal expansion and tensile strength, are also highly anisotropic along the two directions.[183,191,249,265,267] Moulding will tend to align the long axis of the particles in the plane perpendicular to the forming axis. Therefore, these graphites will have two perpendicular WG orientations and one AG orientation. In the case of Gilsocarbon graphite produced by vibration moulding, most of the graphite crystal *c*-axes tend to align perpendicular to the pressing direction and therefore, 90° to the brick axis.[242] The increased isotropic characteristics of Gilsocarbon graphite are consequently derived from its spherical filler particles, which exhibit non-preferential alignment during vibrational moulding.[194,248,258,267,268] Isostatically moulded/pressed graphites have no or minimal preferred orientation of filler grains.[189,242]

Nuclear graphite usually has a bulk density of about 1.7–1.8 g cm^{-3}, lower than the theoretical density of 2.26 g cm^{-3}, due to porosity.[168,172,189,242] The origin, characteristics and role of the various forms of porosity are an important factor in nuclear graphite's microstructural and property changes with irradiation and temperature and hence shall be succinctly outlined.[172,262,264,269,270] There are three main types of porosity existing in nuclear graphite spanning multiple length scales:

(i) Micrometre-sized lenticular cracks and pores inside filler particles due to basal plane shrinkage and delamination, mainly formed during calcination of cokes. In PGA graphite, calcination cracks are formed along crystallographic *a*-axis and hence, in general, parallel to the longer dimension of the needle cokes, *i.e.*, cracks propagate in the WG direction. This is also commonly seen in other nuclear graphites that contain needle-shaped cokes, such as PCEA and PGX.[172,260,271] In Gilsocarbon graphite, lenticular calcination cracks are preferentially aligned with the periphery direction, as a direct consequence of the conchoidal layering structure of its crystallographic basal planes, resulting in the characteristic onion-like layered structures in the filler particles previously mentioned. Similar morphological features of calcination cracks have also been found in NBG-18 and NBG-17 that use spherical filler cokes.[172,259]

(ii) Macro-/millimetre-sized irregularly-shaped and spherical pores in the binder, or at the filler/binder interface, which are usually pore-runs resulting from the bubble percolation of volatile gases during the carbonisation and hardening of the binder phase, as well as impregnant pitch during the baking and impregnation stages.

(iii) Nanoscale lenticular-shaped Mrozowski cracks, as a result of anisotropic thermal shrinkage and inter-platelet shear of graphite crystallites, and nanopores caused by termination of crystals.[262,272,273]

Precise measurement of a full spectrum of these porosities is very challenging and, consequently, a compromise between characterisation resolution and length scale must be made.

1.9.2 Thermomechanical Properties of Nuclear Graphite

This section discusses the physical and thermomechanical properties of several typical grades of nuclear graphite most of which are fine-grained. Many of the properties were summarised in Table 1.1. The properties of some of the medium and medium-fine grained graphites were reported with respect to WG and AG directions especially those with needle-shaped filler particles.[172,260] Mode I fracture toughness values (*i.e.* where the plane of the crack is normal to the tensile stress) are listed in Table 1.1. These include some obtained from sub-sized specimens and differences may exist across the literature due to specimen size,[190,226,274] stress state,[275–277] or depend on the number of samples tested.[278,279]

1.9.2.1 Irradiation Damage Processes

Irradiation damage in graphite materials can be caused by the bombardment of energetic particles displacing carbon atoms from their original lattice sites.[189,252,319,320] A typical process is illustrated in Figure 1.20. The incident energetic particles with kinetic energies exceeding the threshold displacement energy (TDE) of carbon atoms (~24–60 eV[189,252,319,321,322]), displace carbon atoms from their equilibrium lattice positions creating primary knock-on atoms (PKAs) and point defects. These PKAs usually have a kinetic energy in a range of 10^3–10^6 eV and will travel across a distance range of about 100–5000 Å (10–500 nm) before colliding with another lattice carbon atom, to which these PKAs will further transfer their kinetic energy, producing secondary knock-on atoms (SKAs).[167,241,322,323] Displacement events from SKAs and PKAs with energy less than 500 eV will be localised and the cascades of collisions mainly happen within 10 Å (1 nm) and involve small groups of about 5–10 carbon atoms.[189,241,252,319,323] Termination of the passage of PKAs and SKAs due to energy losses makes these displaced carbon atoms interstitials and creates correspondingly vacancies. Most of these localised interstitials will tend to immediately recombine with vacancies when the distance between these short-ranged collision events becomes comparable to graphite atomic spacing, causing the irradiation damage to be partially annealed.[241,252,319,322]

Interstitials are mobile and can diffuse between basal planes for all temperatures that are relevant to the operation of nuclear graphite whereas the

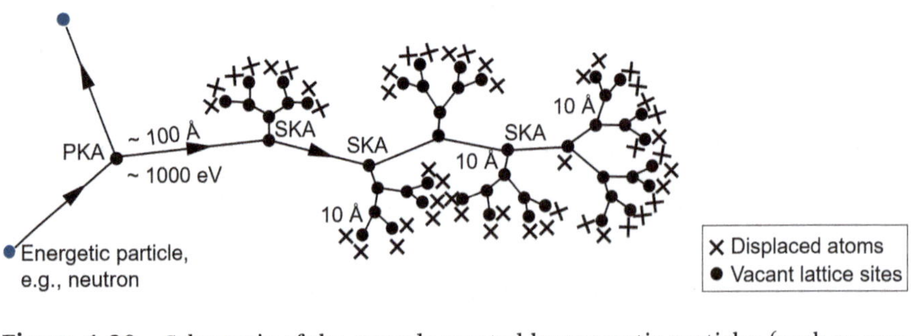

Figure 1.20 Schematic of the cascade created by energetic particles (such as neutrons) interacting with carbon and creating displaced atoms and vacancies. Reproduced from ref. 167, with permission from Elsevier, Copyright 1982. PKA: primary knock-on atom; SKA: secondary knock-on atoms; cross (×): displaced atoms; filled circles (•): vacant lattice sites.

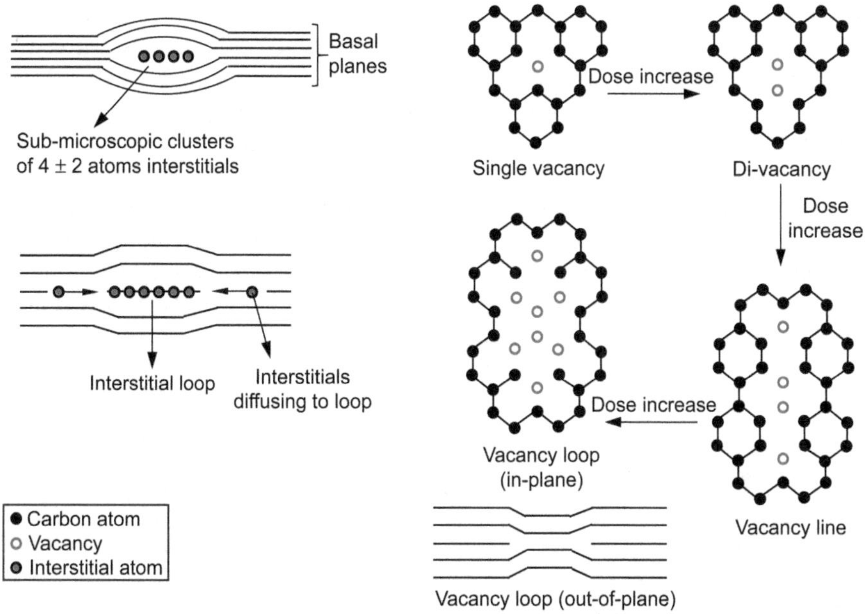

Figure 1.21 The evolution of graphite crystal lattice defects including interstitials, vacancies, loops and extra planes with irradiation. Reproduced from ref. 167, with permission from Elsevier, Copyright 1982.

vacancies are in general immobile especially for irradiation temperature <300 °C and only become more mobile when above 600 °C.[189,323] It has been found that it is easier for interstitials to migrate across basal planes than within basal planes with the opposite trend for vacancies.[189,322,324]

Prismatic edge dislocations can form as a result of the coalescence of interstitials or vacancies, as discussed in Section 1.5.4. The generally accepted irradiation induced defect formation and evolution processes in nuclear graphite are illustrated in Figure 1.21. Firstly, interstitial carbon atoms tend to diffuse between basal planes in two planar dimensions and to coalesce to form interstitial lines of 2, 3 or 4 carbon atoms (sub-microscopic groups of no more than 4 ± 2 atoms each). Secondly, the interstitial lines may either grow in size to form dislocation loops by attracting more diffusing interstitials creating a new atomic plane, or be destroyed by irradiating energetic particles or knock-on carbon atoms (irradiation annealing). Lastly, for vacancies, single point defect vacancies and small groups of vacant sites could be formed and further collapse into vacancy lines and loops, with increasing dose, mainly within basal planes.[167,189,241,252,322,323,325,326]

1.9.2.2 *Crystal and Bulk Dimensional Changes with Irradiation*

With irradiation, the graphite crystal lattice dimensions and properties can be significantly modified. Specifically, interstitials trapped between basal planes could lead to lattice spacing expansion along the *c*-axis, and the migration and coalescence of in-plane vacancies could cause shrinkage along the *a*-axes.[189,319,323,327,328] The lower the irradiation temperature, the greater the crystal dimensional change rate for both *c*- and *a*-plane spacings.[167,168,323,327] The crystal dimensional changes in highly oriented pyrolytic graphite (HOPG) with increasing neutron fluence from 150 °C to 650 °C are shown in Figure 1.22.

It is worth noting that more recent evidence from modelling and experimental work has indicated that other processes, such as breaking, bending and curling of basal planes,[329–331] as well as buckling and ruck and tuck could occur during irradiation.[332–336] Kink band formation[337–339] and a 'ripplocation'

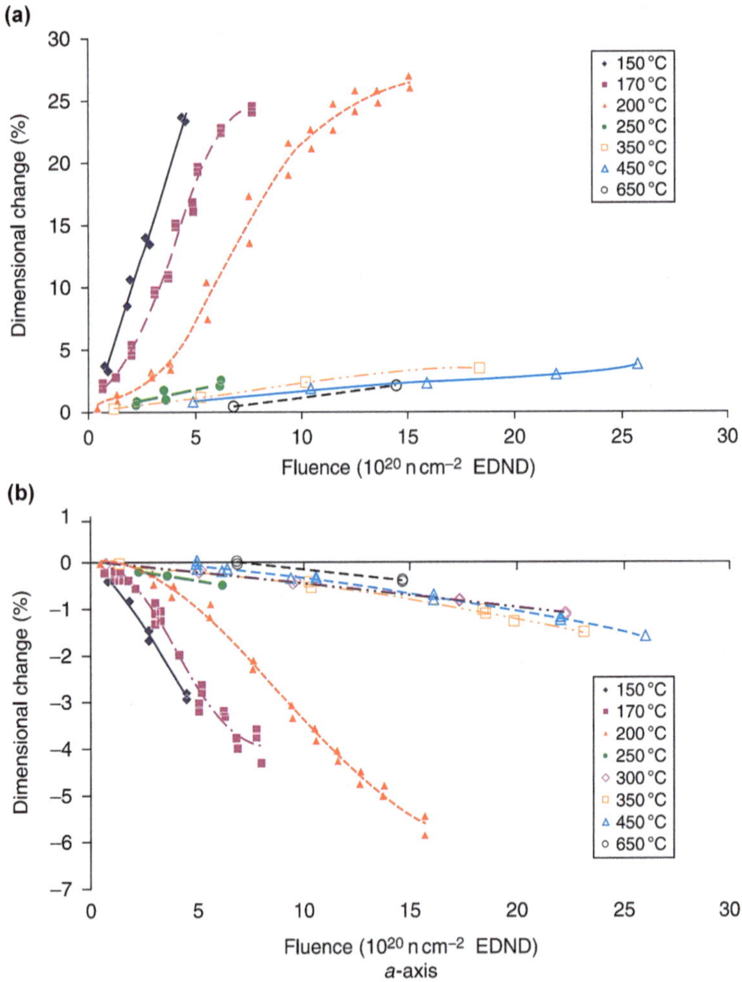

Figure 1.22 Crystal dimensional changes with increasing neutron fluence and temperature in HOPG along (a) *c*-axis and (b) *a*-axis. Reproduced from ref. 168, with permission from Elsevier, Copyright 2020.

model have also been identified as one of the deformation mechanisms of graphite basal planes with irradiation and stress.[340–342]

Irradiation induces bulk dimensional change in nuclear graphite at service temperatures, and it often exhibits shrinkage first, and reaches a 'turn-around' point before volume swelling occurs.[168,189,194,252,322,343] Some results for Gilsocarbon graphite are shown in Figure 1.23. It can be seen that at lower temperatures, *e.g.* 430 °C and 600 °C, the volume shrinkage can be as high as 3%; while at elevated temperatures, 940 °C, 1240 °C and 1430 °C the amount of volume shrinkage is lower at ~2%. The turnaround point is also shifted to lower fluence at these higher temperatures as shown in Figure 1.23.[191] The initial bulk volume shrinkage is usually attributed to the closure of basal plane nanocracks accommodating *c*-plane expansion, therefore the net bulk volume shrinks as a result of *a*-axis contraction.[262,272,344] In addition, Hall *et al.*[267] demonstrated that *c*-axis expansion could also be accommodated by the closure of large lenticular cracks (at hundred micrometre scale) in the Gilsocarbon graphite filler particles, contributing to the bulk dimensional change and affecting the 'turn-around' behaviour of nuclear graphite. Faster 'turn-around' at higher irradiation temperatures was attributed to more accommodation porosity being taken up by thermal expansion in their work.

It is generally believed that the bulk volume expansion starts after the nanoscale basal plane cracks are filled. With further irradiation beyond the turn-around point, the volume expansion creates new voids, which will eventually lead to the structural disintegration of nuclear graphite. It is worth noting that at lower irradiation temperatures, there tends to be an immediate bulk dimensional swelling, as shown in Figure 1.24, where Gilsocarbon swelled at 150 °C. The reason for this remains unclear.

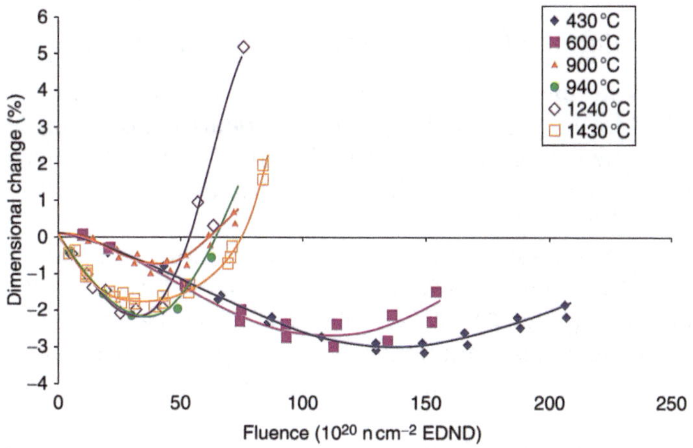

Figure 1.23 Bulk dimensional change curves for near-isotropic, medium-fine grained Gilsocarbon graphite. Reproduced from ref. 168, with permission from Elsevier, Copyright 2020.

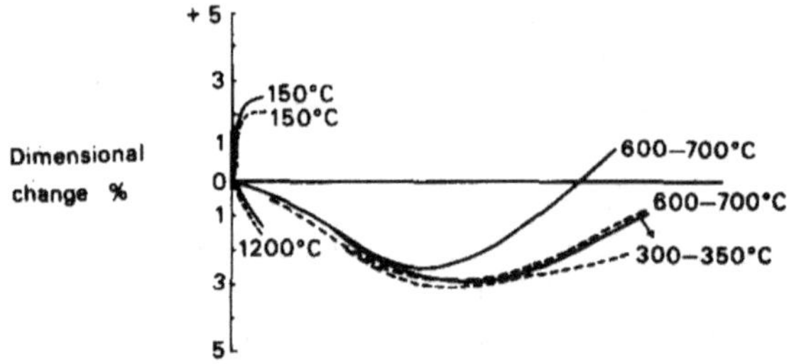

Figure 1.24 Dimensional changes of fast-neutron irradiated Gilsocarbon graphite, showing the initial bulk dimensional swelling at low temperature (150 °C) low dose stage. Original horizontal axis is missing but for reference, the 'turn-around' dose at ~600–700 °C is ~110 × 10^{20} n cm^{-2} Equivalent DIDO Nickel Dose (EDND). Reproduced from ref. 167, with permission from Elsevier, Copyright 1982.

The exact amount of volume change with neutron fluence and temperature varies with different grades of nuclear graphite. For example, for coarser-grained extruded Pile Grade A (PGA) graphite, the bulk dimensional change is different for with grain (WG) and against grain (AG) direction,[168,345] as shown in Figure 1.25. When irradiated by neutrons below 250 °C, PGA graphite directly expands along the AG direction (perpendicular to extrusion); but it shrinks first along the same direction when irradiated above 300 °C. However, for the WG direction (parallel to extrusion) it showed bulk-volume shrinkage at all temperatures below 650 °C. The magnitude of bulk-dimensional shrinkage decreases with increasing temperature.[168,345]

1.9.2.3 *Evolution of Bulk Properties with Irradiation*

Nuclear graphite thermomechanical properties will be modified by irradiation.[224,242,346,347] In general, irradiation introduces more 'brittleness' to nuclear graphite before the 'turn-around' point.[189,195,239,348–351] Both Young's modulus and strength usually increase significantly with increasing neutron fluence.[168,189,239,252,256,308,345]

An example based on Gilsocarbon graphite is shown in Figure 1.26. Specifically, an initial increase in Young's modulus is observed until it reaches a plateau at a slower rate.[168,189] This is followed by a decrease in modulus when dimensional swelling is dominating.[189,308] It has been shown that lower irradiation temperature leads to more rapid and significant increase in Young's modulus.[168]

Nuclear graphite strength usually increases with neutron fluence before the 'turn-around' dose[218,219,248,310,352,353] with similar mechanisms to the increase in Young's modulus.[202,248] Before the 'turn-around' point, Young's modulus and strength have historically been related by the Griffith fracture

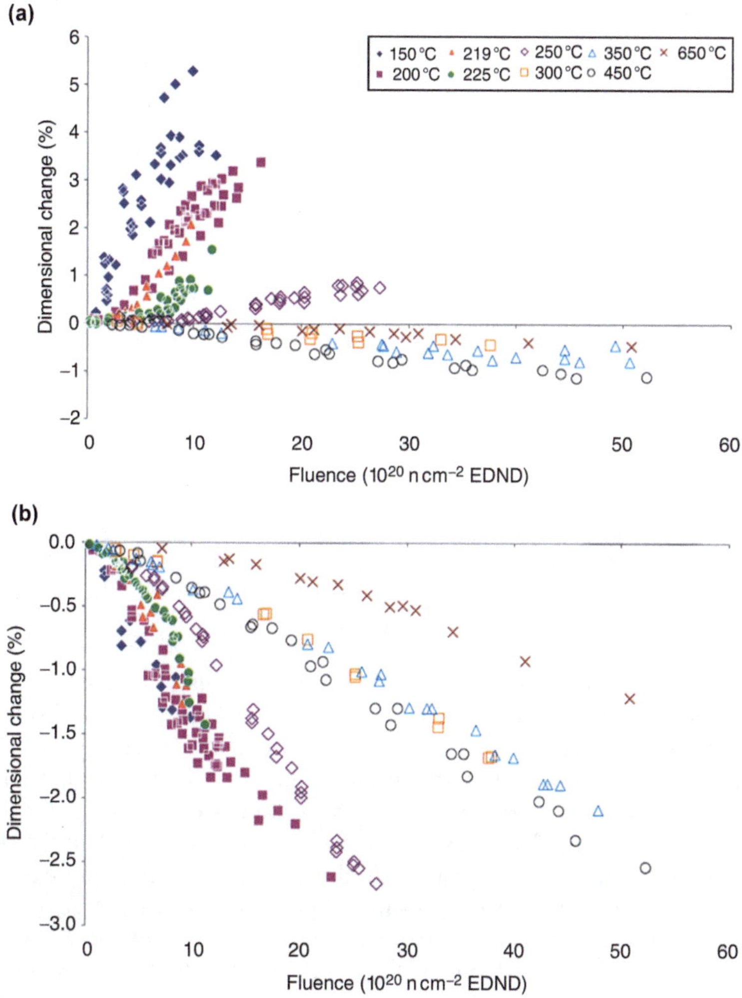

Figure 1.25 Bulk dimensional change in extruded PGA graphite neutron irradiated to low and medium fluence. (a) Perpendicular to extrusion (AG) and, (b) parallel to extrusion (WG). Reproduced from ref. 168, with permission from Elsevier, Copyright 2020.

theory,[354] which is expressed in eqn (1.1), assuming the surface energy per unit area and crack length is invariant to irradiation:

$$\sigma_{\rm f} = \sqrt{\frac{2\gamma E}{\pi c}} \tag{1.1}$$

and therefore,

$$\frac{\sigma}{\sigma_0} = \sqrt{\frac{E}{E_0}} \tag{1.2}$$

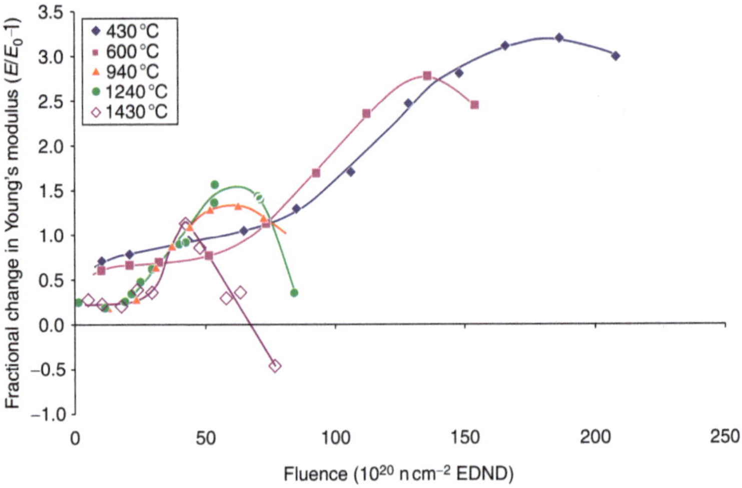

Figure 1.26 Young's modulus change in near-isotropic, medium-grained Gilsocarbon graphite, as a function of fast neutron fluence over a range of temperatures. Reproduced from ref. 168, with permission from Elsevier, Copyright 2020.

where σ_f is the failure strength, E is the Young's modulus, γ is the surface energy per unit area and c is half of a crack length. In eqn (1.2) σ/σ_0 is the ratio of irradiated to un-irradiated strength and E/E_0 is the ratio of irradiated to un-irradiated Young's modulus. This relationship indicates strength increase is directly proportional to modulus increase but it is still subject to discussions.[168,202,355,356] For instance, Marsden *et al.*[356] examined a wide range of strength and modulus data from different grades of graphite and suggested that this square root relationship has only limited use and a value of $m > 0.5$ is more suitable (including $m = 1$ for a linear relationship for oxidised graphite).[352,356] Note that the square root relationship presented in eqn (1.2) is deemed invalid for fluences beyond the 'turn-around' point. For instance, Ishiyama *et al.* showed data on IG-110 graphite[219] and Campbell *et al.* on G347A graphite[189,308] with large deviations from the analytical prediction. This was attributed to a dramatic change in either fracture toughness or critical flaw size that governs the strength at high irradiation doses (Figure 1.27).[189,310]

1.9.2.4 Evolution of Thermal Properties with Irradiation

Heat transfer in graphite crystals mainly occurs by lattice vibration dominated by acoustic phonons above 100 K.[127,357,358] Acoustic phonons are easily scattered by phonons, grain boundaries, impurities, microstructural defects and so on.[127,357,358] A general description of thermal resistivity in graphite can be described as following:[359]

$$K(T) = \beta\left(\frac{1}{K_U} + \frac{1}{K_{GB}} + \frac{1}{K_D}\right)^{-1} \quad (1.3)$$

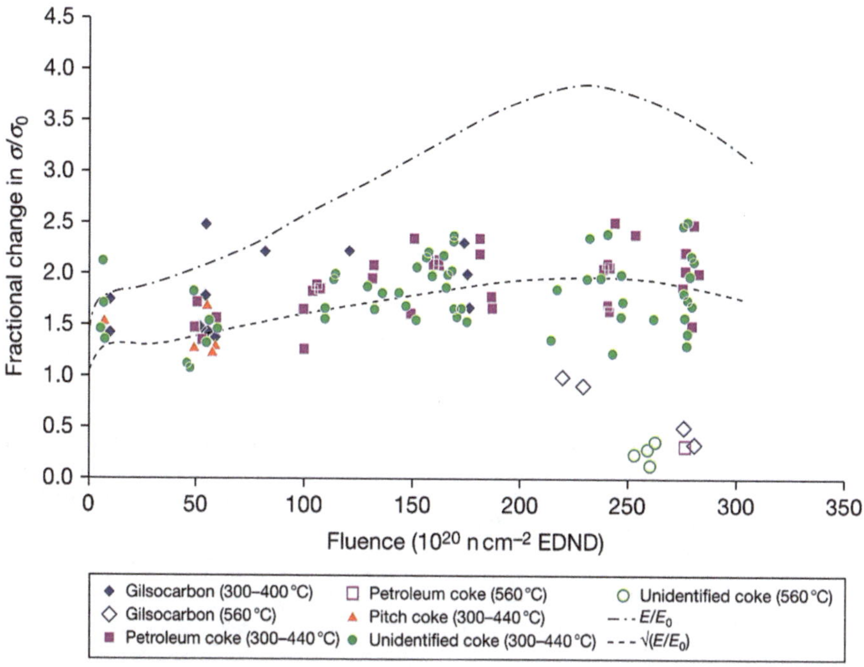

Figure 1.27 Strength change compared to Young's modulus change in Gilsocarbon graphite and other un-identified graphite grades. Due to the scatter in data points, neither square root nor a power law relationship is strictly supported. It was argued by Marsden *et al.* that a direct relationship might be appropriate.[356] Reproduced from ref. 168, with permission from Elsevier, Copyright 2020.

where $K(T)$ is the total thermal conductivity, which is a function of temperature; β is the coefficient representing the contributions from various factors; K_U, K_{GB} and K_D, represents the thermal conductivity contribution from Umklapp scattering (phonon–phonon scattering), grain-boundary scattering and defect scattering, respectively. Introduction of defects into graphite crystal lattice can therefore reduce thermal conductivity. A detailed discussion of the mechanisms responsible for thermal conductivity in graphite can be found in ref. 168, 170, 189, 127, 202, 252, 360–362.

In general, thermal conductivity values of coarser-grained graphite show an initial decrease at low neutron fluences and this was attributed to the formation of interstitial and vacancy defects during irradiation (see example in Figure 1.28 for Gilsocarbon graphite). After the initial decrease in thermal conductivity, a plateau is reached followed by further decrease in thermal conductivity (or increase in thermal resistivity) at high doses. The last stage is attributed to structural disintegration as a result of crystal dimensional swelling due to formation of new cracks/pores.[168,170,173,189,194,202,308,362,363] Note that for fine-grained graphite materials, see example shown in Figure 1.29 for a G347A graphite, there is usually a continued decrease in thermal conductivity (increase in thermal resistivity).[308]

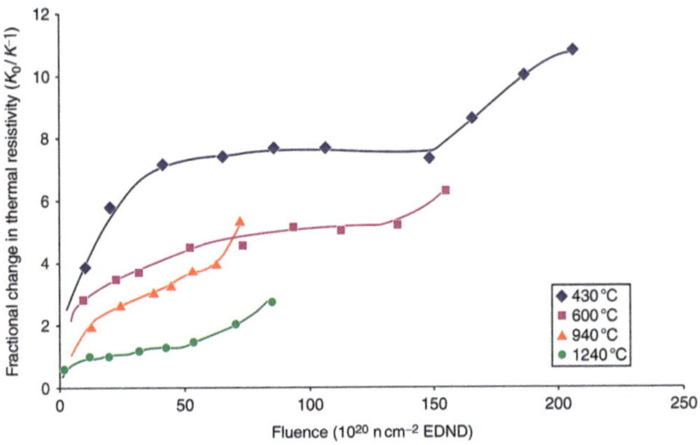

Figure 1.28 Thermal resistivity change in Gilsocarbon graphite due to fast neutron irradiation, K_0 and K are irradiated and un-irradiated thermal conductivity respectively. Reproduced from ref. 168, with permission from Elsevier, Copyright 2020.

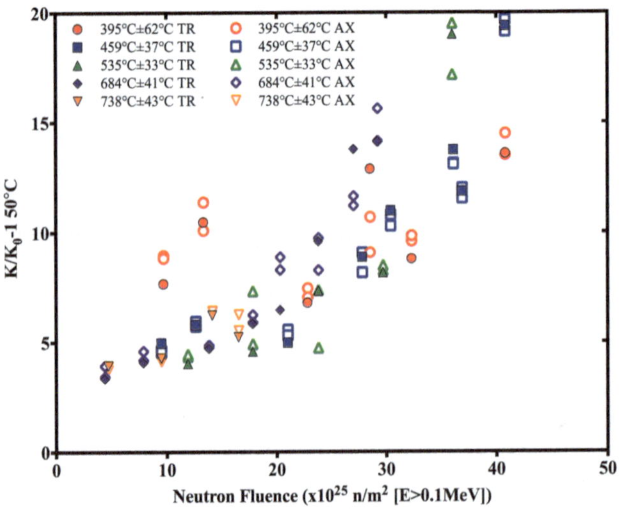

Figure 1.29 Thermal resistivity change in fine-grained G347A graphite as a function of fast neutron irradiation up to ~30 dpa. k_0 is the pristine thermal conductivity and k is irradiated thermal conductivity. TR (transverse) and AX (axial) stand for graphite billet perpendicular and parallel to gravitational force direction during isostatic moulding respectively. Reproduced from ref. 310, with permission from Elsevier, Copyright 2016.

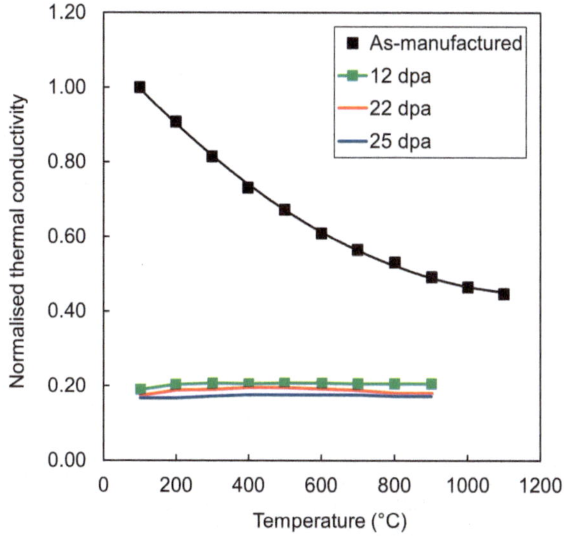

Figure 1.30 Impact of measurement temperature on the thermal conductivity values. Data shown are from IG-110 graphite neutron irradiated at 600 °C to 12 dpa, 22 dpa and 25 dpa. The *y*-axis is normalised to thermal conductivity measured at 100 °C. Pristine thermal conductivity of IG-110 at 100 °C is ~140 W m^{-1} K^{-1}. Reproduced from ref. 219, with permission from Elsevier, Copyright 1996.

The temperature dependence of thermal conductivity in samples irradiated to high neutron fluences will change with irradiation, as shown in Figure 1.30 using IG110 graphite as an example. When measured at increasing temperatures, the thermal conductivity of as-manufactured graphite decreases from 100 °C to 1100 °C. The unirradiated thermal conductivity value is ~140 W m^{-1} K^{-1} at 100 °C, and this value decreased to 20% after irradiation to 12 dpa (displacement per atom). An increase in measurement temperature did not change the thermal conductivity values of the irradiated samples.[219] This implies that thermal conductivity measured in as-manufactured graphite cannot represent the values of irradiated materials.[364]

Lastly, neutron irradiation affects the bulk coefficient of thermal expansion (CTE) of nuclear graphite.[168,170,189,202,242,252] Several examples are shown in Figure 1.31 (Gilsocarbon graphite), Figure 1.32 (GraphNOL N3M graphite) and Figure 1.33 (G347A graphite). For Gilsocarbon graphite, the general trend showed a decrease with increasing neutron fluence. At 940 °C and 1240 °C, the CTE decreased to about 1×10^{-6} K^{-1} at about 10 dpa. The gradient of change become smaller at lower irradiation temperatures (*e.g.* 430 °C and 600 °C). At higher doses, there seems to be a tailing off plateau region where the CTE values remain unchanged with increasing neutron fluences, Figure 1.32. A similar trend applies to GraphNOL N3M graphite and G347A

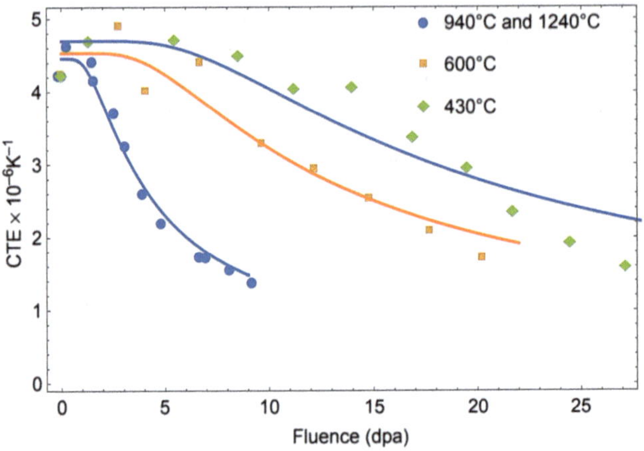

Figure 1.31 Polycrystalline bulk CTE change in Gilsocarbon graphite due to fast neutron irradiation at 430–1240 °C. Reproduced from ref. 202, with permission from Elsevier, Copyright 2017.

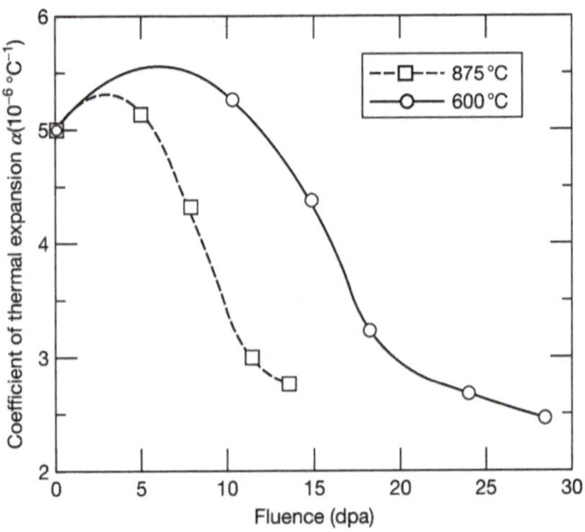

Figure 1.32 Bulk CTE change in GraphNOL N3M graphite due to fast neutron irradiation at 600 °C and 875 °C. Reproduced from ref. 365 from original data in ref. 366, with permission from Elsevier, Copyright 2012.

graphite except there is an obvious increase of bulk CTE at lower fluences, Figures 1.32 and 1.33. More data of various grades of graphite can be found in ref. 168, 170, 189, 202, 242 and 252.

The decrease in CTE is generally attributed to the formation of new cracks and pores,[168,189,202] but Marsden *et al.* pointed out that there is no direct evidence to support this hypothesis. There is also no clear understanding of the correlation between the plateauing of CTE at high doses to the 'turn-around' point.[189]

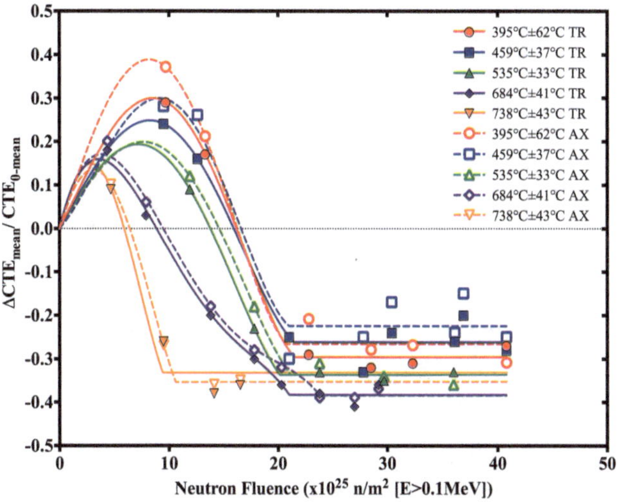

Figure 1.33 Bulk CTE change relative to un-irradiated state in fine-grained G347A graphite due to fast neutron irradiation up to ~30 dpa at a range of irradiation temperatures. Irradiated CTEs are measured at the corresponding irradiation temperatures. TR (transverse) and AX (axial) stand for graphite billet perpendicular and parallel to gravitational force direction during isostatic moulding respectively. Reproduced from ref. 310, with permission from Elsevier, Copyright 2016.

1.10 Graphite in Lithium-ion Batteries

Lithium-ion batteries (LIBs) are ubiquitous in today's world, powering everything from mobile phones and laptops to electric vehicles and satellites. The use of graphite anodes in these batteries is now one of the material's fastest-growing applications, so it is appropriate to include a discussion of LIBs in this chapter. This section begins with a brief description of the lithium-ion battery, followed by a summary of the research which led to the commercialisation of LIBs. Some of the limitations of graphite anodes are then discussed.

1.10.1 Operation of the Lithium-ion Battery

Figure 1.34 illustrates the essential components of a lithium-ion battery. The cathode is a layered transition metal oxide containing lithium, most commonly lithium cobalt oxide ($LiCoO_2$), while graphite is usually the anode material. The electrolyte is a non-aqueous solution, often lithium hexafluorophosphate ($LiPF_6$) dissolved in ethylene carbonate. The polyolefin separator provides a barrier between the cathode and anode while enabling the exchange of Li^+ ions. During charging, lithium ions flow from the cathode through the electrolyte and then intercalate into the graphite, while electrons flow round the external circuit to the anode. When a charged battery is attached to a load, electrons flow from the anode to the cathode, supplying power to the load and the Li^+ ions flow back through the electrolyte. The movement of Li^+

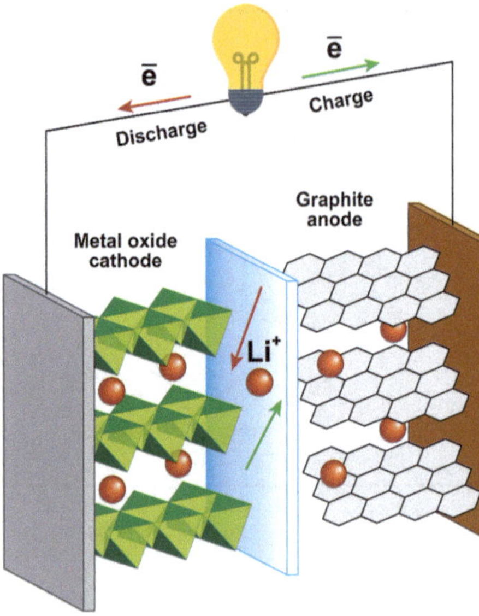

Figure 1.34 Schematic diagram of lithium-ion battery. Reproduced from ref. 369, with permission from MDPI.

ions back and forth between two electrodes led to the devices being dubbed 'rocking-chair' batteries, a concept introduced by M. Armand.[368]

1.10.2 Development of the Lithium-ion Battery

Most of the advances which led to the development of commercial lithium-ion batteries were made by Whittingham, Goodenough and Yoshino, who were jointly awarded the Nobel Prize for Chemistry in 2019. It was Whittingham, working at Exxon in the US in the mid-1970s who demonstrated the first rechargeable Li-ion battery.[370] The battery employed layered titanium(IV) sulfide as the cathode with an Li metal anode and was commercialised by Exxon in the late 1970s. Ultimately, however, the battery was unsuccessful, partly due to dendrite growth on the anode. Goodenough's main contribution was to develop $LiCoO_2$ as a cathode material.[371] Once again this did not lead to a commercial battery. Ultimately, it was work by Yoshino *et al.*, in Japan, that led to the production of the first commercial LIBs in 1991. These batteries initially used coke, as the anode material, although this was later replaced by graphite. Today, as noted above, almost all LIBs employ graphite anodes. Both synthetic and natural graphite are used. Synthetic graphite anodes usually display superior performance, but natural graphite has the advantage of lower cost.[372] Whether synthetic or natural is used, the graphite is normally mechanically processed into spheroidal or 'potato-shaped' particles to allow for more efficient packing in the electrode.[373] Typical spheroidised natural graphite particles are shown in Figure 1.35. The history of the lithium-ion battery is discussed in detail in the literature.[374–376]

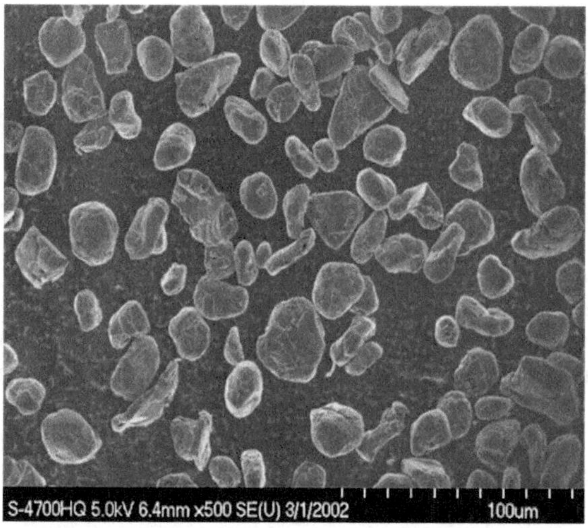

Figure 1.35 SEM image of potato-shaped particles of natural graphite. Reproduced from ref. 377, with permission from Elsevier, Copyright 2003.

1.10.3 Problems with Graphite Anodes

Although lithium-ion batteries have proved to be an extraordinarily successful technology, there remains considerable scope for improving their performance. In particular, graphite anodes have quite serious limitations, specifically with respect to the speed of charging. Some of the issues relate to the solid electrolyte interphase which is generated on the graphite anode during the first few charging cycles.[378] This is a layer, approximately 100 nm thick, which forms when oxidation–reduction reactions occur between electrons, the electrolyte, and the Li^+ ions. It is made up of a mixture of inorganic Li salts, Li alkyl carbonates and other components. The layer plays the crucial role of protecting the Li-intercalated graphite from further reaction with the electrolyte. A consequence of the incorporation of Li^+ ions into the interphase layer is that the capacity of the battery falls by about 10%.[379] Under some circumstances, further growth of the interphase layer may occur, ultimately resulting in battery failure.[380] Another problem can arise due to the deposition of dendritic lithium metal on the graphite anode surface during charging.[381] This issue can be especially acute during rapid charging.

The problems with graphite anodes have prompted a search for alternatives. One of the most promising candidates is hard carbon (*i.e.* non-graphitising carbon[23–25]). Cao *et al.* compared the performances of graphite, soft carbons and hard carbons as LIB anodes and found that hard carbons displayed the highest rate capability and higher charge capacity.[382] However, significant issues need to be overcome before hard carbons become widely used in LIBs.[383] Silicon or SiO_2 also appear to have potential as anode materials[384] and there is renewed interest in lithium-metal anodes of the kind originally used by Whittingham in his pioneering research. There has also been extensive research on the use of graphene and related materials in LIBs, as discussed in Chapter 2, although here the high cost of these materials may limit commercial viability. It seems certain, therefore, that graphite anodes will continue to be used in lithium-ion batteries for many years to come.[385–387]

1.11 Carbon Fibres

The weakness of its interplanar forces means that graphite is of little value as a structural material. However, the sp^2 carbon–carbon bond in the basal plane is among the strongest of all chemical bonds, and this can be exploited through the use of fibres in which all the basal planes run roughly parallel to the axis. Carbon fibres, produced by heat-treating carbonaceous precursors, are one of the most commercially important carbon materials, used in everything from golf clubs to wind turbine blades and aircraft. In these applications the fibres are almost always incorporated in a matrix to produce composite materials. Carbon fibre composites are discussed in Section 1.11.4. Firstly, a brief summary is given of the history, production and properties of carbon fibres. For those who require a more detailed discussion of carbon fibres and their composites than is possible here, several excellent books are available (*e.g.* ref. 388–390).

1.11.1 History

The first carbon fibres were made by Joseph Swan in the UK and Thomas Edison in the USA in the late 19th century.[391,392] The aim was not to produce high strength fibres, but to make filaments for incandescent electric lamps. Edison found that carbonised cotton thread gave the best results, while Swan prepared filaments by carbonising threads extruded from cellulose. Carbon filaments were widely used in electric lights but were eventually replaced by the much longer-lasting tungsten filaments. Serious research into carbon fibres as structural materials began in the 1950s when the advent of jet propulsion led to a demand for high stiffness, low density, materials. This research was given an important boost by the discovery of graphite whiskers in 1958.[393] These whiskers, which were produced in a high-pressure carbon arc, had a scroll-like structure, as illustrated schematically in Figure 1.36. Their diameters ranged from less than 1 μm to >5 μm, with typical lengths of 1–3 cm. Tensile strengths of up to 20 GPa and Young's moduli of ~800 GPa were reported for the whiskers. Scaling up the arc-discharge method for commercial production was not feasible, but the outstanding properties of graphite whiskers demonstrated the enormous potential of graphitic carbon fibres.

Important early work on carbon fibre production was carried out at Union Carbide in the US in the late 1950s. The initial precursor in this work was rayon yarn, which was carbonised and then graphitised at high temperatures, producing fibres in the approximate size range 5–10 μm (much smaller

Figure 1.36 Sketch of scroll-like structure of graphite whiskers. Reproduced from ref. 393 with permission from AIP Publishing.

than the carbon filaments prepared by Swan and Edison) The fibres had tensile strengths of around 1.25 GPa and moduli of *ca.* 170 GPa,[388] and were commercialised in 1964. Rayon-derived carbon fibres were ultimately largely superseded by fibres produced from polyacrylonitrile (PAN). The use of PAN to make carbon fibres was pioneered in Japan[394] and the UK[395] in the 1960s. The PAN-derived fibres had moduli of over 400 GPa.[395]

Today approximately 92% of carbon fibre is produced from PAN, with most of the remainder being made from petroleum pitch. Some of the key research on the production of carbon fibres from pitch was once again carried out at Union Carbide.[396]

1.11.2 Production

Global carbon fibre production currently stands at around 200 000 metric tonnes per year, with the leading producers being Toray Industries and Mitsubishi (Japan), Jilin Tangu (China) and Cytec Industries (USA). A brief outline of the techniques used to produce carbon fibres from PAN and pitch is now given; more details can be found in the literature.[388–390]

Polyacrylonitrile is almost always co-polymerised with a small amount of another monomer, such as vinyl acetate or methyl acrylate. This lowers the glass-transition temperature from ~120 °C to <100 °C, which has the effect of improving the spinning and stretching characteristics. The production process usually involves the following five stages:

1. Spinning the polymer to form a fibre
2. Stretching
3. Stabilisation
4. Carbonisation in an inert atmosphere at ~1600 °C
5. Graphitisation at temperatures up to 3000 °C.

Although it is thermoplastic, PAN degrades before melting under normal conditions. Therefore, in order to spin the polymer into a fibre it is necessary to dissolve it in a solvent, such as dimethyl formamide. A technique known as wet spinning is then normally used to form the fibres. This involves extruding the viscous solution through holes in a spinneret. The spinneret would typically have several thousand holes. The wet-spun fibre is passed through rollers, then dried and stretched. The stretching process promotes the orienting of the PAN polymer. The strands of fibres, known as tows, are then passed into an oven where they are oxidised in air, typically at ~220 °C, under tension. Figure 1.37 presents a simplified view of the reactions that occur during this stabilisation stage, namely cyclisation and oxidation. Carbonisation is then carried out by heating to temperatures between 1000 °C and 1500 °C at a heating rate of ~20 °C min^{-1}, in an inert atmosphere. The final stage involves heating the fibres to very high temperatures, up to 3000

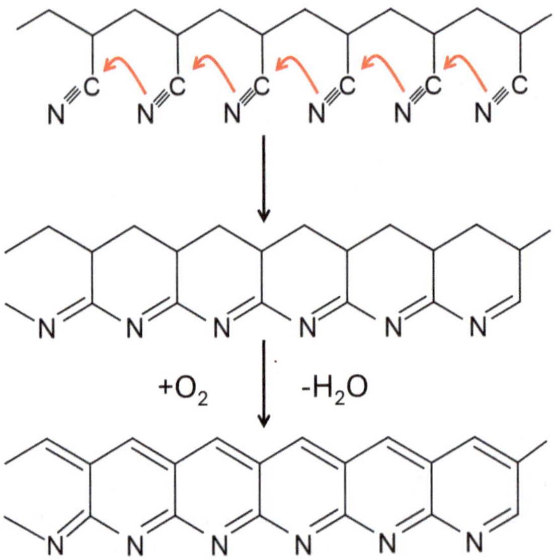

Figure 1.37 Illustration of cyclisation and oxidative carbonisation of polyacrylonitrile (PAN). Adapted from https://en.wikipedia.org/wiki/Carbon_fibers, under the terms of the CC-BY SA 4.0 deed https://creativecommons.org/licenses/by-sa/4.0/.[397]

°C. This results in removal of the nitrogen atoms and the formation of a graphitic structure. However, large-scale graphitisation does not occur since carbon derived from PAN is non-graphitising (see below).

Producing carbon fibres from pitch is somewhat simpler than producing them from PAN in that, liquid crystalline pitch can be directly formed into fine filaments by melt spinning. As the melt cools, the filaments are wound onto a spool. They are then oxidised to convert them from the thermoplastic to the thermoset state to stabilise them, then carbonised at around 1000 °C and graphitised at 3000 °C. Unlike PAN, pitch is a graphitising carbon, so pitch-derived carbon fibres have a higher degree of graphitisation, and higher modulus than PAN-derived fibres (see Section 1.11.3).

In recent years there has been growing interest in producing carbon fibres from electrospun polymers. This topic will not be covered here, but several recent reviews are available (*e.g.* ref. 398 and 399). Catalytically-grown carbon nanofibres also lie outside the scope of the current chapter. Ref. 50, 400 and 401 give comprehensive reviews of the subject.

1.11.3 Properties and Structure

The properties of carbon fibres depend to a large extent on the precursors used for their preparation. Thus PAN-derived fibres tend to have the highest strength, while pitch-derived fibres have the highest modulus. Preparation temperature is also important in determining the mechanical properties of the fibres, with higher temperatures leading to better properties. The differences between the mechanical properties of fibres made from PAN and fibres made from pitch result from the fact that PAN-derived carbon is non-graphitising while pitch-derived carbon is graphitising, as already noted. Thus, pitch-based fibres develop much greater crystallinity than PAN-based ones with high-temperature heat treatment, as illustrated in Figure 1.38.[402]

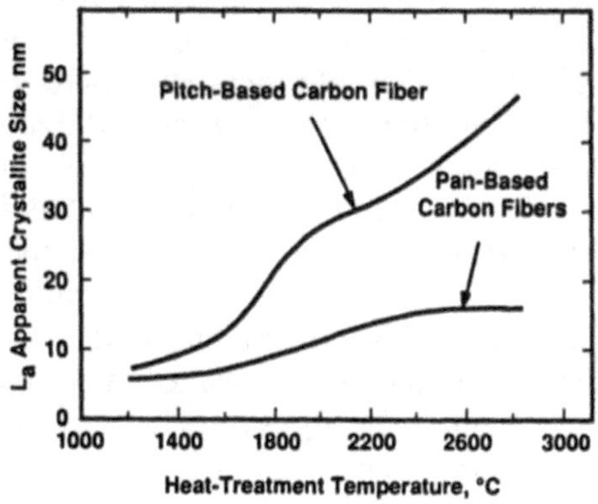

Figure 1.38 Apparent crystallite size *versus* heat-treatment temperature for PAN-based and pitch-based carbon fibres. Reproduced from ref. 402, with permission from Elsevier, Copyright 1989.

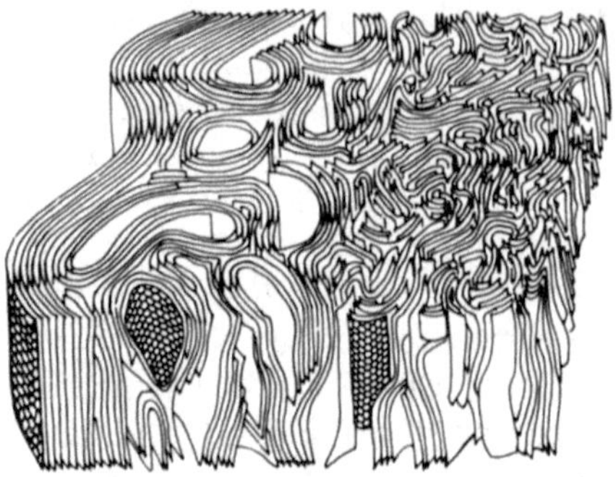

Figure 1.39 Schematic representation of structure in PAN-based carbon fibres. Reproduced from ref. 403 with permission from AIP Publishing.

Electron microscopy and X-ray diffraction studies have shown that PAN-derived fibres only contain small graphitic domains, and have many voids, giving them a low density compared with pitch-derived fibres. A model for the structure of PAN-based carbon fibres put forward by Johnson in 1987 is shown in Figure 1.39.[403] The fact that ex-PAN fibres lack an extended graphitic structure makes them relatively insensitive to flaws, and can have exceptionally high strengths, typically 2–7 GPa.[388–390] Their Young's moduli are generally in the range 250–300 GPa. Pitch-derived fibres have a much more extensive graphitic structure and can have moduli similar to those of graphite whiskers, *i.e.* 800 GPa. This approaches the maximum theoretical modulus for a carbon fibre, namely 1060 GPa.

1.11.4 Carbon Fibre Composites

1.11.4.1 Manufacture of Intermediate Reinforcement Forms

The earliest commercially available carbon fibres were produced in relatively short lengths of ~1 m due to the limitations of the initial manufacturing processes. As the development of processes to deliver continuous fibres greatly increased the availability of the materials, this posed challenges in the manufacture of structures from these fibres.[404] The fibres are supplied in the form of tows of continuous fibres, with 1000–24 000 fibres (Figure 1.40) within the tow for the highest performance applications, with much larger tows being available if the highest mechanical performance is not required. To encourage bonding between the fibres and matrix resins, it is usual to treat the surface of the fibres, and most commercially supplied tows of fibre also have some sizing to protect the fibres from handling damage. In order to facilitate the manufacture of structures more complex than rods and tubes, the tows of fibre need to be assembled into sheets that can be processed into more complex shapes. These sheets are known as intermediate materials. The first such material to become available was known as unidirectional prepreg (short for pre-impregnated). Multiple tows of fibre were passed side by

Figure 1.40 Spools of carbon fibre tows produced by the Mitsubishi Chemical Corporation.[406]

side through a process whereby the fibres became embedded in an uncured or slightly cured resin (generally epoxy) matrix, and were supplied between two sheets of paper or polymer film to protect the tacky resin surface. These material forms could be used to make laminates of the highest quality, as well as lightly curved forms, but were very difficult to use on complex geometries. To overcome this limitation, efforts were put into weaving carbon fibres and today, woven carbon fibre reinforcements are probably the most commonly used. These woven reinforcements, and the increasingly common non-woven reinforcements, are supplied in both dry and pre-impregnated forms depending on the process to be used for manufacturing structures from them. In addition to the reinforcements using continuous fibres there are a range of short fibre reinforcements available, but the continuous fibre variants dominate the market. Ref. 405 gives a much fuller description of the range of intermediate materials available.

1.11.4.2 Manufacture of Carbon Fibre Composite Structure

The process of manufacturing carbon fibre composite structures has essentially three stages. The first is collation (or lay-up) of the required reinforcements onto a prepared tool surface, the second is the application of heat and pressure to cure a thermosetting matrix or consolidate a thermoplastic matrix, the third is the trimming and finishing operations that turn a moulding into a structure like that in Figure 1.41. In practice the procedure is, of course, very much more complicated.

The early development of composite processing, carried out in the 1950s, involved not carbon fibres but fibres such as glass and asbestos. In the UK, this effort was largely led by the Royal Aircraft Establishment and the Bristol Aeroplane Company,[407,408] which developed the two most important

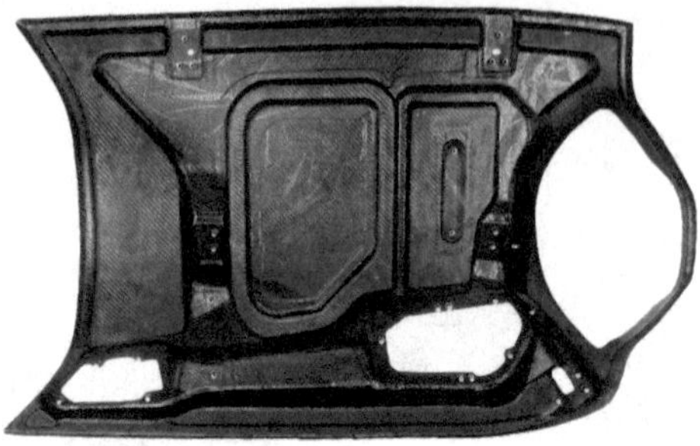

Figure 1.41 Complex autoclave moulding from a supercar door structure, approximately 1 m long. Note all edges have been machined and multiple fasteners have been inserted to support assembly.

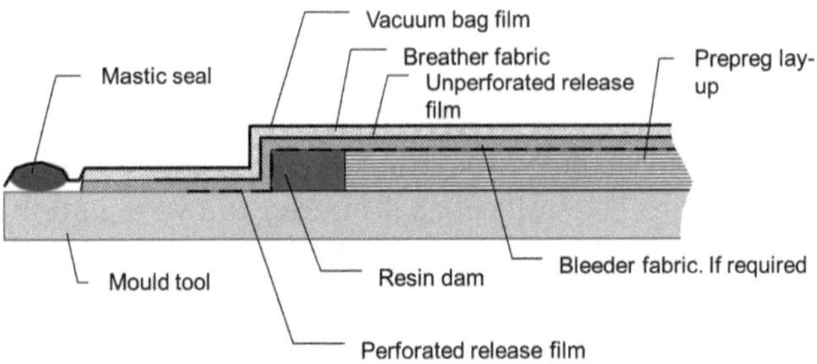

Figure 1.42 Cross-section through the edge of a mould tool, prepared for cure in an autoclave. All the materials shown here apart from the mould tool and the cured prepreg lay-up will be thrown away after use.

processes for high performance composites, autoclave moulding and resin transfer moulding. A wide range of other processes exist,[409,410] but due to limitations of space, will not be considered here.

An autoclave is in essence a heated pressure vessel. Pre-impregnated reinforcement sheets are cut to size and laid onto a suitable tool surface until the required thickness is built up. A non-stick film is then applied to the lay-up surface, a thin polymer film is attached with mastic to the tool surface to form a bag completely enclosing the lay-up and a vacuum is drawn under the bag to extract any air from the lay-up and to act as a separator when pressure (typically 3–7 bar) is applied. The apparatus is illustrated in Figure 1.42.

The autoclave is then taken through a preset programme of temperature and pressure ramps and holds to ensure that the composite is fully cured and of the highest quality in terms of an absence of internal defects. Variants of the process, where an oven replaces the autoclave and the pressure differential is thus limited to 1 bar, have become common to avoid the high capital cost of the autoclave. A critical element of the process is the lay-up activity. Historically this was done by hand and it is still carried out this way for most structures of complex geometry using woven reinforcements,[411] although automated processes[412] have become much more common and are widely used for critical aircraft structures.

The resin transfer moulding (RTM)[413] process, by contrast, uses unimpregnated reinforcements, generally woven or non-woven fabrics which often have a small amount of a binder resin applied to the surface. Whereas a mould in the autoclave is single sided the moulds for RTM are two sided and completely encase the reinforcements controlling all dimensions of the final moulding (Figure 1.43), unlike in the autoclave process where there is limited thickness control. Control of all dimensions allow the manufacture of net-edge components requiring little or no finishing machining (Figure 1.44). Lay-up can be ply-by-ply as in the autoclave, but can also be automated by heating blanks of bound reinforcements to soften the binder and press-forming these into preforms that are handleable and can easily be loaded into the moulding tool. Once the reinforcement has been loaded into the tool, resin can be introduced under pressure to flow through, wet out and encapsulate the reinforcement. To obtain the highest quality, it is necessary to seal the tool and evacuate the tool cavity prior to injection of the resin. This fully tooled version is widely used in the manufacture of structural components for automotive applications, such as the BMW i3 life cell. A variant of the RTM process where the reinforcement is held under an evacuated vacuum

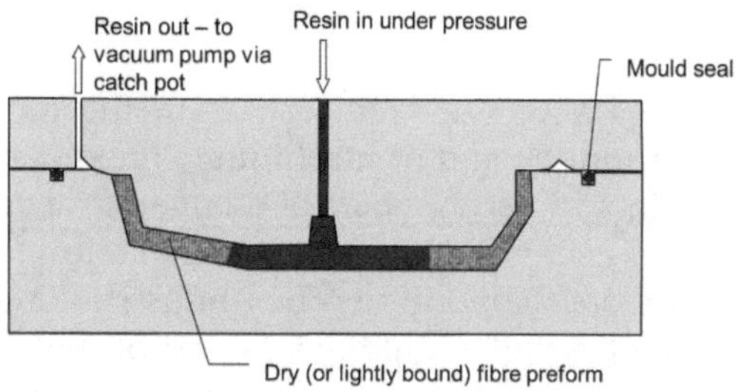

Figure 1.43 Basic layout of the resin transfer moulding process using rigid tooling.

Figure 1.44 Two small complex carbon composite parts made by resin transfer moulding to a net edge without any post-moulding trimming being needed.

bag with resin drawn into the bag by the vacuum is now very widely used especially for large structures, such as yacht hulls and wind turbine blades, and is under study for critical aircraft structures.

Whether the vacuum bag and prepreg or resin infusion families of processing are being used, there are some common problem areas. The first is waste if sheets of reinforcement are being used. When the structure is designed various rules are used. One such rule is that fibres must be applied in multiple orientations to both match the stress fields and avoid any areas of weakness. As with dressmaking a cutting pattern is generated and there is often a very significant waste of reinforcement; up to 40% wastage is by no means unheard of. There is a need to develop processes to recover and reuse this waste as part of the wider targets around carbon fibre recovery and recycling and much of this work is progressing.[414] Automated processes have been developed using strips of unidirectional prepreg to minimise this waste,[412] but at the cost of a much higher intermediate material price. The second problem area is dimensional fidelity. It is usual for the shape of the structure produced to be different from the shape of the tool it was made on. This is largely due to the great difference in thermal expansion coefficient in the plane and through the thickness of the laminate, although other factors, such as aspects of resin cure shrinkage and thermal expansion of the tool, can be important.[415] To an extent these effects are predictable and tooling can be corrected to deliver the correct geometry, but variability in dimensions can be a recurring problem.

Once the mouldings have been removed from the tool, it is generally the case that some trimming and or machining is necessary to finalise the moulding to fit within and become part of a larger structure. In aerospace structures, assembly is generally by the use of mechanical fasteners whereas adhesive bonding is more the norm in other industries. After machining, it is usual to inspect the finished mouldings as part of the quality assurance/quality control system. Inspection will be by a combination of dimensional and visual inspection with additional non-destructive evaluation (generally by ultrasonic probing) required for more critical structures. Due to the combination of variability in the intermediate materials, additive processing with a significant manual input, multistage processing and machining, there are a very significant number of potential defects arising during the manufacturing of composite structures.[416] Great attention to detail is required throughout the design, development and manufacture processes to achieve reliable and high-quality results.

1.11.4.3 Performance of Carbon Fibre Composite Structures

The performance of well-designed and carefully manufactured carbon fibre composites has made them the materials of first choice for a wide range of applications where strength- and stiffness-to-weight ratio are paramount. Mechanical performance data is widely available from the material suppliers in the form of datasheets and typical properties for flat laminates. This data needs to be treated with care as it may not represent the properties achievable in practice for any specific structure and should never be used for design purposes as it represents maximum potential properties and not allowable design properties, which accommodate factors such as the response to environmental conditions, impacts and fatigue loading *etc.* Good design data is much less commonly available but a depository that is excellent in its depth but narrow in its width is maintained in the USA.[417] There are hundreds if not thousands of potential combinations of carbon-fibre type, reinforcement type, and resin type but very good property data is publicly available for only a very few aerospace qualified systems.

When considering the mechanical properties of carbon-fibre composites, it is important to appreciate that the strength and stiffness properties are very dependent on orientation. For a unidirectional laminate, the strength in the direction aligned with the fibres will be of the order of 10 times higher than the strength in the direction aligned at 45° to the fibre direction and 20 times higher than the strength in the direction at 90° to the fibre direction. For complex laminates with multiple fibre directions, the laminate stiffness properties can be estimated quite well using classical laminate theory.[418] For strength properties, this is not the case as multiple additional factors influence the laminate strength.[419] It is commonplace to use laminates, which are balanced in their in-plane properties, and quasi-isotropic laminates (properties essentially equal in all in-plane directions) are perhaps the most commonly used. In this case the through-thickness strength and stiffness will still be very low, in the region of 10% of the in-plane properties. If the through-thickness strength is exceeded, the laminate structure will break down and delaminations will grow between the layers of the structure, which can be very damaging to the overall structural strength.[420] In general the fatigue performance of carbon-fibre composites is very much superior to that of structural light metals such as aluminium.[421] Helicopter rotors use carbon-fibre composites to build heavier rotors than can be safely made in aluminium. This seems counterintuitive for lightweight materials but is a positive feature – as if the engine fails, the kinetic energy carried in the rotor is the only energy available to attempt a controlled landing and can be maximised using carbon composites. Impact on the other hand is more damaging on lightweight composite structures than on metallic structures and the residual strength after an impact often sets the strength available to the structural designer.[422] Structural design of composite structures is more complex than with conventional metals, but the use of appropriate data and analytical methodologies allows robust designs to be produced. Because the composite material's final properties can be significantly affected by the details of the manufacturing processes, the best practice is to use Design for Manufacture concepts to ensure that the designs are fully fit for purpose.

It is important to understand that the non-mechanical properties of carbon-fibre composites are also generally anisotropic and often very different from those of structural metals, such as aluminium or steel.[423] For example carbon-fibre composites will absorb a little moisture and may change shape slightly as a result, but will not corrode. The thermal expansion coefficient of a carbon-fibre laminate will be lower than that of aluminium in the plane of the laminate, but rather higher through the laminate thickness, which can lead to thermally induced stresses in hybrid metal/composite structures. Equally the thermal conductivity of a carbon-fibre laminate is very much lower than that of aluminium or steel, making carbon composite a poor choice for a heat exchanger but excellent if a design calls for a thermal block. In a design environment it is much safer to assume that carbon-fibre composites are different to common metals in all their performance characteristics.

1.11.4.4 Applications of Carbon Fibre Composites

Early applications of carbon fibre composites were largely in the very weight-critical areas of fighter aircraft and space structures. The costs declined as the market capacity grew and composites started to be used in secondary structures on aircraft (structures which will not endanger the aircraft if they fail), especially the leading and trailing edges of wing panels and moving wing structures, such as ailerons and flaps. With increasing confidence and the growth of the industrial capability and capacity, critical aircraft structures, such as pressure bulkheads, inner wing boxes and keel beams, were tackled and have become normal applications on large commercial aircraft even when the majority of the critical structure remains metallic.[424] Currently the majority of the airframe mass in the Boeing 787 and Airbus A350 wide-bodied jets is constructed from carbon-fibre composites. The narrow-bodied jets that form the great bulk of large commercial aircraft have increased their carbon-fibre composites content with each major upgrade. However it is by no means certain that the next generation of these aircraft will follow the A350 and 787, as the build quantities and economics are very different. Alongside military and commercial aircraft, another early adopter of carbon-fibre composites has been the sporting goods industry with multiple applications, such as skis, golf clubs, tennis racquets, bicycle frames and so on, as these applications could tolerate initially high costs.

Currently there are strong markets in aerospace, defence, automotive, marine and renewables (especially ultra-large wind turbines), with growing market potential in the oil and gas, construction and rail sectors.[425] Each of these markets has its own characteristics and shaping new carbon-fibre composite materials, design and manufacturing practices to each market will be necessary as those developed for aerospace applications will usually be completely inappropriate for the emerging applications. Early applications in the automotive sector were in hand-built supercars and race cars, but the technology has developed enough to now be used in high-end sports cars and the demands placed on the structures for battery-powered mass-market cars, present a large potential opportunity if cost-effective manufacturing processes can be demonstrated. At the same time, new classes of carbon-fibre composites with new functionalities, such as structural batteries, are under development[426] and these will trigger new applications.

1.12 Conclusions

In this chapter we have attempted to summarise the most important physical and chemical properties of graphite, as well as describing its structure and the nature of defects that occur in graphite crystals. We have also discussed the occurrence, classification and processing of natural graphite, and the methods used to produce synthetic graphite. As noted at the beginning of the chapter, the applications of graphite date back hundreds, if not thousands, of years. Today there are many more applications than it has been possible to cover in this chapter. For example, graphite is used on a large scale as a refractory material in iron and steel making and other industries, and is also widely used in brushes for electric motors and generators, in electrodes for dry-cell batteries, in resistors and in brake linings. On the whole, these are mature technologies which are no longer the subject of intensive research. For this reason, these applications have not been covered in this chapter. Instead we have focused on three technologically important areas involving graphite, which are still active subjects of research, namely: nuclear power, lithium-ion batteries, and carbon fibres and their composites.

Graphite has been used as a moderator in nuclear reactors since the construction of Chicago Pile 1 in 1942 and, as noted in Section 1.9, is still widely used in today's reactors. It will also be used in some of the planned Generation IV reactors. It is the importance of graphite as a nuclear material that has led to the most intensive programmes of research into its properties. Thus, much of what we know about defects in graphite, discussed in Section 1.5, has been established as a result of nuclear-related research. However there are still unanswered questions concerning the performance of graphite components in the harsh operating conditions of a nuclear reactor. The components will be subject to complex monotonic and/or cyclic loading/stresses,[277,367] which complicate their deformation and fracture behaviours. There is not yet a satisfactory micromechanical model to describe fatigue-crack initiation in nuclear graphite, which emphasises the importance of adopting advanced *in situ* techniques for real-time imaging of the crack tip phenomenon. Further work is also needed to understand graphite deformation under multi-axial stress states. Additionally, irradiation induced dimensional changes including irradiation creep in nuclear graphite is a very important area to investigate. The exact correlation between nanoscale crystallites, Mrozowski cracks, micro-scale pores/microstructures and the resulting bulk dimensional change remains unclear beyond the turn-around point. Temperature and applied stress (tension or compression) play important roles in irradiation creep behaviours of nuclear graphite – graphite with varied raw materials and manufacture conditions will behave differently, which further complicates this topic. Overall, mechanistic understanding of the deformation and fracture of nuclear graphite at high temperature with irradiation is needed to support their safe application in next generation fission reactors.

As noted in Section 1.10, lithium-ion batteries are an essential component of the modern world, and most of these employ graphite anodes.

Graphite has been described as the perfect anode material for LIBs,[385] combining as it does relatively low cost, high-energy density and very long cycle life. This does not mean that there is no room for improvement in graphite's performance, and recent research has explored the possibility of modifying the solid–electrolyte interface through surface modification or electrolyte additives, in order to improve reversible lithium intercalation.[386] The addition of silicon to graphite has also been shown to enhance lithium uptake.[427]

The third major technology to be covered in this chapter is that of carbon fibres and their composites. Here, an important outstanding issue is the lack of commercially viable methods for recovery, recycling and reuse of the composites. This may prove a critical issue for some industries where the use of recycled materials is mandated by law. Most current approaches to the recovery of carbon fibres do so by destruction of the matrix, which is rather wasteful, and recovery of both fibres and matrix would be preferred. Technical solutions are being developed across these requirements, but significant investment will be needed to deliver materials recovery and a circular economy at an industrial scale.[417,428,429]

Acknowledgements

The authors would like to thank the following colleagues for their help and advice in preparing this chapter: Jean-Christophe Charlier, Jean-Marc Leyssale, Hirotsugu Ogi, Ljubisa Radovic, Swati Sharma and Ujjal Tewary.

References

1. V. L. Dumitrescu, *et al.*, in *The Cambridge Ancient History. Vol. III, Part 3*, ed. J. Boardman, *et al.*, Cambridge University Press, 2nd edn, 1982, p. 1.
2. H. Petroski, *The Pencil: A History of Design and Circumstance*, Faber, London, 1989.
3. F. Cirkel, *Graphite: Its Properties, Occurrence, Refining and Uses*, Canada Mines Branch, Ottawa, 1907.
4. E. Ewais, *J. Ceram. Soc. Jpn.*, 2004, **112**, 517.
5. Graphite facts, Government of Canada, 2020, https://www.nrcan.gc.ca/our-natural-resources/minerals-mining/minerals-metals-facts/graphite-facts/24027#L1, accessed 22 January 2024.
6. P. R. Buseck and O. Beyssac, *Elements*, 2014, **10**, 421.
7. G. J. Simandl, S. Paradis and C. Akam, in *Symposium on Critical and Strategic Materials*, British Columbia Geological Survey, 2015, p. 163.
8. G. Erdosh, *Miner. Deposita*, 1970, **5**, 375.
9. P. Touzain, N. Balasooriya, K. Bandaranayake and C. Descolas-Gros, *Can. Mineral.*, 2010, **48**, 1373.
10. J. L. R. Touret, J. M. Huizenga, K. V. W. Kehelpannala and F. Piccoli, *Chem. Geol.*, 2019, **508**, 167.
11. F. J. Luque, J. M. Huizenga, E. Crespo-Feo, H. Wada, L. Ortega and J. F. Barrenechea, *Miner. Deposita*, 2014, **49**, 261.

12. J. Palosaari, R. M. Latonen, J. H. Smått, R. Blomqvist and O. Eklund, *Norw. J. Geol.*, 2016, **96**, 19.
13. Graphite consumption - U.S. Geological Survey, Mineral Commodity Summaries, January 2024, https://pubs.usgs.gov/periodicals/mcs2024/mcs2024-graphite.pdf, accessed 15 May 2024.
14. S. C. Chelgani, M. Rudolph, R. Kratzsch, D. Sandmann and J. Gutzmer, *Miner. Process. Extr. Metall.*, 2016, **37**, 58.
15. A. D. Jara, A. Betemariam, G. Woldetinsae and J. Y. Kim, *Int. J. Min. Sci. Technol.*, 2019, **29**, 671.
16. S. Duan, X. Wu, Y. Wang, J. Feng, S. Hou, Z. Huang, K. Shen, Y. Chen, H. Liu and F. Kang, *New Carbon Mater.*, 2023, **38**, 73.
17. S. Sharma, Lecture: Graphite Ore Processing, https://www.youtube.com/watch?v=FZfqfjyrVvY, accessed 22 January 2024.
18. R. Szymanowitz, *Edward Goodrich Acheson: Inventor - Scientist – Industrialist*, Vantage, 1971.
19. C. M. Despretz, *C. R. Acad. Sci.*, 1850, **30**, 367.
20. E. G. Acheson, Manufacture of Graphite, *US Pat.*, 568323, 1896.
21. R. E. Franklin, *Proc. R. Soc. A*, 1951, **209**, 196.
22. P. J. F. Harris, *Interdiscip. Sci. Rev.*, 2001, **26**, 204.
23. P. J. F. Harris, *J. Mater. Sci.*, 2013, **48**, 565.
24. P. J. F. Harris, *Eurasian Chem.-Technol. J.*, 2019, **21**, 227.
25. C. S. Allen, F. Ghamouss, O. Boujibar and P. J. F. Harris, *Proc. R. Soc. A*, 2022, **478**, 20210580.
26. J. D. Brooks and G. H. Taylor, *Chem. Phys. Carbon*, 1968, **4**, 243.
27. R. H. Hurt and Z. Chen, *Phys. Today*, 2000, **53**(3), 39.
28. T. Q. Li and C. Y. Wang, *New Carbon Mater.*, 2005, **20**, 278.
29. H. Marsh and J. Griffiths, New Processes and New Applications, in *Ext. Abst. of International Symposium on Carbon*, Kagaku Gijutsu-sha, Toyohashi, 1982, p. 81.
30. H. Marsh and I. A. Edwards, *Introduction to Carbon Science*, Butterworth, 1989.
31. A. Oberlin, *Chem. Phys. Carbon*, 1989, **22**, 1.
32. J. W. Martin, J. L. Fogg, G. R. Francas, K. J. Putman, E. P. Turner, I. Suarez-Martinez and N. A. Marks, *Carbon*, 2023, **215**, 118386.
33. S. Liu and C. R. Loper, *Carbon*, 1991, **29**, 1119.
34. M. S. Dresselhaus and G. Dresselhaus, *Adv. Phys.*, 1981, **30**, 139.
35. J.-C. An, H. Kim and I. Hong, *J. Ind. Eng. Chem.*, 2015, **26**, 55.
36. P. L. Walker, *Am. Sci.*, 1962, **50**, 259.
37. H. O. Pierson, *Handbook of Carbon, Graphite, Diamonds and Fullerenes*, Noyes Publications, New Jersey, 1993.
38. J. C. Bokros, *Chem. Phys. Carbon*, 1969, **5**, 3.
39. J. C. Bokros, *Carbon*, 1977, **15**, 353.
40. R. B. More, A. D. Haubold and J. C. Bokros, in *Biomaterials Science - An Introduction to Materials in Medicine*, ed. B. D. Ratner, *et al.*, CRC Press, 3rd edn, 2013, p. 209.

41. L. C. Blackman, G. Saunders and A. R. J. P. Ubbelohde, *Proc. R. Soc. A*, 1961, **264**, 19.
42. A. W. Moore, *Chem. Phys. Carbon*, 1973, **11**, 69.
43. A. W. Moore, *Chem. Phys. Carbon*, 1981, **17**, 233.
44. L. C. F. Blackman and A. R. J. P. Ubbelohde, *Proc. R. Soc. A*, 1962, **266**, 20.
45. H. Legall, H. Stiel, V. Arkadiev and A. A. Bjeoumikhov, *Opt. Express*, 2006, **14**, 4570.
46. A. N. Patel, M. G. Collignon, M. A. O'Connell, W. O. Y. Hung, K. McKelvey, J. V. Macpherson and P. R. Unwin, *J. Am. Chem. Soc.*, 2012, **134**, 20117.
47. C. J. Chen, *Introduction to Scanning Tunneling Microscopy*, Oxford University Press, 3rd edn, 2021.
48. K. S. Novoselov, A. K. Geim, S. V. Morozov, D. Jiang, Y. Zhang, S. V. Dubonos, I. V. Grigorieva and A. A. Firsov, *Science*, 2004, **306**, 666.
49. J. D. Bernal and W. L. Bragg, *Proc. R. Soc. A*, 1924, **106**, 749.
50. P. J. F. Harris, *Carbon Nanotubes and Related Structures: New Materials for the Twenty-first Century*, CUP, Cambridge, 2009.
51. H. Lipson and A. R. Stokes, *Nature*, 1942, **149**, 328.
52. G. E. Bacon, *Acta Crystallogr.*, 1950, **3**, 320.
53. J. P. Nery, M. Calandra and F. Mauri, *Nano Lett.*, 2020, **20**, 5017.
54. F. Laves, *Z. Kristallogr.*, 1956, **107**, 337.
55. Y. Cao, V. Fatemi, A. Demir, S. Fang, S. L. Tomarken, J. Y. Luo, J. D. Sanchez-Yamagishi, K. Watanabe, T. Taniguchi, E. Kaxiras, R. C. Ashoori and P. Jarillo-Herrero, *Nature*, 2018, **556**, 80.
56. Y. Shi, S. Xu, Y. Yang, S. Slizovskiy, S. V. Morozov, S. K. Son, S. Ozdemir, C. Mullan, J. Barrier, J. Yin, A. I. Berdyugin, B. A. Piot, T. Taniguchi, K. Watanabe, V. I. Fal'ko, K. S. Novoselov, A. K. Geim and A. Mishchenko, *Nature*, 2020, **584**, 210.
57. H. Zhou, T. Xie, T. Taniguchi, K. Watanabe and A. F. Young, *Nature*, 2021, **598**, 434.
58. J. C. Charlier, J. P. Michenaud, X. Gonze and J. P. Vigneron, *Phys. Rev. B: Condens. Matter Mater. Phys.*, 1991, **44**, 13237.
59. I. Lobato and B. Partoens, *Phys. Rev. B: Condens. Matter Mater. Phys.*, 2011, **83**, 165429.
60. M. K. Song, S. D. Hong and K. T. No, *J. Electrochem. Soc.*, 2001, **148**, A1159.
61. J. K. Lee, S. C. Lee, J. P. Ahn, S. C. Kim, J. I. B. Wilson and P. John, *J. Chem. Phys.*, 2008, **129**, 234709.
62. Z. Liu, K. Suenaga, P. J. F. Harris and S. Iijima, *Phys. Rev. Lett.*, 2009, **102**, 015501.
63. J. Biscoe and B. E. Warren, *J. Appl. Phys.*, 1942, **13**, 364.
64. V. Jain and B. Kandasubramanian, *J. Mater. Sci.*, 2020, **55**, 1865.
65. O. V. Yazyev and S. G. Louie, *Phys. Rev. B: Condens. Matter Mater. Phys.*, 2010, **81**, 195420.
66. A. A. El-Barbary, R. H. Telling, C. P. Ewels, M. I. Heggie and P. R. Briddon, *Phys. Rev. B: Condens. Matter Mater. Phys.*, 2003, **68**, 144107.

67. K. Izui and F. E. Fujita, *J. Phys. Soc. Jpn.*, 1961, **16**, 1032.
68. G. K. Williamson, *Proc. R. Soc. London, Ser. A*, 1960, **257**, 457.
69. P. Delavignette and S. Amelinckx, *J. Appl. Phys.*, 1960, **31**, 1691.
70. S. Amelinckx and P. Delavignette, *J. Appl. Phys.*, 1960, **31**, 2126.
71. S. Amelinckx, P. Delavignette and M. Heerschap, *Chem. Phys. Carbon*, 1965, **1**, 1.
72. R. H. Telling and M. I. Heggie, *Philos. Mag. Lett.*, 2003, **83**, 411.
73. C. P. Ewels, M. I. Heggie and P. R. Briddon, *Chem. Phys. Lett.*, 2002, **351**, 178.
74. G. R. Hennig, *Appl. Phys. Lett.*, 1964, **4**, 52.
75. G. R. Hennig, *Science*, 1965, **147**, 733.
76. G. R. Hennig, *Chem. Phys. Carbon*, 1966, **2**, 1.
77. J. M. Thomas and E. E. G. Hughes, *Carbon*, 1964, **1**, 209.
78. C. Roscoe and J. M. Thomas, *Proc. R. Soc. London, Ser. A*, 1967, **297**, 397.
79. J. M. Thomas and C. Roscoe, *Chem. Phys. Carbon*, 1968, **3**, 1.
80. I. Suarez-Martinez, G. Savini, G. Haffenden, J.-M. Campanera and M. I. Heggie, *Phys. Status Solidi C*, 2007, **4**, 2958.
81. R. T. Yang and C. Wong, *Science*, 1981, **214**, 437.
82. J. M. Thomas and K. M. Jones, *J. Nucl. Mater.*, 1964, **11**, 236.
83. T. R. Acharya and D. R. Olander, *Carbon*, 1973, **11**, 7.
84. X. Chu and L. D. Schmidt, *Surf. Sci.*, 1992, **268**, 325.
85. F. Stevens, L. A. Kolodny and T. P. Beebe, *J. Phys. Chem. B*, 1998, **102**, 10799.
86. H. Chang and A. J. Bard, *J. Am. Chem. Soc.*, 1991, **113**, 5588.
87. S. Li, M. T. Vahdat, S. Huang, K.-J. Hsu, M. Rezaei, M. Mensi, N. Marzari and K. V. Agrawal, *JACS Au*, 2022, **2**, 723.
88. J. R. Hahn, *Carbon*, 2005, **43**, 1506.
89. M. Levy and P. Wong, *J. Electrochem. Soc.*, 1964, **111**, 1088.
90. S. R. Kelemen and H. Freund, *Carbon*, 1985, **23**, 619.
91. P. L. Walker, F. Rusinko and L. G. Austin, *Adv. Catal.*, 1959, **11**, 133.
92. J. M. Thomas, *Carbon*, 1970, **8**, 413.
93. H. Marsh and A. Foord, *Carbon*, 1973, **11**, 421.
94. I. W. Smith, *Fuel*, 1978, **57**, 409.
95. P. L. Walker, F. J. Vastola and P. J. Hart, in *Fundamentals of Gas-surface Interactions*, ed. H. Saltsburg, et al., Academic Press, 1967, p. 307.
96. P. L. Walker, R. L. Taylor and J. M. Ranish, *Carbon*, 1991, **29**, 411.
97. N. R. Laine, F. J. Vastola and P. L. Walker, *J. Phys. Chem.*, 1963, **67**, 2030.
98. L. R. Radovic and B. Bockrath, *J. Am. Chem. Soc.*, 2005, **127**, 5917.
99. L. R. Radovic, *J. Am. Chem. Soc.*, 2009, **131**, 17166.
100. L. R. Radovic, A. B. Silva-Tapia and F. Vallejos-Burgos, *Carbon*, 2011, **49**, 4218.
101. L. R. Radovic, A. Suarez and F. Vallejos-Burgos, *Carbon*, 2011, **49**, 4226.
102. L. R. Radovic, A. F. Silva-Villalobos, A. B. Silva-Tapia and F. Vallejos-Burgos, *Carbon*, 2011, **49**, 3471.
103. B. C. Brodie, *Philos. Trans. R. Soc. London*, 1859, **149**, 249.
104. W. S. Hummers and R. E. Offeman, *J. Am. Chem. Soc.*, 1958, **80**, 1339.

105. H. He, J. Klinowski, M. Forster and A. Lerf, *Chem. Phys. Lett.*, 1998, **287**, 53.
106. A. Lerf, H. He, M. Forster and J. Klinowski, *J. Phys. Chem. B*, 1998, **102**, 4477.
107. S. K. Srivastava and J. Pionteck, *J. Nanosci. Nanotechnol.*, 2015, **15**, 1984.
108. C. Schafhäutl, *J. Prakt. Chem.*, 1841, **21**, 129.
109. U. Hofmann and W. Rüdorff, *Trans. Faraday Soc.*, 1938, **34**, 1017.
110. L. C. F. Blackman, J. F. Mathews and A. R. J. P. Ubbelohde, *Proc. R. Soc. A*, 1960, **256**, 15.
111. G. R. Hennig, *J. Chem. Phys.*, 1952, **20**, 1443.
112. W. Rüdorff, V. Sils and R. Zeller, *Z. Anorg. Allg. Chem.*, 1956, **283**, 299.
113. L. C. F. Blackman, J. F. Mathews and A. R. J. P. Ubbelohde, *Proc. R. Soc. A*, 1960, **258**, 339.
114. W. Rüdorff and E. Schulze, *Z. Anorg. Allg. Chem.*, 1954, **277**, 156.
115. K. Vilcinskas, F. M. Mulder, S. J. Picken and G. Koper, arXiv Preprint: 1701.06493, 2017, https://arxiv.org/abs/1701.06493, accessed 01 March 2024.
116. J. E. Fischer, *Physica B+C*, 1980, **99**, 383.
117. L. A. Pendrys, R. Wachnik, F. L. Vogel, P. Lagrange, G. Furdin, M. El Makrini and A. Hérold, *Solid State Commun.*, 1981, **38**, 677.
118. R. A. Wachnik, L. A. Pendrys, F. L. Vogel and P. Lagrange, *Solid State Commun.*, 1982, **43**, 5.
119. Y. Li, Y. Lu, P. Adelhelm, M.-M. Titirici and Y.-S. Hu, *Chem. Soc. Rev.*, 2019, **48**, 4655.
120. T. D. Burchell, in *Comprehensive Nuclear Materials*, ed. R. Konings, Elsevier, Oxford, 2012, vol. 2, p. 285.
121. A. Magnus, *Ann. Phys.*, 1915, **48**, 983.
122. A. T. D. Butland and R. J. Maddison, *J. Nucl. Mater.*, 1973, **49**, 45.
123. S. Picard, D. T. Burns and P. Roger, *Metrologia*, 2007, **44**, 294.
124. J. B. Nelson and D. P. Riley, *Proc. Phys. Soc.*, 1945, **57**, 477.
125. E. A. Kellett and B. P. Richards, *J. Nucl. Mater.*, 1964, **12**, 184.
126. D. Yoon, Y.-W. Son and H. Cheong, *Nano Lett.*, 2011, **11**, 3227.
127. B. T. Kelly, *Physics of Graphite*, Applied Science, London, 1989.
128. O. L. Blakslee, D. G. Proctor, E. J. Seldin, G. B. Spence and T. Weng, *J. Appl. Phys.*, 1970, **41**, 3373.
129. A. Bosak, M. Krisch, M. Mohr, J. Maultzsch and C. Thomsen, *Phys. Rev. B: Condens. Matter Mater. Phys.*, 2007, **75**, 153408.
130. M. Grimsditch, *J. Phys. C: Solid State Phys.*, 1983, **16**, L143.
131. C. S. G. Cousins and M. I. Heggie, *Phys. Rev. B: Condens. Matter Mater. Phys.*, 2003, **67**, 024109.
132. D. E. Soule and C. W. Nezbeda, *J. Appl. Phys.*, 1968, **39**, 5122.
133. Z. Liu, S. M. Zhang, J. R. Yang, J. Z. Liu, Y. L. Yang and Q. S. Zheng, *Acta Mech. Sin.*, 2012, **28**, 978.
134. M. F. Yu, O. Lourie, M. J. Dyer, K. Moloni, T. F. Kelly and R. S. Ruoff, *Science*, 2000, **287**, 637.
135. C. Lee, X. Wei, J. W. Kysar and J. Hone, *Science*, 2008, **321**, 385.

136. P. R. Wallace, *Phys. Rev.*, 1947, **71**, 622.
137. A. H. Castro Neto, F. Guinea, N. M. R. Peres, K. S. Novoselov and A. K. Geim, *Rev. Mod. Phys.*, 2009, **81**, 109.
138. T. Ando, *NPG Asia Mater.*, 2009, **1**, 17.
139. S. Reich, C. Thomsen and J. Maultzsch, *Carbon Nanotubes: Basic Concepts and Physical Properties*, Wiley-VCH, Weinheim, 2004.
140. J. C. Slonczewski and P. R. Weiss, *Phys. Rev.*, 1958, **109**, 272.
141. J. W. McClure, *Phys. Rev.*, 1957, **108**, 612.
142. J. W. McClure, *IBM J. Res. Dev.*, 1964, **8**, 255.
143. J. W. McClure, *Carbon*, 1972, **10**, 353.
144. R. W. Newson, Dynamics of carriers and photoinjected currents in carbon nanotubes and graphene, Ph.D. thesis, University of Toronto, 2010.
145. S. Y. Zhou, G. H. Gweon, J. Graf, A. V. Fedorov, C. D. Spataru, R. D. Diehl, Y. Kopelevich, D. H. Lee, S. G. Louie and A. Lanzara, *Nat. Phys.*, 2006, **2**, 595.
146. H. Zhang, T. Pincelli, C. Jozwiak, T. Kondo, R. Ernstorfer, T. Sato and S. Zhou, *Nat. Rev. Methods Primers*, 2022, **2**, 1.
147. M. Orlita, C. Faugeras, G. Martinez, D. K. Maude, M. L. Sadowski and M. Potemski, *Phys. Rev. Lett.*, 2008, **100**, 136403.
148. J. W. McClure, *Carbon*, 1969, **7**, 425.
149. J.-C. Charlier, J.-P. Michenaud and X. Gonze, *Phys. Rev. B: Condens. Matter Mater. Phys.*, 1992, **46**, 4531.
150. J.-C. Charlier, X. Gonze and J.-P. Michenaud, *Carbon*, 1994, **32**, 289.
151. S. Ozdemir, *Electronic Properties of Rhombohedral Graphite*, Springer Theses, 2021.
152. I. L. Spain, *Chem. Phys. Carbon*, 1981, **16**, 119.
153. A. K. Dutta, *Phys. Rev.*, 1953, **90**, 187.
154. I. L. Spain, *Chem. Phys. Carbon*, 1973, **8**, 1.
155. I. L. Spain, A. R. J. P. Ubbelohde and A. D. Young, *Philos. Trans. R. Soc., A*, 1967, **262**, 345.
156. L. A. Pendrys, C. Zeller and F. L. Vogel, *J. Mater. Sci.*, 1980, **15**, 2103.
157. D. D. L. Chung, *Carbon Materials: Science and Applications*, World Scientific, 2019, ch. 2.
158. S. G. Bapat and H. Nickel, *Carbon*, 1973, **11**, 323.
159. G. H. Kinchin, *Proc. R. Soc. London, Ser. A*, 1997, **217**, 9.
160. N. Iwashita, H. Imagawa and W. Nishiumi, *Carbon*, 2013, **61**, 602.
161. P. Esquinazi, *Pap. Phys.*, 2013, **5**, 050007.
162. A. Ballestar, J. Barzola-Quiquia, T. Scheike and P. Esquinazi, *New J. Phys.*, 2013, **15**, 023024.
163. A. Champi, C. E. Precker and P. D. Esquinazi, *New J. Phys.*, 2023, **25**, 093029.
164. The Energy Institute, Statistical Review of World Energy, 2023, https://www.energyinst.org/statistical-review/home, accessed 03 August 2023.
165. Our World in Data, Electricity Production by Source - Stacked Area Chart, https://ourworldindata.org/grapher/electricity-prod-source-stacked, accessed 03 August 2023.

166. World Nuclear Association, Nuclear Power in the United Kingdom, https://www.world-nuclear.org/information-library/country-profiles/countries-t-z/united-kingdom.aspx, accessed 03 August 2023.
167. B. T. Kelly, *Carbon*, 1982, **20**, 3.
168. B. J. Marsden, G. N. Hall and A. N. Jones, *Reference Module in Materials Science and Materials Engineering*, 2020, vol. 14, p. 357.
169. R. E. Nightingale, *Nuclear Graphite: Prepared Under the Auspices of the Division of Technical Information United States Atomic Energy Commission*, Academic Press, 1962.
170. T. D. Burchell, *Carbon Materials for Advanced Technologies*, Elsevier, 1999.
171. W. P. Eatherly, *J. Nucl. Mater.*, 1981, **100**, 55.
172. J. D. Arregui-Mena, R. N. Worth, W. Bodel, B. März, W. Li, A. A. Campbell, E. Cakmak, N. Gallego, C. Contescu and P. D. Edmondson, *Mater. Charact.*, 2022, **46**, 112047.
173. G. Haag, *Properties of ATR-2E Graphite and Property Changes Due to Fast Neutron Irradiation*, Forschungszentrum Jülich in der Helmholtz-Gemeinschaft, 2005.
174. M. Srinivasan, *Nucl. Eng. Des.*, 2021, **381**, 111354.
175. B. Taylor, We're excited to see new concepts emerging, https://nda.blog.gov.uk/were-excited-to-see-new-concepts-emerging/, accessed 01 March 2024.
176. I. R. Cameron, *Nuclear Fission Reactors*, Plenum, 1982.
177. R. Meiswinkel, J. Meyer and J. Schnell, *Design and Construction of Nuclear Power Plants*, Wiley, 2013.
178. International Atomic Energy Agency, *Nuclear Power Reactors in the World*, IAEA, 2021.
179. A. Smaizys, E. Narkunas and P. Poskas, *Radiat. Prot. Dosim.*, 2005, **116**, 270.
180. B. Gurovich, D. Kuleshov, D. Maltsev, O. Chugunov, A. Frolov and Y. I. Shtrombakh, *Procedia Struct. Integr.*, 2019, **23**, 589.
181. A. Pavliuk, S. Kotlyarevsky, E. Bespala and I. Y. Novoselov, *J. Nucl. Mater.*, 2019, **515**, 303.
182. E. Narkunas, S. Norris, P. Poskas and A. Smaizys, *Radiocarbon*, 2018, **60**, 1849.
183. M. Mostafavi, M. Schmidt, B. Marsden and T. Marrow, *Mater. Sci. Eng., A*, 2012, **558**, 265.
184. P. J. Heard, M. R. Wootton, R. Moskovic and P. E. J. Flewitt, *J. Nucl. Mater.*, 2010, **401**, 71.
185. R. Moskovic, P. E. Flewitt, E. Schlangen, G. Smith, A. Crocker, A. Hodgkins, P. Heard and M. Wootton, *Mater. Sci. Technol.*, 2014, **30**, 129.
186. L. Payne, P. J. Heard and T. B. Scott, *PLoS One*, 2015, **10**, e0143041.
187. D. Vulpius, K. Baginski, B. Kraus and B. Thomauske, *Nucl. Eng. Des.*, 2013, **265**, 294.
188. J. Pageot, J. N. Rouzaud, L. Gosmain, A. Duhart-Barone, J. Comte and D. Deldicque, *Prog. Nucl. Energy*, 2018, **105**, 279.

189. A. A. Campbell and T. D. Burchell, in *Comprehensive Nuclear Materials*, ed. R. J. M. Konings, and R. E. Stoller, Elsevier, Oxford, 2nd edn, 2020, p. 398.
190. T. J. Marrow, D. Liu, S. M. Barhli, L. Saucedo Mora, Y. Vertyagina, D. M. Collins, C. Reinhard, S. Kabra, P. E. J. Flewitt and D. J. Smith, *Carbon*, 2016, **96**, 285.
191. D. Liu, T. Zillhardt, P. Earp, S. Kabra, T. Connolley and T. James Marrow, *Carbon*, 2020, **163**, 308.
192. D. Liu, B. Gludovatz, H. S. Barnard, M. Kuball and R. O. Ritchie, *Nat. Commun.*, 2017, **8**, 15942.
193. R. Stainsby, UK Gas Reactors Operational Feedback, https://esfr-smart.eu/wp-content/uploads/2021/04/S52_2_Richard_Stainsby_UK_Gas_Reactors_Stainsby.pdf, accessed 01 August 2023.
194. B. J. Marsden, M. Haverty, W. Bodel, G. N. Hall, A. N. Jones, P. M. Mummery and M. Treifi, *Int. Mater. Rev.*, 2016, **61**, 155.
195. E. Nonbol, Description of the advanced gas cooled type of reactor (AGR), https://web.archive.org/web/20220609170021/https://inis.iaea.org/collection/NCLCollectionStore/_Public/28/028/28028509.pdf, accessed 15 May 2024.
196. A. Fletcher, PhD Thesis, The University of Manchester, 2014.
197. EDF Energy, Hunterston Reactor 3 Graphite Inspections Update, https://www.edfenergy.com/sites/default/files/huntb_reactor_3_graphite_inspections_update_bf.pdf, accessed 01 August 2023.
198. E. E. Lewis, *Fundamentals of Nuclear Reactor Physics*, Elsevier, 2008.
199. M. S. L. Jordan, Doctor of Philosophy, University of Oxford, 2019.
200. U.S. DOE Nuclear Energy Research Advisory Committee and Generation IV International Forum, *A Technology Roadmap for Generation IV Nuclear Energy Systems*, Generation-IV International Forum, 2002.
201. D. Butler, *Nature*, 2004, **429**, 238.
202. B. Marsden, A. Jones, G. Hall, M. Treifi and P. Mummery, in *Structural Materials for Generation IV Nuclear Reactors*, Elsevier, 2017, p. 495.
203. K. Kugeler and Z. Zhang, *Modular High-Temperature Gas-cooled Reactor Power Plant*, Springer, 2018.
204. X. L. Yan, in *Handbook of Generation IV Nuclear Reactors*, Elsevier, 2023, p. 133.
205. M. A. Fütterer, L. Fu, C. Sink, S. de Groot, M. Pouchon, Y. W. Kim, F. Carré and Y. Tachibana, *Prog. Nucl. Energy*, 2014, **77**, 266.
206. *Comprehensive Nuclear Materials*, ed. R. E. Stoller, and R. J. M. Konings, Elsevier, 2nd edn, 2020.
207. M. J. Kania, H. Nabielek, K. Verfondern and H.-J. Allelein, *J. Nucl. Mater.*, 2013, **441**, 545.
208. J. L. Everett and E. J. Kohler, *Ann. Nucl. Energy*, 1978, **5**, 321.
209. H. Gottaut and K. Krüger, *Nucl. Eng. Des.*, 1990, **121**, 143.
210. C. Wallroth, *Feasibility Study of the Dragon Reactor for HTGR Fuel Testing*, General Atomics, San Diego, CA (United States), 1975.
211. G. Engle, *Assessment of Grade H-451 Graphite for Replaceable Fuel and Reflector Elements in HTGR*, General Atomics, San Diego, CA (United States), 1977.

212. Z. Zhang, Y. Dong, F. Li, Z. Zhang, H. Wang, X. Huang, H. Li, B. Liu, X. Wu, H. Wang, X. Diao, H. Zhang and J. Wang, *Engineering*, 2016, **2**, 112.
213. Z. Wu, D. Lin and D. Zhong, *Nucl. Eng. Des.*, 2002, **218**, 25.
214. J. D. Arregui-Mena, D. Griffiths, R. N. Worth, C. E. Torrence, A. Selby, C. Contescu, N. Gallego, P. D. Edmondson, P. M. Mummery and L. Margetts, *Mater. Des.*, 2022, **220**, 110840.
215. R. Bratton, *NGNP Graphite Testing and Qualification Specimen Selection Strategy*, Idaho National Lab.(INL), Idaho Falls, ID (United States), 2005.
216. T. Iyoku, S. Shiozawa, M. Ishihara, T. Arai and T. Oku, *Nucl. Eng. Des.*, 1991, **132**, 23.
217. N. Simos, P. Hurh, N. Mokhov, M. Snead, M. Topsakal, M. Palmer, S. Ghose, H. Zhong, Z. Kotsina and D. Sprouster, *J. Nucl. Mater.*, 2020, **542**, 152438.
218. M. C. R. Heijna, S. de Groot and J. A. Vreeling, *J. Nucl. Mater.*, 2017, **492**, 148.
219. S. Ishiyama, T. D. Burchell, J. Strizak and M. Eto, *J. Nucl. Mater.*, 1996, **230**, 1.
220. T. Shibata, J. Sumita, T. Tada, S. Hanawa, K. Sawa and T. Iyoku, *J. Nucl. Mater.*, 2008, **381**, 165.
221. J. Sumita, T. Shibata, I. Fujita, E. Kunimoto, M. Yamaji, M. Eto, T. Konishi and K. Sawa, *Nucl. Eng. Des.*, 2014, **271**, 314.
222. S.-H. Chi, *J. Nucl. Mater.*, 2016, **476**, 188.
223. M. S. El-Genk and J.-M. P. Tournier, *J. Nucl. Mater.*, 2012, **420**, 141.
224. M. Jiang, A. El-Turke, G. Lolov, K. Ammigan, P. Hurh and D. Liu, *J. Nucl. Mater.*, 2021, **550**, 152876.
225. S.-H. Chi, *Nucl. Eng. Technol.*, 2022, **54**, 2359.
226. A. A. Campbell, A. A. Wereszczak, M. A. Snead and Y. Katoh, *J. Nucl. Mater.*, 2023, **576**, 154263.
227. D. Liu, P. J. Heard, S. Nakhodchi and P. E. J. Flewitt, in *Graphite Testing for Nuclear Applications: The Significance of Test Specimen Volume and Geometry and the Statistical Significance of Test Specimen Population*, 2014, ch. Stp.157820130127, pp. 84–104, DOI: 10.1520/stp157820130127.
228. D. Liu, K. Mingard, O. T. Lord and P. Flewitt, *J. Nucl. Mater.*, 2017, **493**, 246.
229. G. Zheng, P. Xu, K. Sridharan and T. Allen, *J. Nucl. Mater.*, 2014, **446**, 193.
230. C. Karthik, J. Kane, D. P. Butt, W. E. Windes and R. Ubic, *Carbon*, 2015, **86**, 124.
231. K. L. Jones, G. M. Laudone and G. P. Matthews, *Carbon*, 2018, **128**, 1.
232. W. Zhang, B. l. Zhang, J. l. Song, W. Qi, X. J. He, Z. J. Liu, P. F. Lian, Z. T. He, L. N. Gao, H. H. Xia, X. D. Liu, X. T. Zhou, L. B. Sun and X. X. Wu, *New Carbon Mater.*, 2016, **31**, 585.
233. J. J. Lee, J. D. Arregui-Mena, C. I. Contescu, T. D. Burchell, Y. Katoh and S. K. Loyalka, *J. Nucl. Mater.*, 2020, **534**, 152119.

234. H. E. McCoy, R. L. Beatty, W. H. Cook, R. E. Gehlbach, C. R. Kennedy, J. W. Koger, A. P. Litman, C. E. Sessions and J. R. Weir, *Nucl. Appl. Technol.*, 1970, **8**, 156.
235. L. Olson, J. Ambrosek, G. Cao, K. Sridharan, M. Anderson and T. Allen, in *Advances in Materials Science for Environmental and Nuclear Technology*, ed. K. Fox, *et al.*, 2010, p. 145.
236. S. Nagatsuji, New nuclear power plant and high-temperature gas-cooled reactor Hurry up, China looms behind Japan, https://www.sankei.com/article/20200226-5VQ2GC2ZFVJMLOONEUDLC3BAKA/photo/XYD3AWQFQNPYPIYRJ3SX5V45ZM/, accessed 01 August 2023.
237. T. Grebennikova, A. N. Jones and C. A. Sharrad, *Energy Environ. Sci.*, 2021, **14**, 5501.
238. D. A. Petti, *The Next Generation Nuclear Plant*, Idaho National Lab. (INL), Idaho Falls, ID (United States), 2009.
239. J. Wade-Zhu, R. Krishna, A. J. Bodey, M. Davies, N. K. Bourne, C. Rau, B. Davies, A. Tzelepi, A. N. Jones and B. J. Marsden, *Carbon*, 2020, **168**, 230.
240. P. Heard, M. Wootton, R. Moskovic and P. Flewitt, *J. Nucl. Mater.*, 2011, **418**, 223.
241. B. T. Kelly, *Prog. Nucl. Energy*, 1978, **2**, 219.
242. T. D. Burchell and T. R. Pavlov, in *Comprehensive Nuclear Materials*, ed. R. J. M. Konings, and R. E. Stoller, Elsevier, Oxford, 2nd edn, 2020, p. 355.
243. G. B. Neighbour, Microstructural processes leading to fracture in nuclear graphites, PhD thesis, University of Bath, United Kingdom, 1993.
244. S. Ragan and H. Marsh, *J. Mater. Sci.*, 1983, **18**, 3161.
245. T. Boden and B. T. Tripp, *Gilsonite Veins of the Uinta Basin*, Utah Geological Survey, Utah, 2012.
246. M. Mostafavi and T. Marrow, *Fatigue Fract. Eng. Mater. Struct.*, 2012, **35**, 695.
247. M. R. Joyce, T. J. Marrow, P. Mummery and B. J. Marsden, *Eng. Fract. Mech.*, 2008, **75**, 3633.
248. D. Liu and P. E. Flewitt, *Carbon*, 2017, **114**, 261.
249. A. Hodgkins, T. J. Marrow, M. R. Wootton, R. Moskovic and P. E. J. Flewitt, *Mater. Sci. Technol.*, 2013, **26**, 899.
250. P. Béghein, G. Berlioux, B. du Mesnildot, F. Hiltmann and M. Melin, *Nucl. Eng. Des.*, 2012, **251**, 146.
251. J. J. Lee, T. K. Ghosh and S. K. Loyalka, *J. Nucl. Mater.*, 2018, **500**, 64.
252. T. D. Burchell, *MRS Bull.*, 1997, **22**, 29.
253. G. Haag, D. Mindermann, G. Wilhelmi, H. Persicke and W. Ulsamer, *J. Nucl. Mater.*, 1990, **171**, 41.
254. R. M. Paul, J. D. Arregui-Mena, C. I. Contescu and N. C. Gallego, *Carbon*, 2022, **191**, 132.
255. W. R. Hoffmann, *Industrial Carbon and Graphite Materials, Volume I: Raw Materials, Production and Applications*, 2021, vol. 1, p. 338.

256. B. Kelly, *Philos. Mag.*, 1964, **9**, 721.
257. B. Šavija, D. Liu, G. Smith, K. R. Hallam, E. Schlangen and P. E. J. Flewitt, *Eng. Fract. Mech.*, 2016, **153**, 360.
258. K. Shen, S. Yu and F. Kang, *Carbon*, 2019, **153**, 428.
259. K. Shen, X. Zhang, S. Yu and F. Kang, *Carbon*, 2022, **189**, 604.
260. J. Kane, C. Karthik, D. P. Butt, W. E. Windes and R. Ubic, *J. Nucl. Mater.*, 2011, **415**, 189.
261. K. Y. Wen, T. J. Marrow and B. J. Marsden, *Carbon*, 2008, **46**, 62.
262. D. Liu and D. Cherns, *Philos. Mag.*, 2018, **98**, 1272.
263. C. I. Contescu, J. D. Arregui-Mena, A. A. Campbell, P. D. Edmondson, N. C. Gallego, K. Takizawa and Y. Katoh, *Carbon*, 2019, **141**, 663.
264. C. Karthik, J. Kane, D. P. Butt, W. E. Windes and R. Ubic, *Microsc. Microanal.*, 2012, **18**, 272.
265. P. J. Hacker, G. B. Neighbour and B. McEnaney, *J. Phys. D: Appl. Phys.*, 2000, **33**, 991.
266. L. Babout, B. Marsden, P. Mummery and T. Marrow, *Acta Mater.*, 2008, **56**, 4242.
267. G. Hall, B. J. Marsden and S. L. Fok, *J. Nucl. Mater.*, 2006, **353**, 12.
268. Y. Vertyagina and T. J. Marrow, *Carbon*, 2016, **109**, 711.
269. H. M. Freeman, A. N. Jones, M. B. Ward, F. S. Hage, N. Tzelepi, Q. M. Ramasse, A. J. Scott and R. M. D. Brydson, *Carbon*, 2016, **103**, 45.
270. D. Baker, *Nucl. Eng. Des.*, 1971, **14**, 413.
271. H. M. Freeman, B. Mironov, W. Windes, M. Alnairi, A. Scott, A. Westwood and R. Brydson, *J. Nucl. Mater.*, 2017, **491**, 221.
272. S. Mrozowski, *Proceedings of the 2nd Conference on Carbon*, Buffalo Univ., 1954, p. 31.
273. A. Sutton and V. Howard, *J. Nucl. Mater.*, 1962, **7**, 58.
274. M. Metcalfe, N. Tzelepi and D. Wilde, in *Graphite Testing for Nuclear Applications: The Significance of Test Specimen Volume and Geometry and the Statistical Significance of Test Specimen Population*, ASTM International, 2014.
275. M. Mostafavi, S. A. McDonald, H. Çetinel, P. M. Mummery and T. J. Marrow, *Carbon*, 2013, **59**, 325.
276. F. Berto, S. M. J. Razavi and M. R. Ayatollahi, *Procedia Struct. Integr.*, 2017, **3**, 68.
277. T. D. Burchell, T. Yahr and R. Battiste, *Carbon*, 2007, **45**, 2570.
278. M. O. Tucker and N. McLachlan, *J. Phys. D: Appl. Phys.*, 1993, **26**, 893.
279. M. O. Tucker, A. Rose and T. D. Burchell, *Carbon*, 1986, **24**, 581.
280. N. Simos, P. Nocera, Z. Zhong, R. Zwaska, N. Mokhov, J. Misek, K. Ammigan, P. Hurh and Z. Kotsina, *Phys. Rev. Accel. Beams*, 2017, **20**, 071002.
281. A. A. Campbell, C. I. Contescu, E. Cakmak, N. C. Gallego and T. D. Burchell, *Effects of High-Temperature Irradiation on the Microstructure of Nuclear Graphite*, Oak Ridge National Lab.(ORNL), Oak Ridge, TN (United States), 2019.
282. Z. Zhou, W. Bouwman, H. Schut and C. Pappas, *Carbon*, 2014, **69**, 17.

283. C. Paraskevoulakos, B. Roebuck, K. R. Hallam and P. Flewitt, *SN Appl. Sci.*, 2023, **5**, 28.
284. A. A. Campbell and G. S. Was, *Carbon*, 2014, **77**, 993.
285. A. A. Campbell, Doctor of Philosophy, University of Michigan, 2014.
286. POCO Graphite, POCO Graphite manufacturer datasheet, https://poco.entegris.com/content/dam/poco/resources/reference-materials/brochures/brochure-graphite-properties-and-characteristics-11043.pdf, accessed 22 February 2024.
287. N. Simos, *Long Baseline Neutrino Experiment (LBNE) Target Material Radiation Damage from Energetic Protons of the Brookhaven Linear Isotope Production (BLIP) Facility*, Brookhaven National Lab. (BNL), Upton, NY (United States), 2016.
288. S.-H. Chi and G.-C. Kim, *J. Nucl. Mater.*, 2008, **381**, 98.
289. TOYO TANSO, Special Isotropic Graphite, https://www.toyotanso.co.jp/Products/Special_graphite/data_en.html, accessed 22 February 2024.
290. T. Yamada, Y. Matsushima, M. Kuroda, J. Sumita, T. Shibata, I. Fujita and K. Sawa, *Nucl. Eng. Des.*, 2014, **271**, 323.
291. A. A. Campbell and Y. Katoh, *Report on Effects of Irradiation on Material IG-110-Prepared for Toyo Tanso Co., Ltd*, Oak Ridge National Lab.(ORNL), Oak Ridge, TN (United States), 2017.
292. T. D. Burchell, D. Erdmann III, R. R. Lowden, J. Hunter and C. Hannel, *The Fracture Toughness of Nuclear Graphites*, ORNL/TM-2016/678, Oak Ridge National Laboratory, 2016.
293. B. März, K. Jolley, T. J. Marrow, Z. Zhou, M. I. Heggie, R. Smith and H. Wu, *Mater. Des.*, 2018, **142**, 268.
294. B. März, K. Jolley, T. J. Marrow, Z. Zhou, M. I. Heggie, R. Smith and H. Wu, *Mater. Des.*, 2018, **159**, 103.
295. B. März, K. Jolley, T. J. Marrow, Z. Zhou, M. I. Heggie, R. Smith and H. Wu, *Data Brief*, 2018, **19**, 651.
296. X. Jin, T. J. Marrow, J. Wang, Y. Chen, H. Chen, D. Scotson, B. Wang, H. Wu and X. Fan, *Carbon*, 2022, **193**, 356.
297. H. X. Hui Yanga, D. Huanga and H. Wu, *Presented in Part at the International Nuclear Graphite Specialist Meeting*, INGSM-19, Bruges, Belgium, 16–19 Sept. 2019.
298. C. Paraskevoulakos, B. Roebuck, K. R. Hallam and P. Flewitt, *SN Appl. Sci.*, 2023, **5**, 28.
299. J. Shen, T. J. Marrow, D. Scotson, X. Jin, H. Wu and H. Chen, *J. Nucl. Mater.*, 2022, **563**, 153642.
300. Y. Gao and D. K. Tsang, *Nucl. Eng. Technol.*, 2020, **52**, 206.
301. Q. Huang and H. Tang, *Mater. Sci. Technol.*, 2019, **35**, 962.
302. X. Zhou, Z. Li, Y. Lu, H. Huang, Z. He, Z. Dai and H. Xu, *Strategic Study Chin. Acad. Eng.*, 2019, **21**, 29.
303. Z. He, Z. Liu, T. J. Marrow and J. Song, *J. Nucl. Mater.*, 2021, **550**, 152921.
304. R. Gundlapalli and S. Jayanti, *J. Power Sources*, 2020, **467**, 228225.
305. R. Gundlapalli and S. Jayanti, *J. Power Sources*, 2019, **427**, 231.
306. J. Yoon, T. Byun, J. Strizak and L. Snead, *J. Nucl. Mater.*, 2011, **412**, 315.

307. G. Liu, Y. Tang, K. Hattar, Y. Wang, W. Windes, A. Haque and J. Du, *J. Mater. Res.*, 2023, **38**, 1984.
308. S.-H. Chi and G. C. Kim, *J. Nucl. Mater.*, 2017, **491**, 37.
309. M. Jiang, K. Ammigan, G. Lolov, F. Pellemoine and D. Liu, *Carbon*, 2023, **213**, 118181.
310. A. A. Campbell, Y. Katoh, M. A. Snead and K. Takizawa, *Carbon*, 2016, **109**, 860.
311. A. A. Campbell, M. A. Snead and Y. Katoh, *J. Nucl. Mater.*, 2023, **576**, 154269.
312. M. Li, C. Shi, H. Schut, Z. Zhang and Z. Li, *J. Nucl. Sci. Technol.*, 2019, **56**, 744.
313. C. Goodwin, A. Barkatt and M. Al-Sheikhly, *Radiat. Phys. Chem.*, 2014, **97**, 38.
314. T. D. Burchell, *Grade 2114: Flexure Strength and Elastic Properties*, Oak Ridge National Lab.(ORNL), Oak Ridge, TN (United States), 2019.
315. T. D. Burchell, *Mersen Grade 2114: A Comparison of Tensile Strength Data*, Oak Ridge National Lab.(ORNL), Oak Ridge, TN (United States), 2020.
316. Mersen, Speciality Graphite Materials for Continuous Casting, https://www.graphite-eng.com/uploads/downloads/continuous_casting_brochure, accessed 22 August 2023.
317. T. Becker, T. Marrow and R. Tait, *J. Nucl. Mater.*, 2011, **414**, 32.
318. M. Roedig, G. Kleist and H. Nickel, Fracture mechanics applied to a reactor graphite, 1980. https://inis.iaea.org/collection/NCLCollectionStore/_Public/12/632/12632925.pdf, accessed 20 May 2024.
319. B. Kelly, B. Marsden, K. Hall, D. Martin, A. Harper and A. Blanchard, *IAEA Tecdoc*, 2000, p. 1154.
320. G. S. Was, *Fundamentals of Radiation Materials Science: Metals and Alloys*, Springer, 2016.
321. A. McKenna, T. Trevethan, C. Latham, P. Young and M. I. Heggie, *Carbon*, 2016, **99**, 71.
322. R. Telling and M. I. Heggie, *Philos. Mag.*, 2007, **87**, 4797.
323. B. T. Kelly, *J. Vac. Sci. Technol., A*, 1986, **4**, 1171.
324. P. Thrower and R. Mayer, *Phys. Status Solidi A*, 1978, **47**, 11.
325. B. Kelly, W. Martin, A. Price and J. Bland, *Philos. Mag.*, 1966, **14**, 343.
326. B. T. Kelly, *Carbon*, 1977, **15**, 117.
327. B. T. Kelly, W. H. Martin and P. Nettley, *Philos. Trans. R. Soc. London, Ser. A*, 1966, **260**, 37.
328. J. H. W. Simmons, *Radiation Damage in Graphite*, Pergamon Press, Oxford, 1st edn, 1965.
329. S. Johns, L. He, K. Bustillo, W. E. Windes, R. Ubic and C. Karthik, *Carbon*, 2020, **166**, 113.
330. K. Niwase, *Phys. Rev. B: Condens. Matter Mater. Phys.*, 1995, **52**, 15785.
331. T. Tanabe, S. Muto and K. Niwase, *Appl. Phys. Lett.*, 1992, **61**, 1638.
332. M. I. Heggie, I. Suarez-Martinez, C. Davidson and G. Haffenden, *J. Nucl. Mater.*, 2011, **413**, 150.
333. S. Johns, L. He, J. J. Kane, W. E. Windes, R. Ubic and C. Karthik, *Carbon*, 2020, **159**, 119.

334. D. Liu, D. Cherns, S. Johns, Y. Zhou, J. Liu, W.-Y. Chen, I. Griffiths, C. Karthik, M. Li and M. Kuball, *Carbon*, 2021, **173**, 215.
335. K. Niwase and T. Tanabe, *Mater. Trans., JIM*, 1993, **34**, 1111.
336. B. K. Annis, D. F. Pedraza and S. P. Withrow, *J. Mater. Res.*, 1993, **8**, 2587.
337. G. Plummer, H. Rathod, A. Srivastava, M. Radovic, T. Ouisse, M. Yildizhan, P. O. Å. Persson, K. Lambrinou, M. W. Barsoum and G. J. Tucker, *Mater. Today*, 2021, **43**, 45.
338. M. W. Barsoum, A. Murugaiah, S. R. Kalidindi, T. Zhen and Y. Gogotsi, *Carbon*, 2004, **42**, 1435.
339. S. Basu, A. Zhou and M. W. Barsoum, *J. Struct. Geol.*, 2009, **31**, 791.
340. M. W. Barsoum, X. Zhao, S. Shanazarov, A. Romanchuk, S. Koumlis, S. J. Pagano, L. Lamberson and G. J. Tucker, *Phys. Rev. Mater.*, 2019, **3**, 013602.
341. J. Gruber, A. C. Lang, J. Griggs, M. L. Taheri, G. J. Tucker and M. W. Barsoum, *Sci. Rep.*, 2016, **6**, 33451.
342. H. O. Badr and M. W. Barsoum, *Carbon*, 2023, **201**, 599.
343. J. Brocklehurst and B. Kelly, *Carbon*, 1993, **31**, 155.
344. B. N. Gareth, *J. Phys. D: Appl. Phys.*, 2000, **33**, 2966.
345. H. Bridge, B. Kelly and P. Nettley, *Carbon*, 1964, **2**, 83.
346. N. Simos, H. G. Kirk, P. Thieberger, H. Ludewig, J. O. Conor, L. Mausner, P. T. Trung, K. T. McDonald, K. Yoshimura and J. R. J. Bennett, *J. Nucl. Mater.*, 2008, **377**, 41.
347. F. Pellemoine, K. Ammigan, C. Barbier and K. Yonehara, *J. Instrum.*, 2023, **18**, T07006.
348. T. J. Marrow, I. Šulak, B. S. Li, M. Vukšić, M. Williamson and D. E. J. Armstrong, *Carbon*, 2022, **191**, 236.
349. B. Kelly, *Fusion Technol.*, 1989, **16**, 96.
350. M. Birch and D. J. Bacon, *Carbon*, 1983, **21**, 491.
351. M. Mostafavi, S. McDonald, P. Mummery and T. Marrow, *Eng. Fract. Mech.*, 2013, **110**, 410.
352. R. Price, Strength of irradiated graphite: A review, https://inis.iaea.org/collection/NCLCollectionStore/_Public/11/499/11499311.pdf, accessed 09 September 2023.
353. T. D. Burchell and L. L. Snead, *J. Nucl. Mater.*, 2007, **371**, 18.
354. A. A. Griffith, *Philos. Trans. R. Soc. London, Ser. A*, 1921, **221**, 163.
355. E. D. Eason, G. N. Hall, B. J. Marsden and G. B. Heys, *J. Nucl. Mater.*, 2013, **436**, 208.
356. B. Marsden, S. Fok, T. Marrow and P. Mummery, *Proceedings of HTR-2004 Beijing*, 2004.
357. A. A. Balandin, *Nat. Mater.*, 2011, **10**, 569.
358. D. T. Morelli and G. A. Slack, in *High Thermal Conductivity Materials*, ed. S. L. Shindé, and J. S. Goela, Springer New York, New York, NY, 2006, p. 37, DOI: 10.1007/0-387-25100-6_2.
359. L. L. Snead, in *Carbon Materials for Advanced Technologies*, ed. T. D. Burchell, Elsevier Science Ltd, Oxford, 1999, p. 389.
360. R. Taylor, B. T. Kelly and K. E. Gilchrist, *J. Phys. Chem. Solids*, 1969, **30**, 2251.

361. D. B. Timothy, *Phys. Scr.*, 1996, **1996**, 17.
362. G. Engle and W. Eatherly, *High Temp. – High Pressures*, 1972, **4**(2), 119.
363. L. L. Snead and T. D. Burchell, *J. Nucl. Mater.*, 1995, **224**, 222.
364. M. S. L. Jordan, D. Liu, M. Jiang and A. Tzelepi, in *Graphite Testing for Nuclear Applications: The Validity and Extension of Test Methods for Material Exposed to Operating Reactor Environments*, ed. A. Tzelepi, and M. Metcalfe, ASTM International, West Conshohocken, PA, 2022, p. 80, DOI: 10.1520/STP163920220037.
365. T. D. Burchell, in *Comprehensive Nuclear Materials*, ed. R. J. M. Konings, Elsevier, Oxford, 2012, p. 299.
366. T. D. Burchell and W. P. Eatherly, *J. Nucl. Mater.*, 1991, **179–181**, 205.
367. D. Liu, M. Mostafavi, P. E. Flewitt, T. J. Marrow and D. J. Smith, *Key Eng. Mater.*, 2014, **577**, 485.
368. A. Mauger, C. M. Julien, J. B. Goodenough and K. Zaghib, *J. Electrochem. Soc.*, 2019, **167**, 070507.
369. N. Akhmetov, A. Manakhov and A. S. Al-Qasim, *Electronics*, 2023, **12**, 5.
370. M. S. Whittingham, *Science*, 1976, **192**, 1126.
371. K. Mizushima, P. C. Jones, P. J. Wiseman and J. B. Goodenough, *Mater. Res. Bull.*, 1980, **15**, 783.
372. Fastmarkets, Synthetic versus natural graphite debate rages on: 2023 preview, https://www.fastmarkets.com/insights/synthetic-versus-natural-graphite-%09debate#:~:text=Natural%20graphite%20anode%20has%20the,charge%20turnaround %20and%20battery%20longevity, accessed 12 January 2024.
373. M. Mundszinger, S. Farsi and M. Rapp, *et al.*, Morphology and texture of spheroidized natural and synthetic graphites, *Carbon*, 2017, **111**, 764.
374. M. Li, J. Lu, Z. Chen and K. Amine, *Adv. Mater.*, 2018, **30**, 1800561.
375. M. V. Reddy, A. Mauger and C. M. Julien, *et al.*, Brief history of early lithium-battery development, *Materials*, 2020, **13**, 8.
376. C. J. Murray, *Long Hard Road: The Lithium-Ion Battery and the Electric Car*, Purdue University Press, 2022.
377. K. Zaghib, X. Song, A. Guerfi, R. Rioux and K. Kinoshita, *J. Power Sources*, 2003, **119–121**, 8.
378. S. K. Heiskanen, J. Kim and B. L. Lucht, *Joule*, 2019, **3**, 2322.
379. J. P. Pender, G. Jha and D. H. Youn, Electrode degradation in lithium-ion batteries, *ACS Nano*, 2020, **14**, 1243.
380. S. J. An, J. Li and C. Daniel, *et al.*, *Carbon*, 2016, **105**, 52.
381. A. S. Ho, D. Y. Parkinson and D. P. Finegan, *et al.*, *ACS Nano*, 2021, **15**, 10480.
382. W. Cao, J. Zheng, D. Adams and J. P. Zheng, *ECS Trans.*, 2014, **61**, 37.
383. N. LeGe, X.-X. He, Y.-X. Wang, Y. Lei, Y.-X. Yang, J.-T. Xu, M. Liu, X. Wu, W.-H. Lai and S.-L. Chou, *Energy Environ. Sci.*, 2023, **16**, 5688.
384. C. P. Grey and D. S. Hall, *Nat. Commun.*, 2020, **11**, 6279.
385. H. Zhang, Y. Yang, D. Ren, L. Wang and X. He, *Energy Storage Mater.*, 2021, **36**, 147.

386. J. Asenbauer, T. Eisenmann, M. Kuenzel, A. Kazzazi, Z. Chen and D. Bresser, *Sustainable Energy Fuels*, 2020, **4**, 5387.
387. J. Wang, B. Ge, H. Li, M. Yang, J. Wang, D. Liu, C. Fernandez, X. Chen and Q. Peng, *Chem. Eng. J.*, 2021, **420**, 129739.
388. J.-B. Donnet and R. C. Bansal, *Carbon Fibers*, CRC Press, Boca Raton, 3rd edn, 1998.
389. P. Morgan, *Carbon Fibers and Their Composites*, CRC Press, Boca Raton, 2005.
390. H. Lengsfeld, H. Mainka and V. Altstädt, *Carbon Fibers: Production, Applications, Processing*, Hanser Publications, 2nd edn, 2021.
391. M. E. Swan and K. R. Swan, *Sir Joseph Wilson Swan FRS, Inventor and Scientist*, Oriel Press, 1968.
392. R. W. Clark, *Edison: The Man Who Made the Future*, Putnam, New York, 1977.
393. R. Bacon, *J. Appl. Phys.*, 1960, **31**, 283.
394. A. Shindo, *Osaka Kogyo Gijitsu Shikenjo Koho*, 1961, **12**, 110.
395. R. Moreton, W. Watt and W. Johnson, *Nature*, 1967, **213**, 690.
396. L. S. Singer, *Fuel*, 1981, **60**, 839.
397. Wikipedia, Carbon fibers, https://en.wikipedia.org/wiki/Carbon_fibers, accessed 24 January 2024.
398. X. Yang, Y. Chen, C. Zhang, G. Duan and S. Jiang, *Composites, Part B*, 2023, **249**, 110386.
399. T. Wang, Z. Chen, W. Gong, F. Xu, X. Song, X. He and M. Fan, *ACS Omega*, 2023, **8**, 22316.
400. M. Endo, Y. A. Kim, T. Hayashi, K. Nishimura, T. Matusita, K. Miyashita and M. S. Dresselhaus, *Carbon*, 2001, **39**, 1287.
401. J. W. Seo, A. Magrez, M. Milas, K. Lee, V. Lukovac and L. Forró, *J. Phys. D: Appl. Phys.*, 2007, **40**, R109.
402. E. Fitzer, *Carbon*, 1989, **27**, 621.
403. D. J. Johnson, *J. Phys. D: Appl. Phys.*, 1987, **20**, 286.
404. K. Potter, in *The Structural Integrity of Carbon Fibre Composites*, ed. P. W. R Beaumont, and C. Soutis, Springer International Publishing AG, 2016, p. 29.
405. K. Potter, in *Design and Manufacture of Structural Composites*, ed. L. Harper, and M. Clifford, Woodhead Publishing, 2023, p. 83.
406. Mitsubishi Chemical Corporation, https://www.m-chemical.co.jp/carbon-fiber/en/, accessed 24 January 2024.
407. P. McMullen, *Composites*, 1984, **15**, 222.
408. K. Potter, *Composites, Part A*, 1999, **30**, 619.
409. M. Volk, O. Yuksel, I. Baran, J. Hattel, J. Spangenberg and M. Sandberg, *Composites, Part B*, 2022, **246**, 110135.
410. M. Azeem, H. Ya, M. Alam, M. Kumar, P. Stabla, M. Smolnicki and M. Mustapha, *J. Energy Storage*, 2022, **49**, 103468.
411. M. Elkington, D. Bloom, C. Ward and K. Potter, *Adv. Manuf.: Polym. Compos. Sci.*, 2015, **1**, 138.

412. K. Potter, in *Design and Manufacture of Structural Composites*, ed. L. Harper, and M. Clifford, Woodhead Publishing, 2023, p. 125.
413. C. Rudd, A. Long, K. Kendall and C. Mangin, *Liquid Moulding Technologies: Resin Transfer Moulding, Structural Reaction Injection Moulding and Related Processing Techniques*, Elsevier, 1997.
414. E. Pakdel, S. Kashi, R. Varley and X. Wang, *Conserv. Recycl.*, 2021, **166**, 105340.
415. G. Fernlund, N. Rahman, R. Courdji, M. Bresslauer, A. Poursartip, K. Willden and K. Nelson, *Composites, Part A*, 2002, **33**, 341.
416. K. Potter, B. Khan, M. Wisnom, T. Bell and J. Stevens, *Composites, Part A*, 2008, **39**, 1343.
417. National Centre for Advanced Materials Performance, Wichita State University, USA, https://www.wichita.edu/industry_and_defense/NIAR/Research/ncamp.php, accessed 22 January 2024.
418. R. Khandan, S. Noroozi, P. Sewell and J. Vinney, *J. Mater. Sci.*, 2012, **47**, 5901.
419. B. G. Green, M. R. Wisnom and S. R. Hallett, *Composites, Part A*, 2007, **38**, 867.
420. S. Sridharan, in *Delamination behaviour of composites*, Elsevier, 2008.
421. O. Konur and F. Matthews, *Composites*, 1989, **20**, 317.
422. S. Shah, S. Karuppanan, P. Megat-Yusoff and Z. Sajid, *Compos. Struct.*, 2019, **217**, 100.
423. N. Forintos and T. Czigany, *Composites, Part B*, 2019, **162**, 331.
424. P. Jerome, *Proc. 13th Int. Conf. Compos. Mater.*, Beijing, 2001, p. 1.
425. Composites Leadership Forum, The 2016 UK Composites Strategy, https://compositesuk.co.uk/wp-content/uploads/2021/09/Strategy-final-version_1.pdf, accessed 22 January 2024.
426. F. Danzi, R. Salgado, J. Oliveira, A. Arteiro and P. Camanho, *Molecules*, 2021, **11**, 2203.
427. Z. Yan, S. Yi, X. Li, J. Jiang, D. Yang and N. Du, *Mater. Today Energy*, 2023, **31**, 101225.
428. M. Longana, N. Ong, H. Yu and K. Potter, *Compos. Struct.*, 2016, **153**, 271.
429. B. Wang, S. Ma, S. Yan and J. Zhu, *Green Chem.*, 2019, **21**, 5781.

CHAPTER 2

Synthesis, Properties and Applications of Graphene and Related Materials

NICHOLAS YUNG LI LOH[a,b], WAN TING TEE[a], BILLIE YAN ZHANG HIEW[c], SVENJA HANSON[b], SUYIN GAN[a] AND LAI YEE LEE[a*]

[a]Department of Chemical and Environmental Engineering, University of Nottingham Malaysia, Jalan Broga, 43500 Semenyih, Selangor, Malaysia; [b]Department of Chemical and Environmental Engineering, University of Nottingham Ningbo China, Ningbo, 315100, China; [c]School of Engineering and Physical Sciences, Heriot-Watt University Malaysia, 62200 Putrajaya, Malaysia

*E-mail: lai-yee.lee@nottingham.edu.my

2.1 Introduction

Carbon is one of the most abundant elements and can exist in many fascinating structures, such as graphite, graphene, fullerenes and carbon nanotubes. The subject of this chapter, graphene, is a single layer of hexagonally packed carbon atoms in an sp^2 hybridisation state. Graphene was first isolated and characterised in 2004 by Andre Geim and Konstantin Novoselov, and they were awarded the Nobel Prize in Physics for this work in 2010.[1] Graphene has demonstrated exceptional properties including adjustable bandgap, high Young's modulus (~1 TPa), outstanding thermal and electrical conductivities,

Inorganic Materials Series No. 15
Elemental Carbon
Edited by Peter J. F. Harris

Published by the Royal Society of Chemistry, www.rsc.org

and ultrahigh specific surface area (~2630 m^2 g^{-1}).[2] Because of these properties, graphene has received worldwide recognition in various research fields including hydrogen storage,[3] biosensors[4] and supercapacitors,[5] as evident from the increasing number of publications displayed in Figure 2.1.

Different forms of graphene, specifically graphene oxide (GO) and reduced graphene oxide (rGO), also have potential in various applications due to the existence of aromatic rings, free π–π electron mobility and reactive functional groups on the basal plane.[6] The main research domains involving graphene materials are illustrated in Figure 2.2.

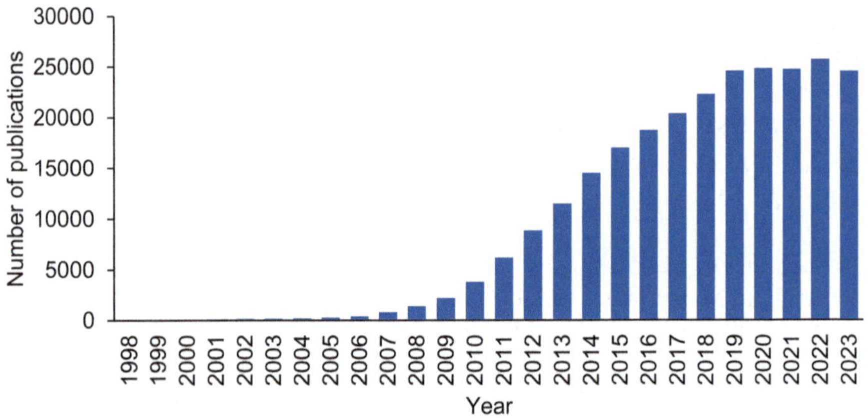

Figure 2.1 Trend in graphene-related publications. Extracted from Scopus on 13 February 2024.

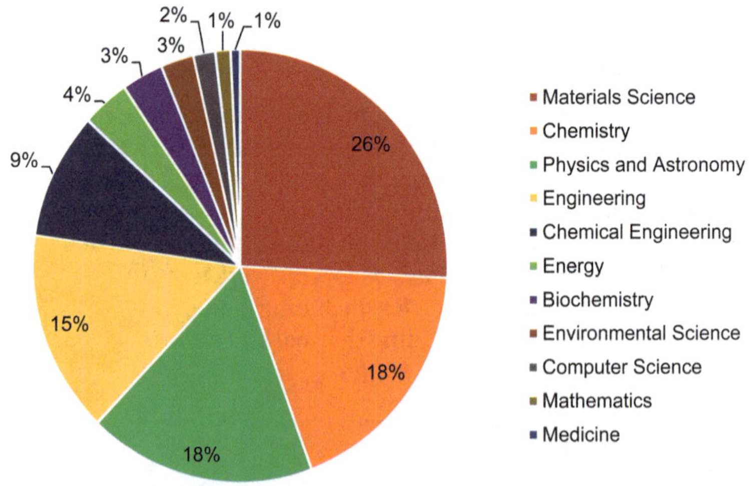

Figure 2.2 Research domain coverage of graphene materials. Extracted from Scopus on 1 May 2023.

Notably, materials science (26%) is the most widely explored area for graphene, followed by engineering (24%, including chemical engineering), chemistry (18%), physics and astronomy (18%), and others (13%). This trend could be attributed to graphene's ease of modification, enabling infinite possibilities for functionalisation of the carbon backbone. Oxygenation, fluorination and hydrogenation of graphene can produce GO, fluorographene and graphane, respectively, and these processes bestow interesting properties not available in pristine graphene. Moreover, the insertion of carbyne chains between the carbon hexagons of graphene yields graphyne and related forms, which possess similar properties to graphene. In the search for new properties and applications, special attention has been focused on designing graphene-derived materials to introduce desired attributes or even surpass the properties of pristine graphene.

Numerous publications have documented the remarkable performance of graphene materials in a wide range of applications. To take a random example, a sulfur-doped graphene sponge has exhibited a copper(II) ion (Cu^{2+}) adsorption capacity of 228 mg g^{-1}, which was 40 times greater than activated carbon.[7] In energy storage, Pilathottathil *et al.* (2021) have compared the performance of supercapacitor devices fabricated from different carbon allotropes and molybdenum disulfide.[8] The study revealed that the graphene-based device had a relatively high specific capacitance (202 F g^{-1}) and energy density (5.68 W h kg^{-1}).[8] Graphene and its derivatives can also significantly enhance the mechanical properties of materials. Thus, Naseem *et al.* (2022) demonstrated that the tensile and compressive strengths of polymer modified cement improved by 59% and 33%, respectively, with the incorporation of GO.[9] This was due to GO in the hardened cement paste structure inhibiting the formation of air bubbles during the cement mixing.[9] In these, and many other areas, graphene materials are considered to have great potential for technological development.

This chapter aims to provide an overview of graphene and its alternative forms, such as GO and rGO. Additionally, emerging graphene-related materials including halogenated graphene, hydrogenated graphene and graphyne are also discussed. The structures of these nanomaterials are presented first, followed by a discussion of their synthesis methods and a summary of their applications in various fields.

2.2 Structures of Graphene Materials

2.2.1 Graphene

Over the years, literature data on two-dimensional (2D) carbons, including graphene and its spin-off forms, has accumulated significantly, and many definitions have been used to describe the graphene products. The commonly used terminologies for graphene materials are listed in Table 2.1 along with the respective definitions.

Table 2.1 Different types of graphene and related materials with their respective definitions. Adapted from ref. 10 with permission from Elsevier, Copyright 2013.

Graphene material	Definition
Graphene	• An atom-thick sheet of sp^2-bonded carbon atoms in hexagonal arrangement which is not an integral part of a carbon material but is freely suspended or adhered on foreign substrate. • Lateral dimension ranging from several nanometres to macroscale.
Turbostratic carbon	• Three-dimensional sp^2-bonded carbon material where there is no defined orientation of the layers.
Bilayer, trilayer graphene	• A 2D sheet of free-standing films/flakes or a substrate-bound coating consisting of 2 or 3 well-defined, countable, stacked graphene layers of extended lateral dimension.
Multilayer graphene (MLG)	• A 2D sheet of free-standing films/flakes or a substrate-bound coating consisting of a small number (2 to 10) of well-defined, countable, stacked graphene layers of extended lateral dimension.
Few-layer graphene (FLG)	• A subset of MLG with 2 to 5 layers of graphene.
Graphene nanosheet	• A graphene sheet with lateral dimension less than 100 nm.
Graphene microsheet	• A graphene sheet with lateral dimension ranging between 100 nm and 100 μm.
Graphene nanoribbon	• A graphene sheet in which the longer lateral dimension exceeds the shorter lateral dimension by at least an order of magnitude.
Graphene oxide (GO)	• Graphene modified *via* oxidation. • Monolayer material with high oxygen content and C/O ratio <3.
Graphite oxide	• Bulk solid from oxidation of graphite with basal plane functionalisation and increased interlayer spaces. • Can be exfoliated in solution to form GO or few-layer GO.
Reduced graphene oxide (rGO)	• GO reduced by chemical, thermal, microwave, photochemical, photo-thermal or biological methods to decrease its oxygen content.

In graphene, each carbon atom shares the sp^2 electrons with three other carbon atoms forming a planar structure with a bond angle of 120° between the carbon–carbon bonds. As such, the carbon atoms are arranged in a hexagonal lattice. Figure 2.3 illustrates the sp^2 hybridisation of the carbon atoms whereby each atom has a π-orbital from the remaining unhybridised 2p orbital, thus contributing to the network of delocalised electrons.

A graphical illustration of the graphene structure is shown in Figure 2.4. The carbon–carbon σ-bonds are 0.142 nm in length; hence, the unit cells of graphene have a lattice constant, a, of 0.246 nm.[12] The thickness of graphene is 0.43 nm, comparable to its theoretical thickness of 0.34 nm.[13]

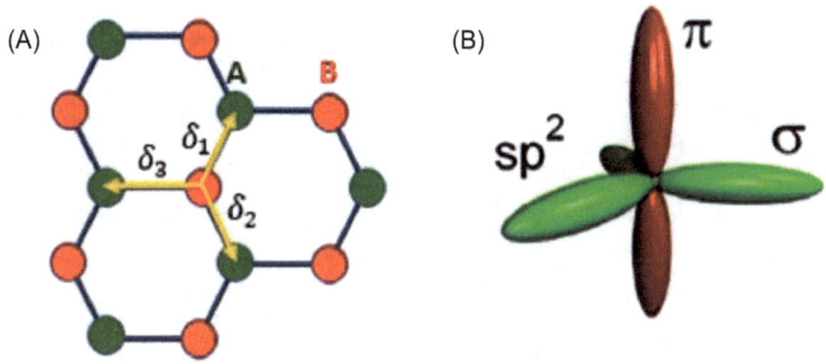

Figure 2.3 (A) 3σ bonds (δ_1, δ_2 and δ_3) in sp^2 hybridised graphene (green dots = Carbon A species; red dots = Carbon B species). (B) 3σ bonds in-plane and 2π bonds perpendicular to 3σ bonds. Reproduced from ref. 11 with permission from Elsevier, Copyright 2021.

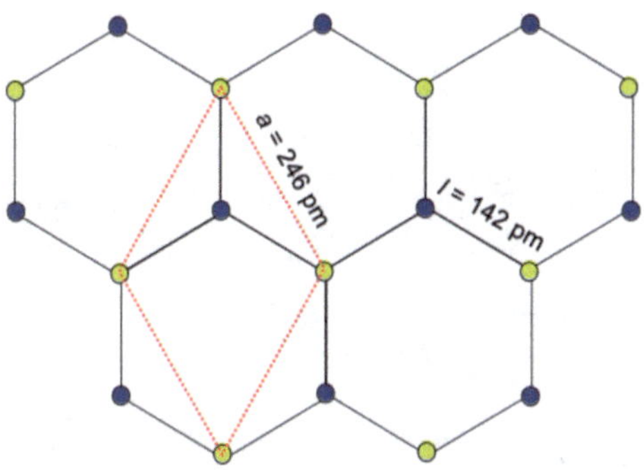

Figure 2.4 Hexagonal lattice of graphene with its unit cell (red dotted lines) and lattice constant, *a*. Reproduced from ref. 12 with permission from Taylor & Francis, Copyright 2018.

Figure 2.5 shows the edge configurations of graphene known as *armchair* and *zigzag*. The edges of graphene monolayers exhibit a higher chemical reactivity than at the basal planes, as well as a higher electronic density of states. The distribution of electrons is influenced by the edge configuration of graphene, hence the importance of crystallographic orientation in controlling the electronic properties of graphene.

The underlying source of the electronic properties of graphene lies in its band structure, which features Dirac cones and Dirac points (Figure 2.6). The Dirac points are points at which the valence and conduction bands of graphene coincide at a single point at the Fermi level. Consequently, graphene can be considered as a zero-band gap semiconductor or semi-metal with zero density of states at the Fermi level. Comparatively, typical semiconductors have band gaps ranging from 1 to 2 eV, but graphene is an exception with a band gap of zero.[15] This allows the free movement of electrons, giving

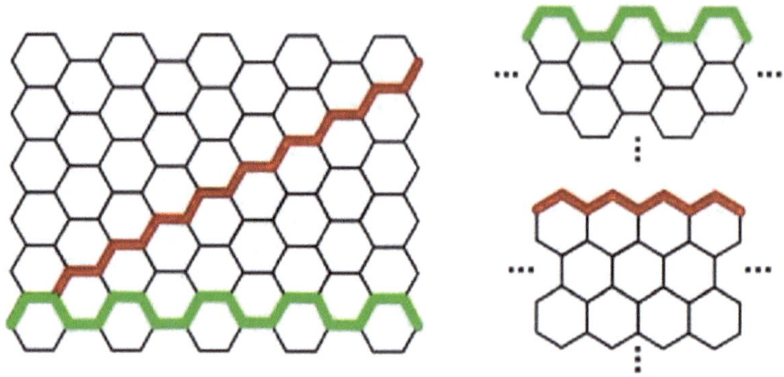

Figure 2.5 Edge configurations of graphene: armchair (green) and zigzag (red). Reproduced from ref. 14 with permission from Springer Nature, Copyright 2009.

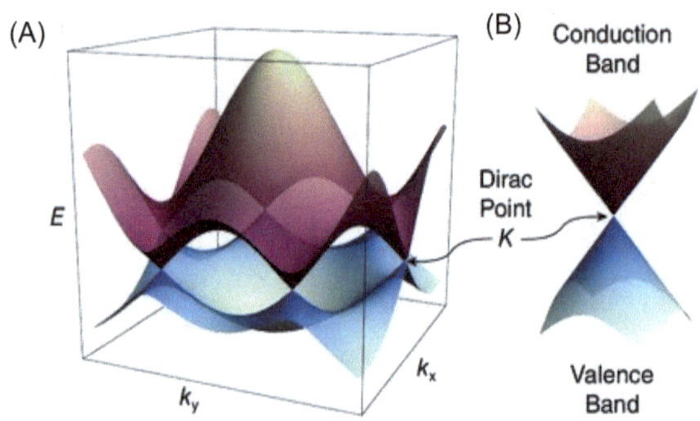

Figure 2.6 (A) Dirac cones and (B) Dirac point of graphene. Reproduced from ref. 18 with permission from John Wiley and Sons, Copyright 2012.

graphene extraordinary electronic properties that can be exploited in numerous applications.

One of the most important discoveries in the field of graphene electronics is the observation of superconductivity in twisted bilayer graphene. Twisting of bilayer graphene involves rotating two layers of graphene stacked on each other at a specific angle (typically around 1.1°) and this can modify the interactions between energy bands to create flat bands near the Fermi level. The presence of these flat bands enables the twisted bilayer graphene to act as a superconductor with electric current flowing in the absence of resistance.[16]

The cones which surround the Dirac point are known as the Dirac cones and are linear toward the Dirac point. This has important implications with regard to the mobility of electrons in graphene. The intersection of the conduction and valence bands at the Dirac points enable electrons to move easily between the two bands, and as such, the application of a voltage allows current to flow through graphene. The linear bands near the Dirac point influence the effective mass of electrons where electrons possess mass-less behaviour in these regions of zero curvature. This means that electrons are able to move rapidly, which imparts the high electron mobility of graphene – as high as 2.5×10^5 cm^2 V^{-1} s^{-1}.[17]

2.2.1.1 Defective Graphene Structures

As already noted, the ideal structure of graphene exhibits zero band gap, high tensile strength and thermal conductivity. However, various graphene structural defects, such as surface corrugations, vacancies, adatoms, substitutions, dislocations, grain boundaries and other deformations (Figure 2.7), have been observed in the planar sheet of graphene, which adversely impact the ideal properties of graphene. Table 2.2 illustrates some of the impacts of imperfections on the properties of graphene. These defects can occur from the synthesis of graphene, such as chemical vapour deposition and oxidation–reduction reactions of graphite. In addition, energetic particle irradiation or chemical treatments can also introduce various deformity effects in the graphene layer.

Graphene layers can exhibit corrugative and wave-like structures as a consequence of structural deformation which affect its intrinsic optical and

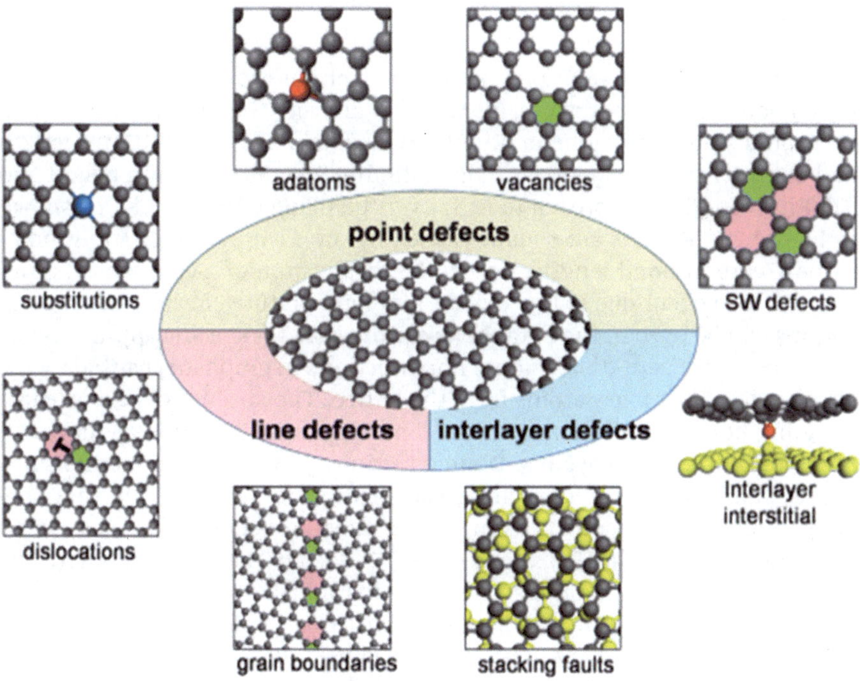

Figure 2.7 Various defects of graphene framework. Reproduced from ref. 19 with permission from Elsevier, Copyright 2018.

Table 2.2 Comparison of properties between pristine graphene and defective graphene. Reproduced from ref. 19 with permission from Elsevier, Copyright 2018.

Properties	Pristine graphene	Defective graphene
Young's modulus (TPa)	~1	0.15–0.95
Fracture strength (GPa)	90–130	47–117
Electrical conductivity (S m^{-1})	10^8	$1–10^5$
Electron mobility (cm^2 V^{-1} s^{-1})	$10^5–10^6$	$1–10^4$
Thermal conductivity (W m^{-1} K^{-1})	$10^3–10^4$	$10–10^3$

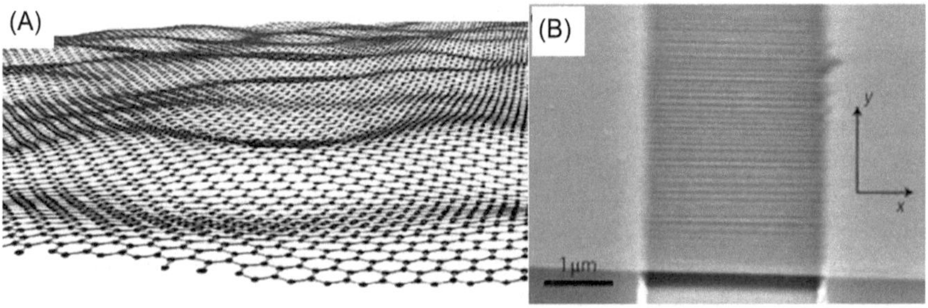

Figure 2.8 (A) Ripples and (B) wrinkles on graphene. Reproduced from ref. 21 with permission from Elsevier, Copyright 2016.

electronic properties. Examples of the corrugations are ripples and wrinkles, which can induce magnetic fields possibly altering the local potentials and electronic states of graphene. Ripples with dimensions of 8–10 nm length and 0.7–1.0 nm height, and wrinkles with dimensions of <15 nm height and >100 nm length were proven to exist over the plane of graphene[20] as shown in Figure 2.8. Ripples are usually caused by the asymmetric distribution of carbon–carbon bond lengths due to the localisation of π-electrons.[12] Wrinkles can also occur due to line defects and temperature effects. These corrugations can be used to tune graphene's properties for specific applications.

Vacancies are zero-dimensional imperfections in graphene, caused by the absence of one or a few atoms from the lattice. The absence of atoms leaves dangling bonds in adjacent atoms and induces reconstruction as they rearrange into different polygons. Double vacancies (two missing atoms) show higher stability compared to single vacancies as there are no residual dangling bonds after reconstruction.[19] The most frequently observed defect configurations are the 5–8–5 and 555–777 defects (Figure 2.9), where the former represents two opposing pentagons and a shared octagon, while the latter represents three pentagons and three heptagons. The dynamic transformation among configurations have been observed by transmission electron microscopy (TEM), for instance, the migration of double vacancies from 5–8–5 to 555–777 and *vice versa*.[22] Larger densities of vacancies in graphene may lead to the formation of carbon allotropes, such as 2D carbon

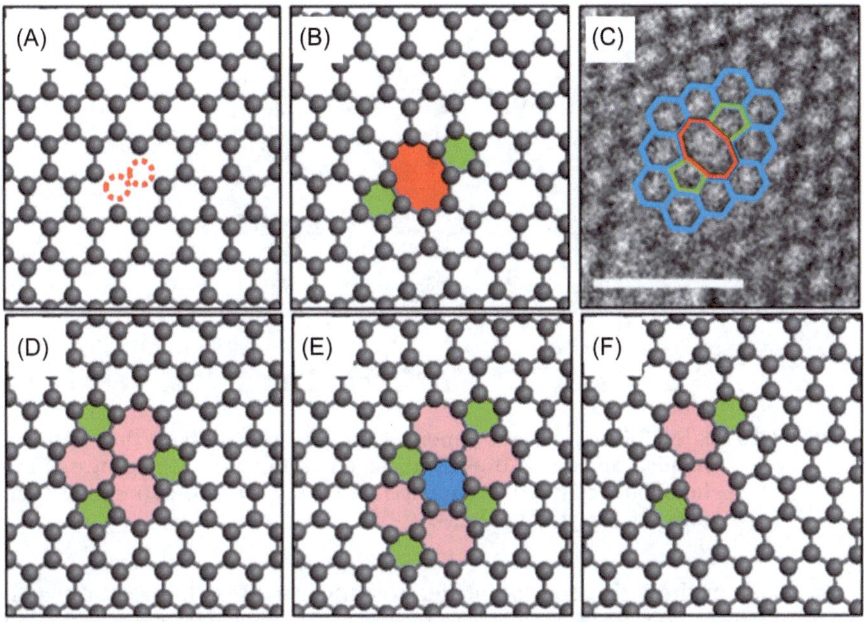

Figure 2.9 Atomic arrangements for (A) and (B) 5–8–5 defect from double vacancies, (C) TEM micrograph of 5–8–5 defect, (D) 555–777, (E) 5555–6–7777, and (F) 5–7–7–5 defects. Reproduced from ref. 19 with permission from Elsevier, Copyright 2018.

glass, where a random arrangement of polygons is formed from sp^2 carbons (Haeckelite). Functional groups, such as hydroxyl and carboxyl groups, can also be attached to the dangling bonds of vacancies. In this manner, the reactivity of graphene may be enhanced with the presence of sufficient defects and this may be useful for applications, such as hydrogen storage,[23] catalysis,[24] polymer reinforcement[25] and environmental pollution remediation.[26]

Inserting additional carbon atoms into the graphene lattice, for instance, out-of-plane atoms or adatoms, can strain the local 2D graphene structure. Adatoms can be situated at different sites, such as a bridge site (B) above the centre of bonds, top site (T) directly above an atom and in the centre of the hexagonal ring (H) (Figure 2.10). Some degree of local sp^3 hybridisation may occur as adatoms are bonded to the underlying atoms in the graphene layer, particularly in the B and T sites, as new covalent bonds are formed between the adatoms and the carbon atoms of the graphene framework. In the H site, the adatoms can easily migrate over the graphitic sheet at room temperature and may combine with vacancies upon contact. The adatoms are not limited to carbon atoms: foreign atoms, such as metallic elements, may also be incorporated into the graphene lattices through chemisorption or physisorption. Specifically, metallic elements and partially filled d-shell transition elements are favoured at the H site, whereas non-metallic elements and filled d-shell transition elements are preferred at the B site. Hydrogen and halogens are

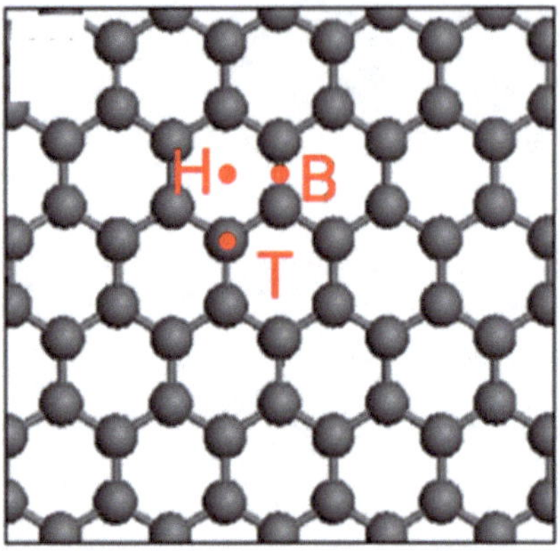

Figure 2.10 Sites for adatom attachment – bridge site (B) above the centre of bonds, top site (T) directly above an atom, and in the centre of the hexagonal ring (H). Reproduced from ref. 19 with permission from Elsevier, Copyright 2018.

most stable at the T site due to the electrostatic attraction between the adatoms and the graphene surface. Note that the sp^3 characteristic in the local graphene regions due to the presence of adatoms can lead to distortion of the planar structure.

Foreign atoms can also be substituted into the graphene lattice replacing one or two carbon atoms based on their atomic radius. Boron and nitrogen atoms are common replacement atoms due to their similar atomic radius to carbon. The difference in the number of electrons in boron and nitrogen, as compared to carbon, endows the doped graphene with semiconducting behaviour. Thus, the additional electron in nitrogen is donated to the graphene layer, which transforms it into an n-type semiconductor, whereas a p-type is observed with boron doping.[27] Figure 2.11 illustrates the various sites in which nitrogen can be doped onto the graphene network.

Dislocations can be initiated by the presence of a vacancy chain in the *armchair* or *zigzag* direction, whereby the carbon atoms surrounding the chain are reconstructed (Figure 2.12). Specifically, the dangling bonds from the edges of the vacancy chain are reconnected to recover the graphene sheet forming a 5–7 pair at both ends of the strip. To accommodate such strains, rippling of the planar structure occurs as the dislocations are introduced into the graphene sheets. One way of inducing dislocations is the use of electron beam irradiation to sputter carbon atoms out of the graphene sheet. In such a way, the electronic transport in graphene can be altered as leakage currents are introduced through periodic dislocations.

Grain boundaries are line defects separating two independent grains tilted relative to each other. These imperfections are often observed in graphene synthesised by chemical vapour deposition in which simultaneous nucleation at different points occurs causing a mismatch of lattice orientations (Figure 2.13). Grain boundaries exist where the different orientations meet

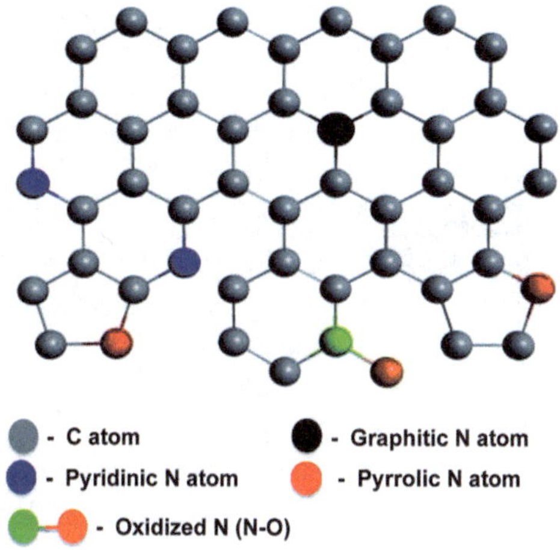

Figure 2.11 Doping sites for graphitic, pyridinic and pyrrolic nitrogen. Reproduced from ref. 28 with permission from the Royal Society of Chemistry.

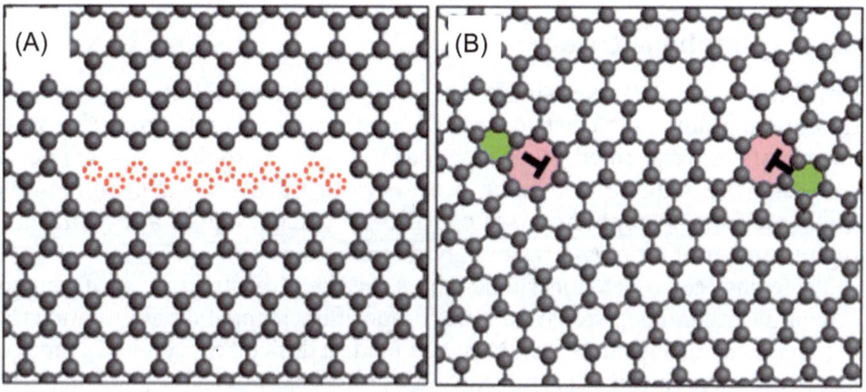

Figure 2.12 Dislocation defects due to (A) vacancy chain and (B) pairs of 5–7 defects formation. Reproduced from ref. 19 with permission from Elsevier, Copyright 2018.

and the lattices are tilted relative to each other by a misorientation angle, θ. Small θ defects (<20°) cause strong rippling in the graphene sheet, whereas graphene sheets with larger θ (20–30°) are flat. Grain boundaries can enhance the conductivity of graphene along the line defects as the pentagon and octagon carbon rings become conducting metallic wires.

While defects can be intentionally introduced into the graphene lattice with the goal of adjusting the graphene properties, some defects are naturally formed and have been observed in graphene synthesised *via* different methods, as already noted. In some applications that require pristine graphene properties, there is a requirement to prevent these defects forming.[19]

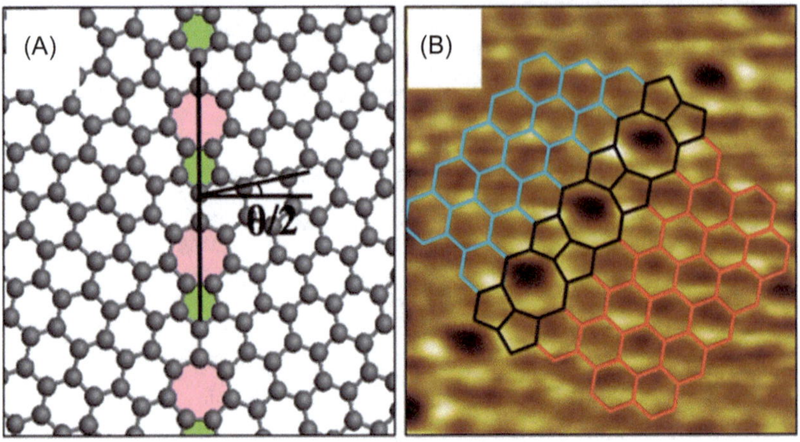

Figure 2.13 Grain boundaries in graphene: (A) Line of 5–7 pairs formed by two grains coalescing, and (B) alternating line of pentagons and octagons (black) in graphene. Reproduced from ref. 19 with permission from Elsevier, Copyright 2018.

2.2.2 Graphene Oxide

Graphene oxide (GO) is a single layer of graphene containing oxygen rich functional groups, such as hydroxyl, carbonyl, carboxyl and epoxy. The presence of these oxygenated groups imbues GO with hydrophilicity and chemical reactivity, making it a useful precursor for the synthesis of numerous graphene forms. Graphene oxide has also been suggested as a very promising raw material for mass production of graphene.[2]

There has been much uncertainty about the chemical structure of GO due to its complexity and non-stoichiometric composition. Sample variability and lack of precise characterisation tools have also made it difficult to establish a precise structure.[29] The early models mostly proposed a regular GO lattice with discrete repeating units. Hofmann and Holst (1939) put forward a model incorporating only epoxy groups forming an ideal net with a molecular formula of C_2O.[30] However, this model does not contain any hydrogen atoms, which are known to exist in GO. To account for the hydrogen content in GO, hydroxyl and 1,3-ether groups were incorporated into the model by Ruess (1947).[31] In this model, carbons were formed into a sp^3-hybridised carbon grid with *chair*-like conformation instead of the sp^2-hybridised model of Hofmann and Holst. Scholz and Boehm (1969) proposed a corrugated backbone model without the presence of epoxy and 1,3-ether groups.[32] Nakajima *et al.* (1988) suggested a new structure model of double carbon layers linked by perpendicular sp^3 bonds where the hydroxyl and carbonyl groups were combined into double layers.[33] The lattice-based atomic structures of the GO models are displayed in Figure 2.14.

Later models have taken into account the non-stoichiometric nature of GO and have eliminated the lattice-based concepts, notably the well-known model by Lerf *et al.* (1998) (Figure 2.15).[34] The use of solid-state nuclear magnetic

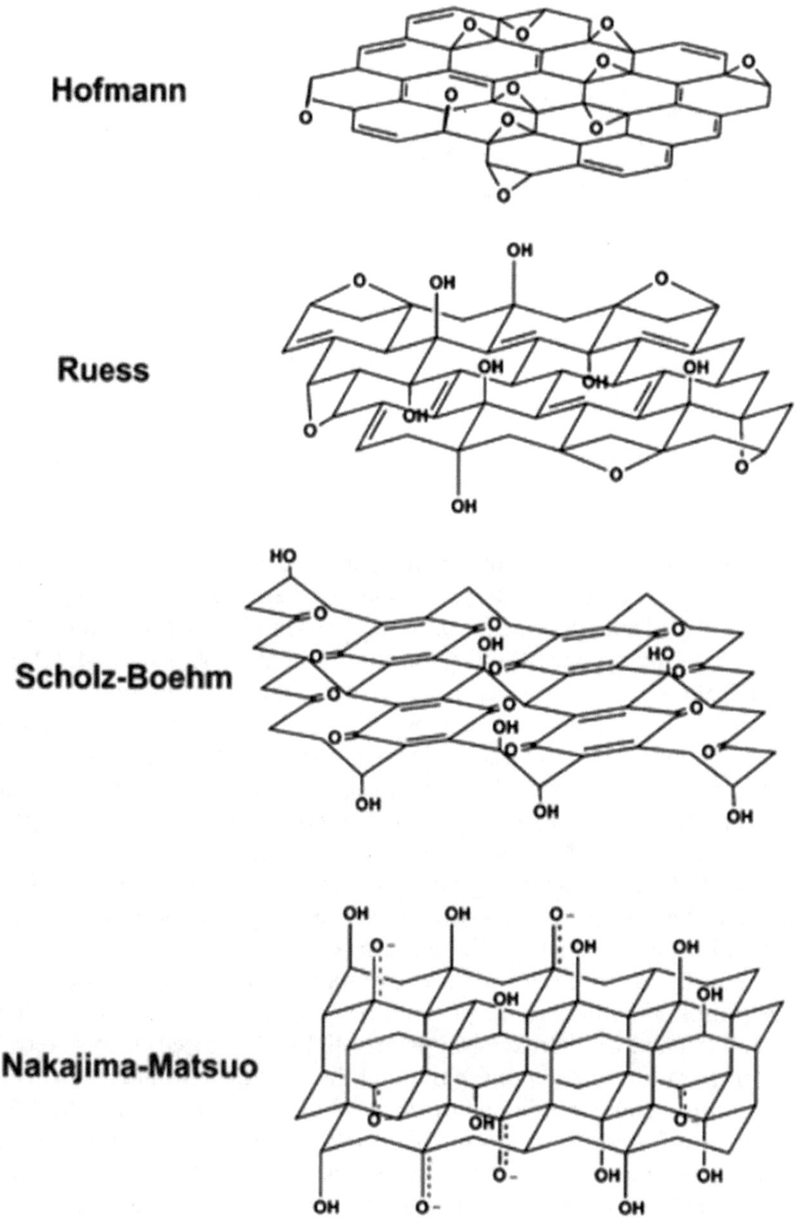

Figure 2.14 Various lattice-based atomic structure models of GO. Reproduced from ref. 34 with permission from American Chemical Society, Copyright 1998.

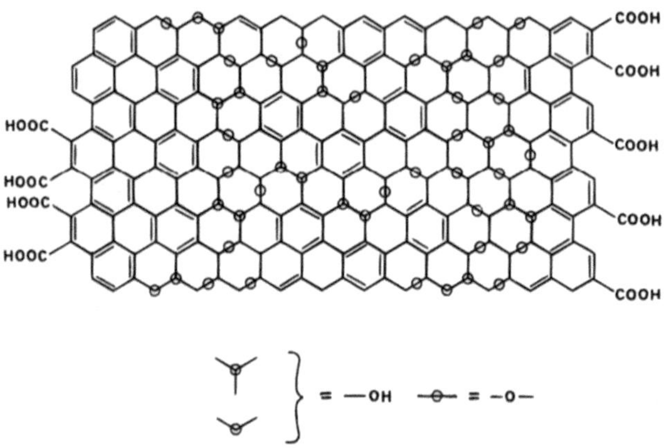

Figure 2.15 Lerf–Klinowski model of GO. Reproduced from ref. 34 with permission from American Chemical Society, Copyright 1998.

resonance (ssNMR) to detect functional groups of GO was essential to this model. Lerf *et al.* (1998)[34] noted that the formation of 1,3-ether groups in GO proposed by Ruess (1947)[31] and Mermoux *et al.* (1991)[35] was unlikely in the synthesis using potassium permanganate ($KMnO_4$) and sulfuric acid (H_2SO_4). Instead, 1,2-ether (epoxy) groups were incorporated. Additionally, the concentration of isolated double bonds in the model was very low compared to aromatic double bonds as the former bonds were easily oxidised in potassium permanganate/sulfuric acid.

The Lerf–Klinowski model showing the location of the oxygen-containing groups is depicted in Figure 2.15. The epoxy and hydroxyl groups are attached above and below the graphitic sheet, whereas carboxyl (comprising of carbonyl and hydroxyl) groups are usually located at the edges. Lerf *et al.* concluded that GO comprised two regions, namely aromatic sites (benzene rings) and oxygenated sites (aliphatic rings). The extent of oxidation in GO determines the relative sizes of these two regions. The aromatic regions and epoxy groups contribute to the planar structure of the carbon grid, while the carbons attached to hydroxyl groups are slightly distorted in their tetrahedral configuration causing unevenness of the graphitic surfaces.

Szabó *et al.* (2006) utilised analytical tools, such as TEM, ssNMR, diffuse reflectance infrared Fourier transform (DRIFT), X-ray photoelectron (XPS) and electron spin resonance (ESR) spectroscopies, to highlight the merits and drawbacks of previous models of GO.[36] It was clear that each of these models featured structural characteristics that could not be accounted for in the observed properties of GO. To explain the experimental results, the authors modified the Scholz–Boehm model resulting in a mixture of Ruess and Scholz–Boehm models, as shown in Figure 2.16. The model featured two regions including corrugated carbon ribbons and cyclohexane *chairs*. A slight tilting angle existed between the boundary of these regions contributing to the unevenness of the GO sheet as observed in the TEM images.

Aberration-corrected TEM images of GO by Erickson *et al.* (2010) supported the distinction of aromatic (graphitic) regions and oxygenated

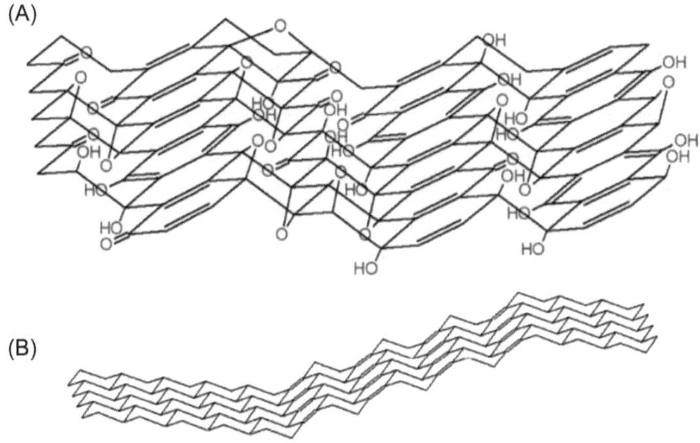

Figure 2.16 Szabó's model of GO – (A) combination of Ruess and Scholz–Boehm models and (B) unevenness of GO sheet. Reproduced from ref. 36 with permission from American Chemical Society, Copyright 2006.

(disordered) regions[37] similar to the Lerf–Klinowski model (Figure 2.15). In addition to these regions, the presence of holes in the GO layer was detected and their formation was attributed to the aggressive oxidation of graphene leading to the release of carbon monoxide and carbon dioxide. Carbonyl groups existed along the edges of the holes, whereas hydroxyl and 1,2-ether (epoxy) groups were the predominant functionalities in the oxygenated region. Interestingly, the authors noted multiple instances of mobile oxygenated groups with time. This dynamic motion was due to the incident 80 keV electron bombardment during the TEM analysis, which was greater than the bond energies of hydroxyl and epoxy, inducing bond breakage and mobility.

2.2.3 Reduced Graphene Oxide

The presence of oxygenated functional groups in GO disrupts the sp^2 bonding networks leading to changes in the properties of pristine graphene. For instance, pure graphene is an electrical conductor whereas GO is electrically insulating depending on the extent of oxidation. Hence, the restoration of a π-network to partially recover graphene is one of the most important reactions of GO. Table 2.3 summarises the key properties of graphene, GO and reduced graphene oxide (rGO). In general, rGO exhibits properties between those of GO and graphene, with high dependence on the extent of reduction process experienced.

Establishing the detailed structure of rGO is challenging despite the numerous spectroscopic and microscopic analyses performed, owing to its partially amorphous nature. High resolution TEM revealed distinct areas of graphitic and defective regions which supported the Lerf–Klinowski model.[39] A large portion of the layer contained graphitic regions where the well-known hexagonal lattice of graphene was observed, as depicted in Figure 2.17. The dark grey regions represent the areas where carbonaceous

Table 2.3 Comparison of graphene, GO and rGO. Reproduced from ref. 38 with permission from Elsevier, Copyright 2018.

Material	Graphene	GO	rGO
Atomic structure	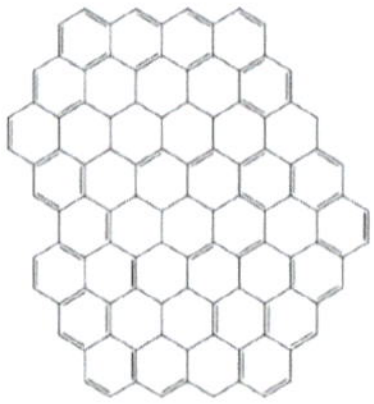	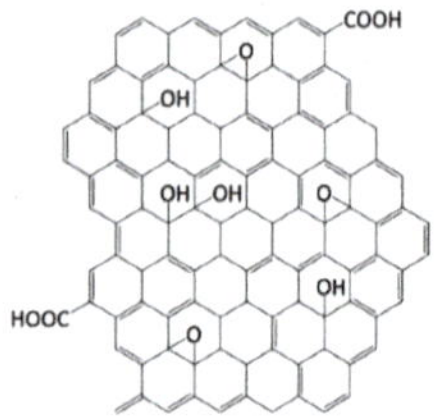	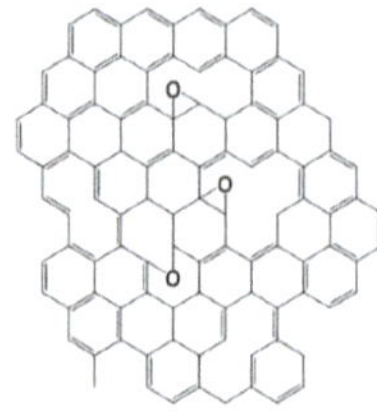
Carbon : oxygen ratio	No oxygen	2 : 4	8 : 246
Young's modulus (GPa)	1 000	200	250
Electron mobility ($cm^2 V^{-1} s^{-1}$)	10 000–50 000	Insulator	0.05–200

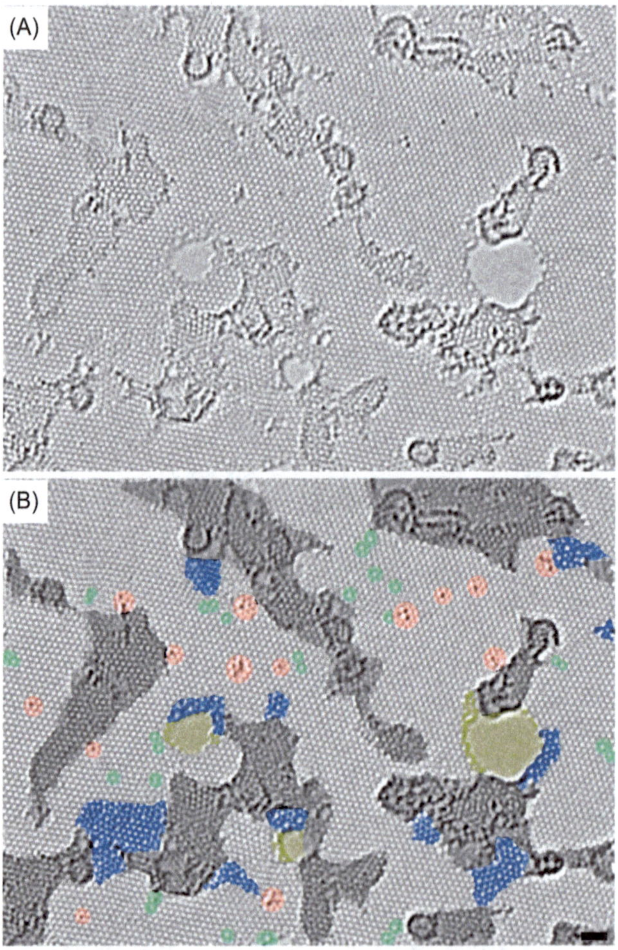

Figure 2.17 TEM images of (A) rGO and (B) shaded images – defective area (dark grey), disordered single-layer carbon network (blue), individual adatom (red), topological defect (green) and holes (yellow). Reproduced from ref. 39 with permission from American Chemical Society, Copyright 2018.

adsorbates and heavier atoms were trapped, while the green and blue regions indicate topological defects that were isolated and clustered, respectively. Isolated topological defects were made of carbon pentagon–heptagon pairs and the extended defects predominantly included a larger number of carbon pentagons, heptagons and rotated hexagons. These defects were only observable after the oxidation–reduction process and were remnants of the reactions. Although these carbon networks have not fully restored the hexagonal graphene framework, all carbon atoms in these networks were bonded to three adjacent atoms, forming a planar geometry with strong sp^2 characteristics.

Erickson *et al.* (2010) compared aberration-corrected TEM images of pristine graphene, GO and thermally rGO to determine the local chemical structures of these materials,[37] and some of their TEM images are shown in Figure 2.18. Reduced graphene oxide was found to be structurally similar to GO, with the presence of holes, graphitic regions and disordered regions (Figure 2.18(B) and (C)). However, an increase in the area of holes was observed in rGO due to the release of CO and CO_2 gases during annealing. Naturally, the areas containing oxygenated groups are much smaller in rGO compared to GO. In terms of graphitic sheet restoration, rGO exhibited substantial differences compared to graphene, such as the presence of remaining functional groups and

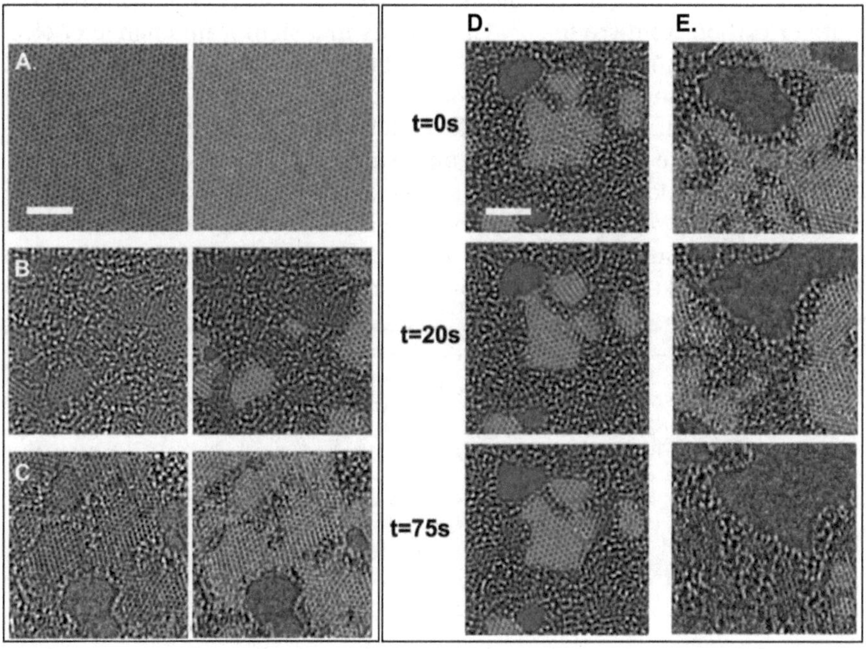

Figure 2.18 (Left) Aberration-corrected TEM images of (A) graphene, (B) GO and (C) rGO with graphitic areas (yellow), holes (blue), disordered regions/oxygenated groups (red). (Right) Degradation of graphitic areas with time for: (D) GO and (E) rGO. Reproduced from ref. 37 with permission from John Wiley and Sons, Copyright 2010.

holes. Notably, rGO was found to be highly sensitive to the incident electron beams of TEM and thus, experienced significant degradation over time (Figure 2.18(E)).

2.2.4 Other Graphene Materials

2.2.4.1 Halogenated Graphene

Graphene materials can be further modified into different structures by incorporating heteroatoms such as halogens into the graphitic lattice. For instance, graphite fluoride is a well-known derivative of graphite, used in lubrication and as an electrode material in batteries. Recent work has established that the building block of graphite fluoride, namely graphene fluoride (fluorographene), has promising electronic properties. This graphene derivative will now be briefly discussed.

Through quantum mechanics calculations, a comparison of thermodynamic stability revealed that fluorographene was the most stable among all graphene halides with their stability decreasing in the order F > Cl > Br > I; it was also more stable than graphane.[40] Unlike fluorographene, other fully halogenated counterparts (chlorographene, bromographene and iodographene) have not yet been synthesised; only theoretical studies on these counterparts have been reported.[41–43]

The attachment of fluorine atoms to graphene changes the hybridisation state of carbon from sp^2 to sp^3. This results in a significant change of electronic properties and structure of the material including buckling of the graphene sheet and loss of π-electron clouds above and below the graphitic plane. As a result, fluorographene has a lower charge carrier mobility compared to graphene and behaves like an insulator.

As graphite fluoride comprises stacks of graphene fluoride layers, the structure of fluorographene can be determined from measurements on bulk graphite fluoride. The structure of fluorographene is represented in Figure 2.19. With the change in hybridisation state of carbon from sp^2 to sp^3

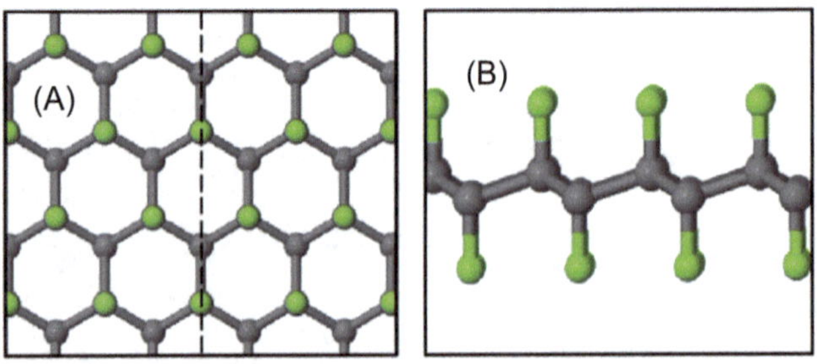

Figure 2.19 Structure of fluorographene: (A) hexagonal lattice and (B) buckled structure (side view along the armchair direction marked by the dashed line in (A)). Reproduced from ref. 44 with permission from Springer Nature, Copyright 2011.

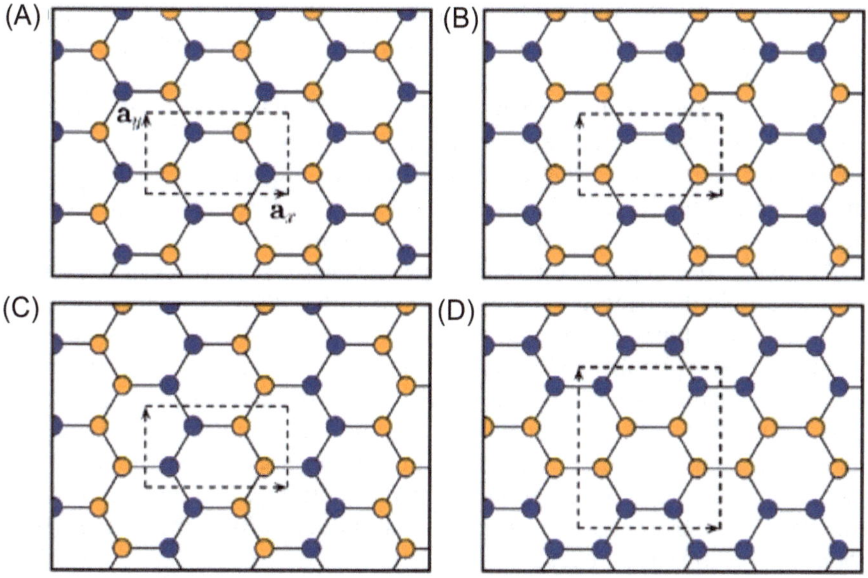

Figure 2.20 Configurations of fluorographene with fluorine atoms above (yellow) and below (blue) the graphene plane: (A) chair, (B) boat, (C) zigzag and (D) armchair. Reproduced from ref. 45 with permission from American Chemical Society, Copyright 2013.

owing to the attachment of halogens, the halogenated graphene layer is no longer flat but has a buckled structure, as already noted.

Figure 2.20 depicts the different structural configurations of fluorographene. The structure of a fully halogenated graphene can take several configurations, such as the *chair*, *boat*, *zigzag* and *armchair* forms. Predictions using density functional theory calculations found that the *armchair* conformation was most stable, followed by the *zigzag* conformation.[45] The *boat* configuration is deemed improbable as the pairs of halogen atoms alternating on one side of the carbon layer experience strong repulsion forces.[40] Nevertheless, experimental observations confirmed a wide range of lattice spacings, in contrast to only one or two distinct lattice spacings, which should be expected if the crystal contains only a single form. This suggests that there are other locally stable configurations such as the *twist-boat-chair* form.[45] The structures of chlorinated, brominated and iodinated graphene are expected to be similar to that of fluorographene.

2.2.4.2 Graphane, Graphone and Hydrogenated Graphene

Graphane is a variant of graphene in which each C atom is bound to a H atom and thus, all carbon atoms are fully saturated. Graphane lacks the double bonds observed in graphene, and as a result, the carbon atoms in graphane are sp^3 rather than sp^2 hybridised. This gives rise to a rearrangement of inter-atomic bonds and angles. In addition, the electronic properties of graphane are altered due to the removal of conducting π-bands and opening of band gaps.

Graphane can be obtained by hydrogenation of graphene as shown in Figure 2.21. However, current attempts at producing graphane have led to only partial success. Hereafter, such structures are referred to as hydrogenated graphene, instead of graphane. The various synthesis methods of hydrogenated graphene are discussed in Section 3.6.

The basic unit of graphane is a cyclohexane with each carbon atom bound to three neighbouring carbon atoms and one hydrogen atom. Just as in cyclohexane, when compared to its counterpart benzene, graphane does not conform to the planarity of graphene: the sp^3 hybridisation of carbon atoms transforms it into an out-of-plane buckled conformation. Depending on how the hydrogen atoms are arranged on both sides of the carbon layer, graphane is known in two distinct configurations – *chair* graphane with alternating hydrogen atoms and *boat* graphane with alternating pair of hydrogen atoms. Intermediate structures of graphane exist as the *twist-boat* and *twist-boat-chair* as shown in Figure 2.22.[47]

The transition of sp^2 to sp^3 in graphane results in an increase in C–C bond length from 0.142 nm to 0.152–0.156 nm.[48] A range of bond lengths are observed in graphane due to its different configurations. Note that the tetrahedral bond arrangement in sp^3 entails dihedral angles between C–C bonds of 109.5° whereas planar sp^2 bonds possess dihedral angles of 120° producing three-dimensionality in graphane.

The *chair* form of graphane is the most stable as the structure does not exhibit steric strain from the interaction of parallel hydrogen atoms. Therefore, the C–C bond length in *chair* graphane is uniform across the whole structure, at 0.152 nm. On the other hand, two C–C bond lengths, 0.152 and 0.156 nm, are observed in the *boat* isomer of graphane. The interactions of pairs of hydrogen atoms (flagpole hydrogens) impose steric strain on the C–C bond resulting in an increased bond length. In comparison to the *chair* and *boat* forms, the *twist-boat* conformation possesses higher in-plane shrinkage and is less stable. The twisting of C–C bonds induces higher bond energies

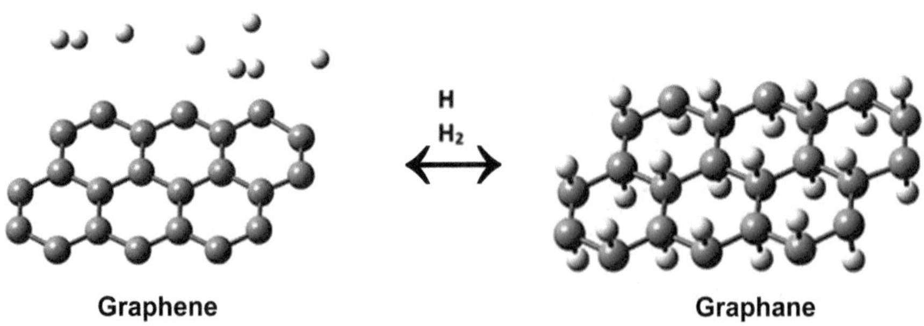

Figure 2.21 Reversible hydrogenation of graphene to graphane. Reproduced from ref. 46 with permission from Elsevier, Copyright 2017.

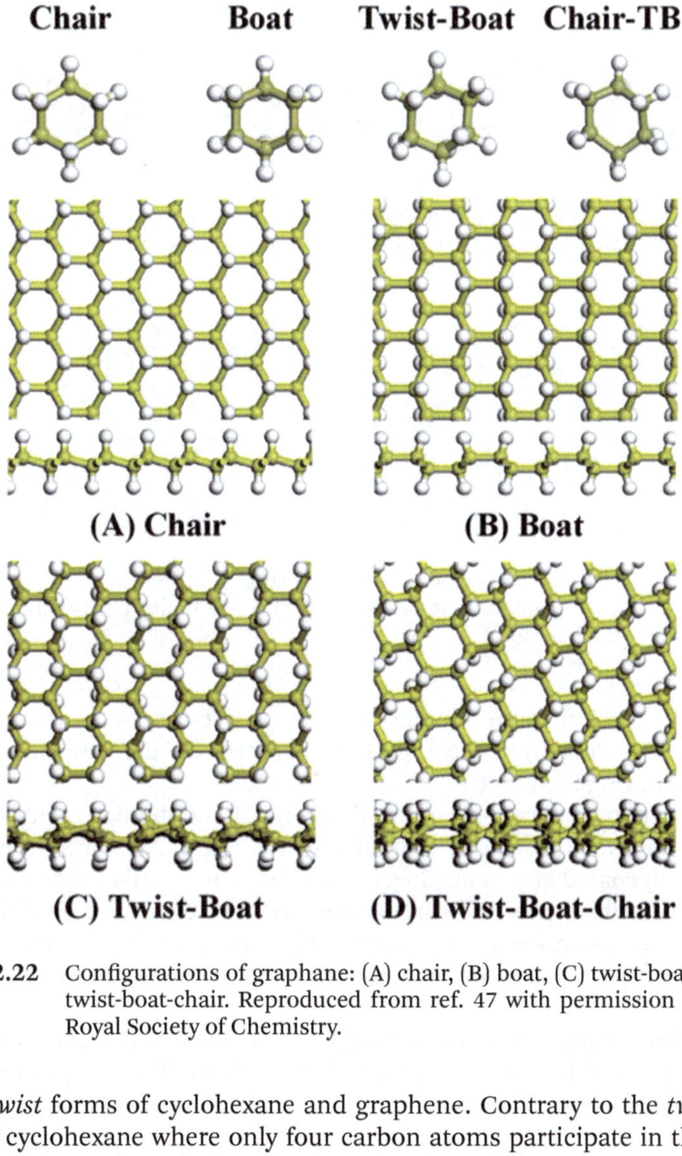

Figure 2.22 Configurations of graphane: (A) chair, (B) boat, (C) twist-boat and (D) twist-boat-chair. Reproduced from ref. 47 with permission from the Royal Society of Chemistry.

in the *twist* forms of cyclohexane and graphene. Contrary to the *twist-boat* form of cyclohexane where only four carbon atoms participate in the bond twisting process, the *twist-boat* graphane involves all six carbon atoms in this process, hence the additional energy required for the C–C bonds.[49] Experimentally, a wide range of different graphane configurations are likely to be observed. Samarakoon and Wang (2011) suggested the *twist-boat-chair* configuration to be a stable structure consisting of adjacent *twist-boat* and *chair* structures, thus supporting the existence of locally stable *twist-boat* configurations in graphane.[50]

Transformation between the different graphane configurations would entail the flipping or switching of positions of hydrogen atoms, similar to the conformational change in cyclohexane. Such a phenomenon is highly

Figure 2.23 Graphone: one-sided (50%) hydrogenation of graphene with hydrogen atoms (white). Reproduced from ref. 51 with permission from MDPI.

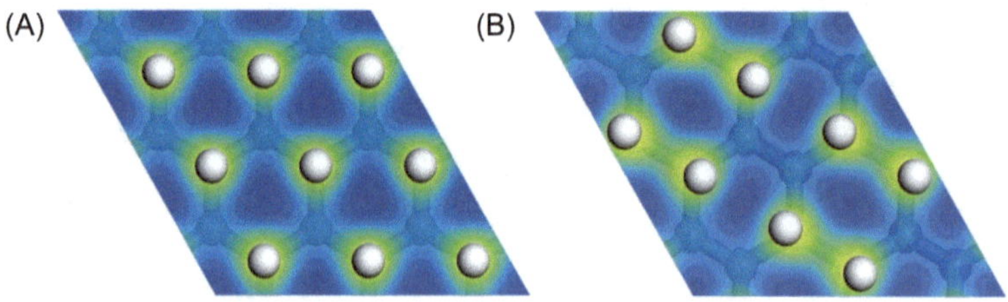

Figure 2.24 Hydrogen atoms (white) on (A) triangular graphone and (B) rectangular graphone. Reproduced from ref. 52 with permission from AIP Publishing.

unlikely as it requires the flipping of atoms to occur between the graphane plane and the elongation of carbon bonds to accommodate the said atoms in the plane during the process.[49]

The reversible conversion of graphene–graphane implies many forms of possible intermediates. The midpoint of such a conversion, graphone, is 50% hydrogenated graphene. Specifically, the term 'graphone' refers to one-sided hydrogenation of the carbon sheet (Figure 2.23). The partial saturation of carbon atoms in the structures of graphone and other hydrogenated graphenes results in a mixture of sp^2 and sp^3 hybridised carbon atoms in the plane.

Feng and Zhang (2012) investigated the stability of different structures of graphone and found that 'rectangular' graphone (Figure 2.24) was more stable compared to 'triangular' graphone.[52] Furthermore, calculations conducted by the authors determined the anti-ferromagnetic property of rectangular graphone, whereas ferromagnetism was observed for triangular graphone.

2.2.4.3 Graphyne

Graphyne is a two-dimensional planar-carbon form similar to graphene. The structural transformation from graphene to graphyne is represented in Figure 2.25. In graphyne, carbon hexagons are linked by a linear acetylenic or carbyne chains (sp hybridised carbon). The term graphyne arises from its chemical structure where acetylenic linkages replace one-third of the C–C

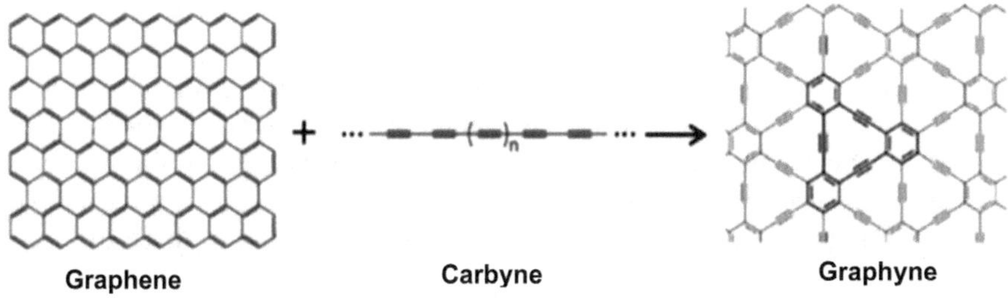

Figure 2.25 Chemical structure transformation of graphene to graphyne. Reproduced from ref. 54 with permission from Elsevier, Copyright 2021.

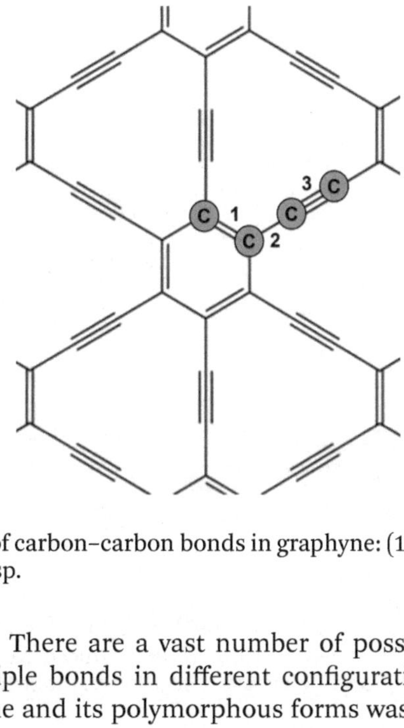

Figure 2.26 Types of carbon–carbon bonds in graphyne: (1) sp^2–sp^2, (2) sp^2–sp and (3) sp–sp.

bonds in graphene. There are a vast number of possible graphynes, each with double and triple bonds in different configurations. The theoretical proposal of graphyne and its polymorphous forms was made by Baughman *et al.* (1987).[53]

Owing to its smaller number of coordination bonds, graphyne is predicted to be softer than graphene. In addition, graphyne materials are expected to possess high thermal stability with similar mechanical properties to graphite.[55]

Unlike graphene, there are three types of C–C bonds in graphyne, as shown in Figure 2.26: (1) C(sp^2)–C(sp^2) forming the central ring, (2) C(sp^2)–C(sp) connecting adjacent double and triple bonds, and (3) C(sp)–C(sp) for the linkages.

Owing to the different hybridisation states of carbon and their configurations in the graphyne sheet, a great number of graphyne polymorphs are possible. Their structures can be derived from four basic polymorphs of graphene (Figure 2.27), which are hexagonal graphene (L_6), graphene 4–8 ($L_{4\text{-}8}$), graphene 3–12 ($L_{3\text{-}12}$), and graphene 4–6–12 ($L_{4\text{-}6\text{-}12}$).[56]

The various forms of graphyne can be classified into three groups with regard to the gradual conversion of graphene into graphyne, *i.e.* the substitution of sp^2 hybridised carbon atoms in graphene by carbyne chains.

In the first group, the carbyne chain substitutes only one of the three sp^2 hybridised atoms of graphene (Figure 2.28(B)), resulting in three different

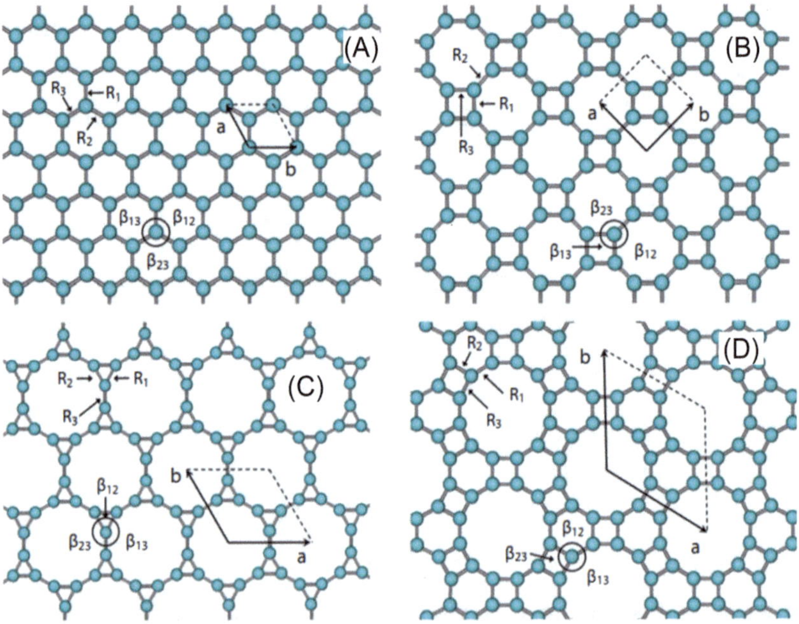

Figure 2.27 Polymorphs of graphene for derivation of graphynes: (A) hexagonal graphene (L_6), (B) graphene 4–8 ($L_{4\text{-}8}$), (C) graphene 3–12 ($L_{3\text{-}12}$) and (D) graphene 4–6–12 ($L_{4\text{-}6\text{-}12}$). Reproduced from ref. 56 with permission from John Wiley and Sons, Copyright 2019.

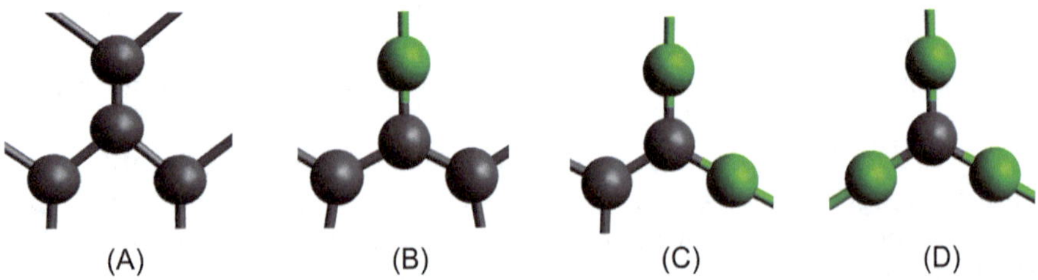

Figure 2.28 (A) Three-coordinated atoms in graphene and substitution of (B) 1, (C) 2 and (D) 3 carbon atoms by carbyne chains (two-coordinated atoms). Reproduced from ref. 56 with permission from John Wiley and Sons, Copyright 2019.

structures (Table 2.4). These graphyne polymorphs are designated as γ-graphyne. The most stable carbon structure is a hexagonal layer of graphene where all the atoms are three-coordinated (Figure 2.28(A)). Of the various graphyne structures, γ-graphyne is the most stable, with a minimum proportion of just one two-coordinated atom per three-coordinated atom.

The additional substitution of the C–C bonds with a carbyne chain gives rise to the second group – β-graphyne. Three different forms can be obtained through this substitution, namely, β_1-, β_2-, β_3-graphyne (Table 2.4). These graphyne polymorphs contain two two-coordinated atoms and one three-coordinated atom.

In the last group, all three adjacent atoms are replaced with carbyne chains resulting in α-graphyne. Unlike the various polymorphs of γ- and β-, α-graphyne formed from hexagonal graphene has only one form (Table 2.4).

Table 2.4 Atomic structures of hexagonal graphene, L_6, and polymorphs of graphyne derived from hexagonal graphene, L_6 (γ_1-, γ_2-, γ_3-, β_1-, β_2-, β_3- and α-graphyne) with carbyne chains (green). Reproduced from ref. 56 with permission from John Wiley and Sons, Copyright 2019.

Graphyne polymorph	Number of carbyne chain substitutions	Forms
L_6-graphene	0	
γ-L_6-graphyne	1	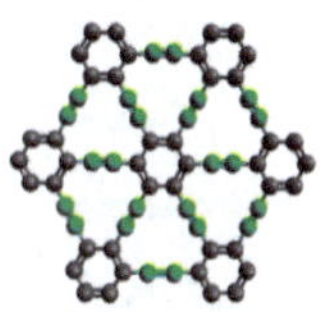 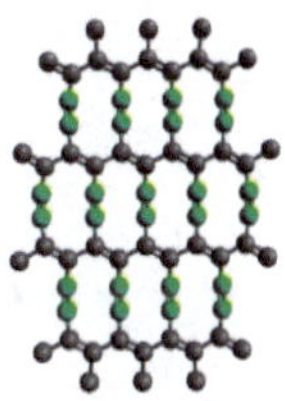 
β-L_6-graphyne	2	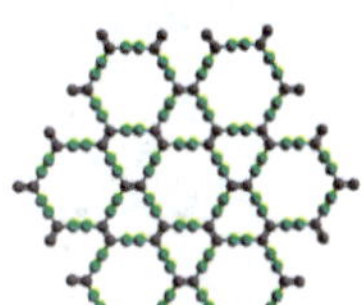 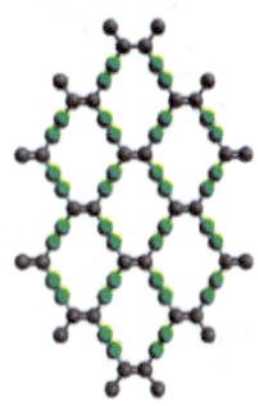 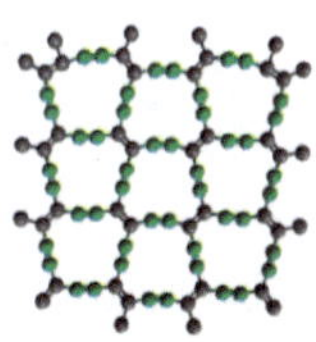
α-L_6-graphyne	3	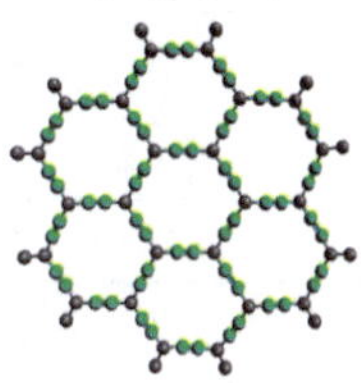

Therefore, graphyne derived from hexagonal graphene (L_6) may contain seven structural forms, which are the α-, β_1-, β_2-, β_3-, γ_1-, γ_2- and γ_3-graphyne. It should be noted that graphyne may also be obtained from the combination of these seven basic forms in various proportions, thus leading to a great variety of combined structures.

Similar to graphynes derived from hexagonal graphene, the other graphene polymorphs ($L_{4\text{-}8}$, $L_{3\text{-}12}$, $L_{4\text{-}6\text{-}12}$) could be used to obtain a series of graphyne structural forms. Tables 2.5, 2.6 and 2.7 present the basic structures of graphyne obtained from $L_{4\text{-}8}$, $L_{3\text{-}12}$, and $L_{4\text{-}6\text{-}12}$, respectively.

Graphynes based on graphene 4–8 and graphene 4–6–12 form seven possible basic structures: one α-, three β- and three γ-structures (Tables 2.5 and 2.7). However, with graphene 3–12, only three basic structures may be derived: one for each of the α-, β- and γ-structures (Table 2.6).

Table 2.5 Atomic structures of polymorphs of graphyne derived from graphene 4–8, $L_{4\text{-}8}$ (γ_1-, γ_2-, γ_3-, β_1-, β_2-, β_3- and α-graphyne) with carbyne chains (green). Reproduced from ref. 56 with permission from John Wiley and Sons, Copyright 2019.

Graphyne polymorph	Number of carbyne chain substitutions	Forms
γ-$L_{4\text{-}8}$-graphyne	1	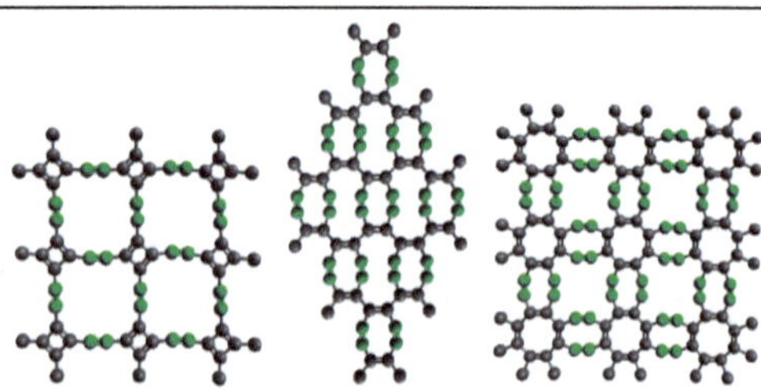
β-$L_{4\text{-}8}$-graphyne	2	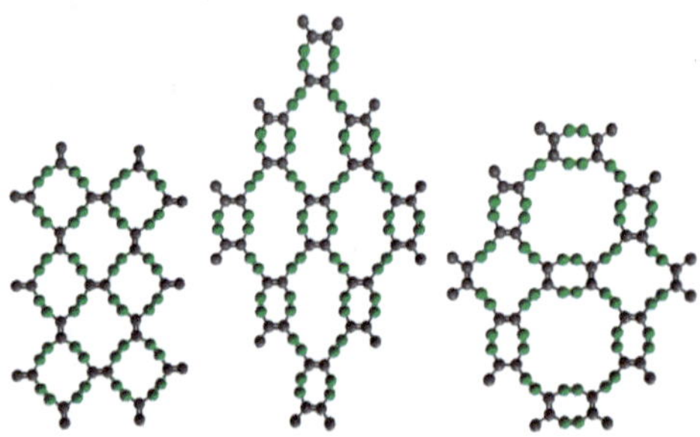
α-$L_{4\text{-}8}$-graphyne	3	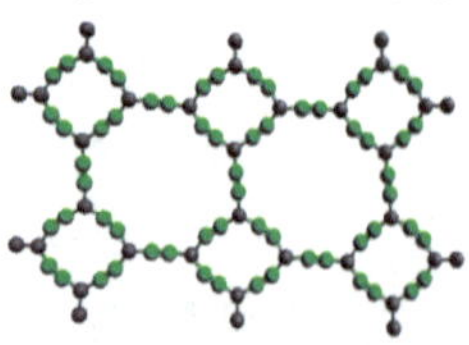

Table 2.6 Atomic structures of polymorphs of graphyne derived from graphene 3–12, L_{3-12} (γ-, β- and α-graphyne) with carbyne chains (green). Reproduced from ref. 56 with permission from John Wiley and Sons, Copyright 2019.

Graphyne polymorph	Number of carbyne chain substitutions	Forms
γ-L_{3-12}-graphyne	1	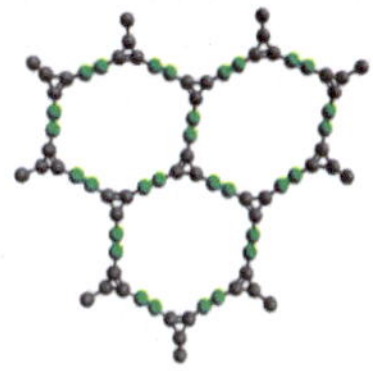
β-L_{3-12}-graphyne	2	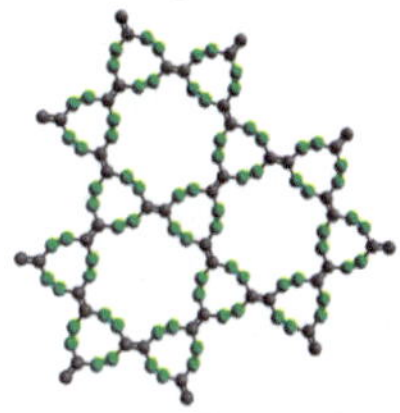
α-L_{3-12}-graphyne	3	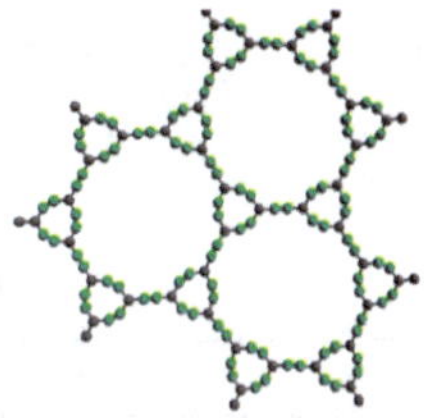

Note that graphynes obtained from the graphene polymorphs (L_{4-8}, L_{3-12}, L_{4-6-12}) are less stable than those from hexagonal graphene due to the greater number of different atomic positions. L_6-based graphynes have only two different atomic positions whereas the other graphynes have three or four different atomic positions.[56] Furthermore, the highest sublimation energies and stability are expected from graphynes formed from L_6, followed by L_{4-8} graphene as they most resemble hexagonal graphene (all atoms are three-coordinated).

Graphyne structures can be expanded through the extension of the carbyne chains, replacing the two-atom carbyne chains with longer chains containing an even number of atoms. For instance, in γ_1-graphyne, substitution with fragments of carbyne chain of four atoms results in γ_1-graphyne-2 or graphdiyne as it has been termed.[40,54,57–63] Table 2.8 shows the structures of L_6-based graphyne with four atom-carbyne chain substitutions.

Therefore, considering the various classifications of graphyne, a general designation for graphyne may be presented as X_m-L-graphyne-*n*, where X = α, β, or γ; *m* = 1, 2, or 3; L = L_6, L_{4-8}, L_{3-12}, or L_{4-6-12}; *n* = 1, 2, 3, ... depending on the number of pairs of two-coordinated atoms in the carbyne chains.

Table 2.7 Atomic structures of polymorphs of graphyne derived from graphene 4–6–12, $L_{4\text{-}6\text{-}12}$ (γ_1-, γ_2-, γ_3-, β_1-, β_2-, β_3- and α-graphyne) with carbyne chains (green). Reproduced from ref. 56 with permission from John Wiley and Sons, Copyright 2019.

Graphyne polymorph	Number of carbyne chain substitutions	Forms
γ-$L_{4\text{-}6\text{-}12}$-graphyne	1	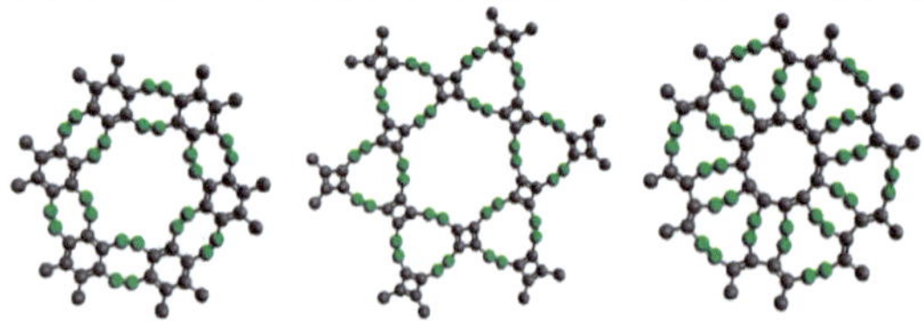
β-$L_{4\text{-}6\text{-}12}$-graphyne	2	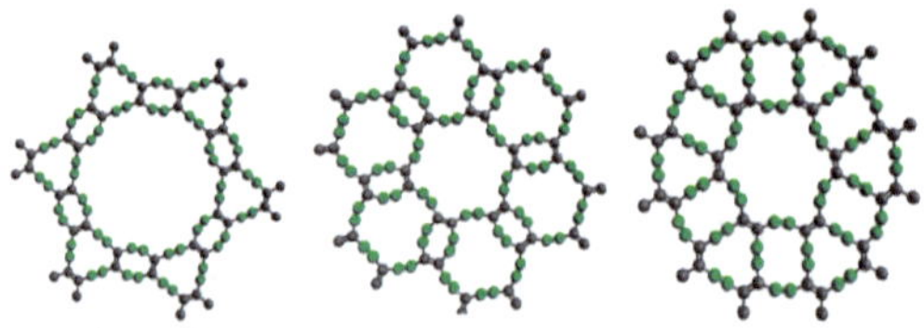
α-$L_{4\text{-}6\text{-}12}$-graphyne	3	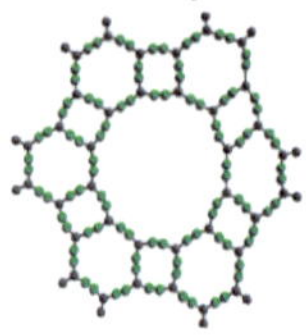

Table 2.8 Atomic structures of polymorphs of graphyne derived from hexagonal graphene, L_6 with four atom-carbyne chain substitutions (γ_1-, γ_2-, γ_3-, β_1-, β_2-, β_3- and α-graphyne-2) with carbyne chains (green). Reproduced from ref. 56 with permission from John Wiley and Sons, Copyright 2019.

Graphyne polymorph	Forms
γ-L_6-graphyne-2	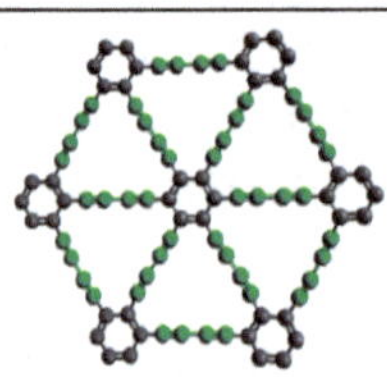
β-L_6-graphyne-2	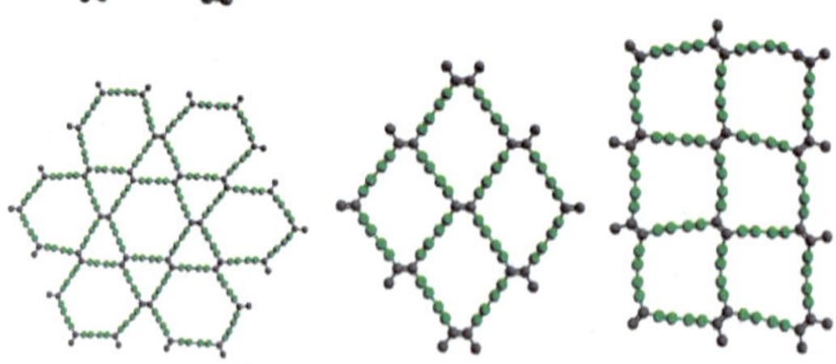
α-L_6-graphyne-2	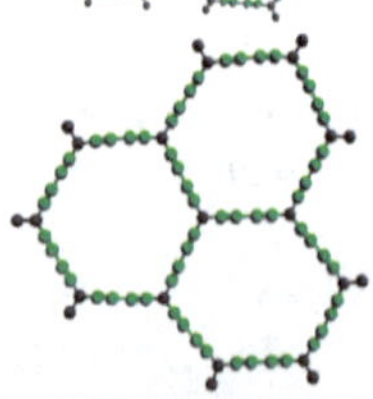

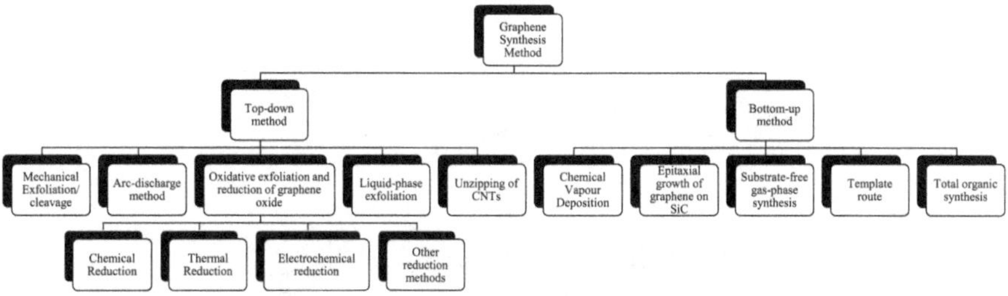

Figure 2.29 Top-down and bottom-up approaches for graphene synthesis. Reproduced from ref. 2 with permission from Elsevier, Copyright 2019.

2.3 Synthesis of Graphene

Graphene can be prepared *via* top-down or bottom-up synthesis methods, as set out in Figure 2.29. The top-down approach involves the breakdown of a precursor, for instance, separation of a graphite interlayer to form graphene sheets. Mechanical exfoliation, liquid-phase exfoliation, electrochemical exfoliation and arc-discharge are among the well-known top-down methods. Conversely, bottom-up approaches involve the use of a carbon source for the build-up of graphene with techniques, such as chemical vapour deposition, templating, epitaxial growth and total organic synthesis.

2.3.1 Top-down Synthesis of Graphene

In general, the top-down methods for producing graphene are advantageous in the sense of being highly scalable for industrial applications and yielding products of high quality. Nonetheless, these methods are usually limited by the extent of control over the properties of the final product and difficulty in maintaining consistent properties and quality of products. The most popular of these techniques will now be outlined.

2.3.1.1 Mechanical Exfoliation

The mechanical exfoliation of graphite can be performed based on two directional forces – normal force and lateral force.[64] The exertion of a normal force aims to overcome the van der Waals attraction between the individual graphite layers, whereas lateral forces exploit the self-lubricating ability of graphite to induce motion between the individual graphite layers. An example of the normal force approach is the famous Scotch tape method,[1] where repetitive cleavage of the graphitic layers reduces graphite to a single graphene layer (Figure 2.30). Although graphene sheets prepared through this method are of high quality and large areas, the process is labour intensive, time consuming and not feasible for industrial scale production. In an attempt to improve the Scotch tape method, a three-roll mill machine with polymer adhesive has been explored to derive a continuous exfoliation process.[65] However, this approach experiences difficulties in relation to the complex separation methods of isolating graphene from the polymer adhesives. Jayasena and Subbiah (2011) described

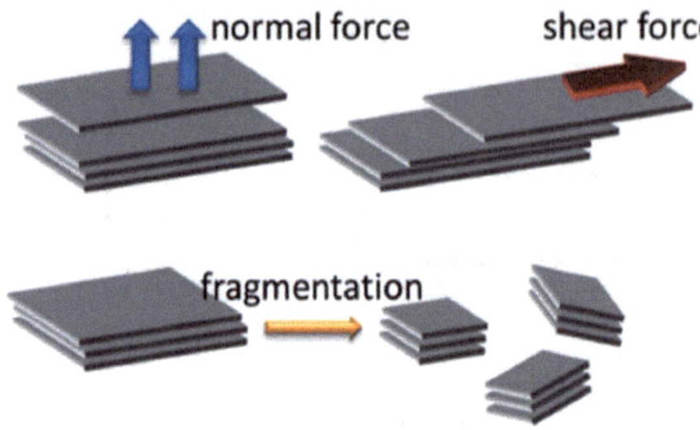

Figure 2.30 Mechanical exfoliation, as a result of normal force and shear force. Also shown is the fragmentation which can occur during exfoliation. Reproduced from ref. 64 with permission from the Royal Society of Chemistry.

an ultrasharp single-crystal diamond wedge to cleave the graphene layers of graphite with the aid of ultrasonic oscillations along the wedge.[66] Although the technique was able to generate graphene layers of thickness around 10 nm, it cannot be used for large-scale production of graphene due to the intensive operational time required to generate a small amount of graphene.

Another technique utilising shear force for exfoliating graphite is ball milling. This technique is conventionally used for grinding materials where mechanical force is generated through the rotation of the mill and grinding balls impact onto the material. In this process, both normal and shear forces are exerted onto the material. However, the normal forces may destroy the crystalline structure of the product and thus, there is a need to reduce these forces while ensuring that the shear forces remain strong enough to produce exfoliation.[64]

2.3.1.2 Liquid-phase Exfoliation

In liquid-phase exfoliation, a sample of graphite/graphite oxide is first dispersed in a suitable solvent. Exfoliation can then be achieved through techniques, such as sonication, jet cavitation, microfluidisation and high-shear mixing. These techniques offer the advantages of not requiring laborious chemical oxidation steps or high temperatures and vacuum environments. However, the graphene layers produced by liquid-phase exfoliation are of average quality and are often defective. Nevertheless, the graphene products are still valuable for certain applications, such as catalysis and energy storage, which benefit from the presence of edge defects.[67]

The most commonly used technique in liquid-phase exfoliation is sonication. Liquid cavitation is the underlying mechanism behind this technique (Figure 2.31). Cavitation bubbles are created through sonication, which, upon their collapse, create strong shockwaves and microjets, as well as the subsequent generation of extreme pressures and temperatures. Under these intense conditions, edge and void defects are introduced into graphene. Sonication can be performed *via* bath sonication or probe/tip sonication. However, probe sonication can only be used for limited amounts of solutions and produces low yields with long exfoliation durations.[67]

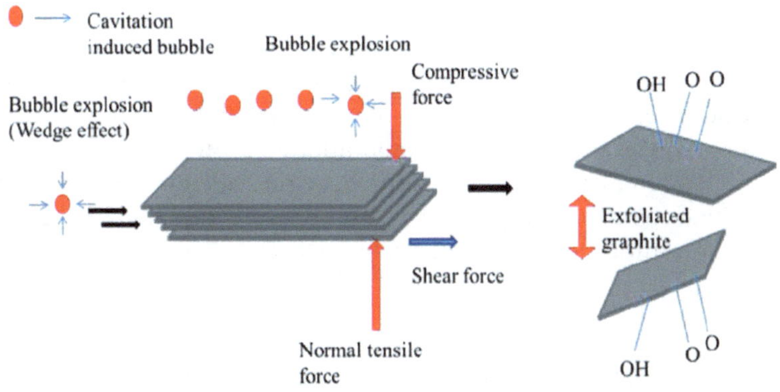

Figure 2.31 Mechanism of liquid-phase graphite exfoliation. Reproduced from ref. 68 with permission from MDPI.

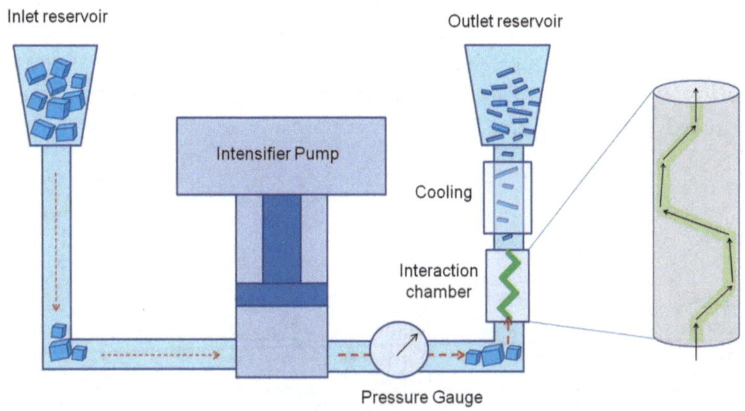

Figure 2.32 Microfluidisation of graphite. Reproduced from ref. 70 with permission from American Chemical Society, Copyright 2017.

Compared to sonication, jet cavitation is more efficient as cavitations produced through this approach are more uniform throughout the solution. Exfoliation can also be achieved in a shorter time with this technique. Jet cavitation utilises both the normal and shear directional forces to exfoliate. Cavitation and pressure difference produces the normal forces whereas turbulence and velocity gradient in the flow channel produce the shear forces.[64] Liang *et al.* (2015) investigated the effects of initial concentration of graphite, jet pressure and treatment time on the yield of graphene.[69] An exfoliation fraction of ~12 wt% was obtained by treating 10 L of graphite dispersion under 20 MPa for 8 hours. Their findings further indicated that high graphene yield was favoured at appropriate initial graphite concentration, high jet pressure and long exfoliation time.[69]

Microfluidisation is similar to jet cavitation, but higher pressures (207 MPa) are used in much narrower channels (<100 μm) (Figure 2.32).[67] It has been established that exfoliation through microfluidisation is only efficient with multiple cycles, with the optimum number of cycles being 70 for producing high-quality graphene.[70] Due to the small dimensions of the channels, much smaller initial sizes of graphite are required to prevent blockage. In addition, there is a need for a cooling system to achieve temperatures as low as −20 °C to avoid solvent evaporation. Despite these requirements, microfluidisation has the potential for industrial scale production due to the high production rate of 9.2 g h^{-1} for a single channel and high throughput of 12 L min^{-1} for multiple parallel microchannels.[70]

2.3.1.3 Electrochemical Exfoliation

In electrochemical exfoliation, ionic species are intercalated between the graphitic layers, which then produce gaseous molecules that inflate and separate the graphene layers. The setup comprises an electrochemical cell containing an aqueous or solid electrolyte and the use of electric currents to induce exfoliation at the graphite anode or cathode. Various forms of graphite can be used as the working electrodes, including graphite powder, flakes and rods. The type of power supply to the electrodes determines the type of exfoliation mechanism for the production of graphene, namely, anodic or cathodic exfoliation (Figure 2.33).

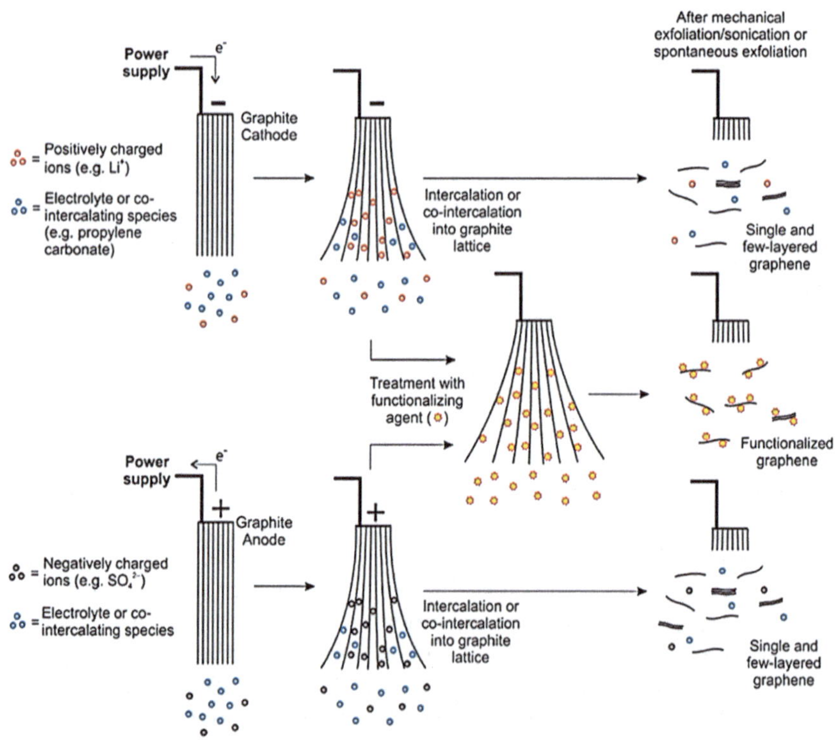

Figure 2.33 Mechanism of electrochemical exfoliation cathodic (top) and anodic (bottom). Reproduced from ref. 72 with permission from John Wiley and Sons, Copyright 2019.

A positive bias on the electrode creates anodic exfoliation in which bulky anions are intercalated into graphite to facilitate the exfoliation of the sheets. Electrolytes, such as sulfuric, nitric and perchloric acids, or salts, such as sodium benzoate and sodium citrate, can be used to provide these anionic species.[67] Water plays a significant role in the electrochemical exfoliation of graphite, as it generates oxygen and hydroxyl radicals, which can aid in intercalation and exfoliation.[71] Cathodic exfoliation is induced when a negative current is passed through the graphite electrode. Another significant difference between the anodic and cathodic exfoliations is that the electrolytes used in cathodic exfoliation are organic chemicals, such as dimethylformamide (DMF) and *N*-methyl-2-pyrrolidone (NMP).[67]

Generally, anodic electrochemical exfoliation requires large potentials ranging from 5 to 10 V, which hasten the intercalation of ions and the electrochemical reactions. This can cause the undesirable oxidation of graphene and the introduction of numerous hole defects. Cathodic exfoliation avoids these issues, but the application of a cathodic potential is associated with low efficiencies, as this exfoliation process is much slower than anodic exfoliation. One improvement is to combine shear-induced effects in tandem with the electrochemical exfoliation process. This integrated technique allows lower potentials to be used and hence reduces unwanted oxidation of graphene sheets. This can be achieved since shear forces are the dominant exfoliating mechanism in the intercalation of graphite layers through electrochemical means.[73]

Despite being a relatively easy one-step method, the use of electrochemical exfoliation to produce graphene is challenging on a large scale as this entails the design of large electrodes and electrochemical cells.

2.3.1.4 Arc Discharge

The arc-discharge technique is carried out in an electric arc chamber or oven, which consists of a sacrificial graphite rod anode and cathode in a water-cooled steel vacuum chamber. A constant direct current or alternating current is applied to produce a plasma and induce the vaporisation of the graphite anode. Graphene is formed in the plasma and deposited on the walls of the chamber. Gas atmospheres, such as helium, hydrogen, argon, ammonia and carbon dioxide, or mixtures of these gases, have been used in the arc-discharge chamber. The type of atmosphere in which discharge occurs influences the quality of graphene synthesised. For instance, graphene produced in ammonia and helium gas mixtures is reported to be doped with nitrogen atoms,[74] whereas hydrogen-inert gas mixtures yielded graphene with high electrical conductivities, as hydrogen apparently inhibited the rolling and closing of graphene sheets.[75] The arc-discharge method shows potential in terms of scalability and ease of operation. Nonetheless, it is also associated with a lack of control over the formation of graphene and the unwanted doping of atoms onto the graphene sheet.

2.3.2 Bottom-up Synthesis of Graphene

2.3.2.1 *Chemical Vapour Deposition*

The most popular bottom-up method is chemical vapour deposition (CVD) as it offers simplicity and good control over the product. In this process, carbon precursors are decomposed at elevated temperatures which upon contact with a catalyst form graphene. Various kinds of carbon precursors (solids, liquids or gases) have been investigated for the deposition of graphene including methane,[76] propane,[77] hexane,[78] and propanol.[79] The quality of graphene is dependent on the carbon source, mass transfer of gas and type of substrate. Substrates, such as metals, alloys and oxides, have been reported. Although CVD produces graphene of comparable quality to mechanical exfoliation, this technique overcomes the scalability issues associated with the top-down method. Nevertheless, CVD preparation of graphene is costly due to the complex and energy-intensive apparatus. In addition, further processing is required to remove residual catalysts and to transfer the graphene sheets to other substrates.

Some interesting studies have reported the continuous production of chemical vapour deposited graphene by an automated roll-to-roll process (Figure 2.34). Kidambi *et al.* (2018) designed a high-speed graphene manufacturing process with a 5 cm min^{-1} production rate.[80] Although the prepared graphene was not ideal for electronic applications, it showed great potential in preparing nanoporous atomically thin membranes with size-selective molecular transport performance. In other work, Kobayashi *et al.* (2013) fabricated a 100 m long graphene sheet through low pressure thermal CVD synthesis.[81] The graphene sheet synthesised was of high quality with a sheet resistance of only 150 Ω sq^{-1}, deemed suitable for use in semiconductor devices.

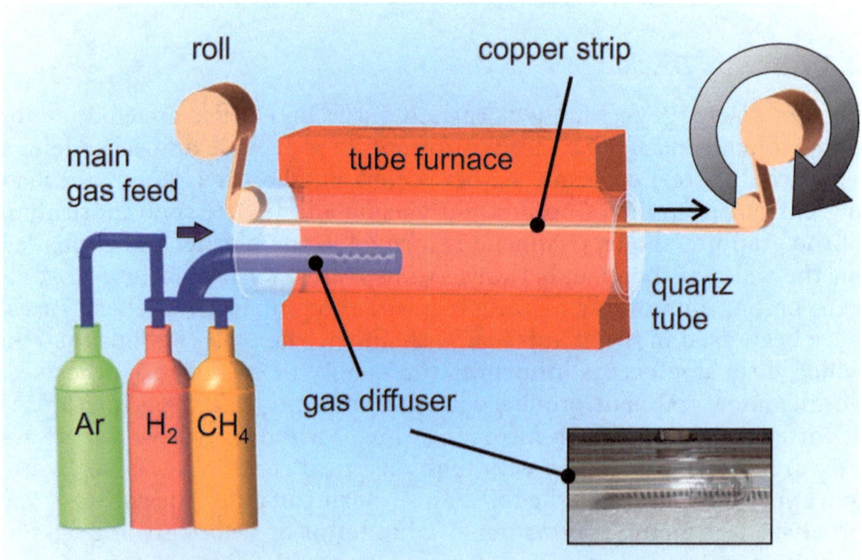

Figure 2.34 Roll-to-roll chemical vapour deposition of graphene onto copper using methane precursor. Reproduced from ref. 82 with permission from AIP Publishing.

2.3.2.2 Epitaxial Growth

The epitaxial growth of graphene promises great potential for electronic applications due to its capability of producing high quality graphene with minimal defects. Graphene is grown *via* thermal decomposition of silicon carbide, where simultaneous evaporation of silicon atoms and aggregation of remaining carbon atoms leads to the formation of sp^2 carbon-rich layers. Temperatures as high as 1200–1600 °C have been used to initiate the thermal decomposition of silicon carbide where graphene was found to form at 1100 °C.[2,83] This method of producing graphene is expensive due to the energy-intensive process and cost of substrates. Moreover, the high growth rate leads to difficulty in obtaining thin and homogenous graphene layers.

Molecular beam epitaxy offers an alternative to the decomposition of silicon carbide. Graphene layers produced using molecular beam epitaxy technology possess very high quality and can be achieved at much lower temperatures than with CVD.[84] In this technique, graphene growth occurs in an ultrahigh vacuum and in the absence of any carrier gases resulting in a highly pure graphene crystalline structure (Figure 2.35). Zheng *et al.* (2017) investigated the growth of graphene films on iron through low temperature molecular beam epitaxy.[85] The

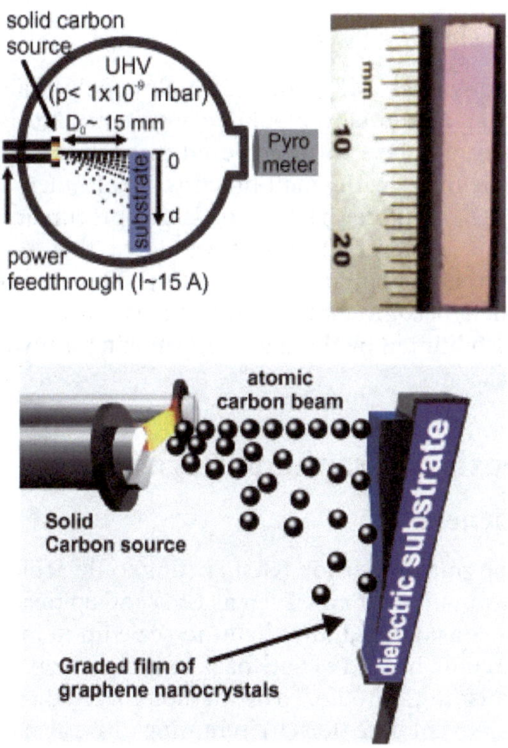

Figure 2.35 Molecular beam epitaxy of graphene under ultrahigh vacuum. Reproduced from ref. 86 with permission from Elsevier, Copyright 2012.

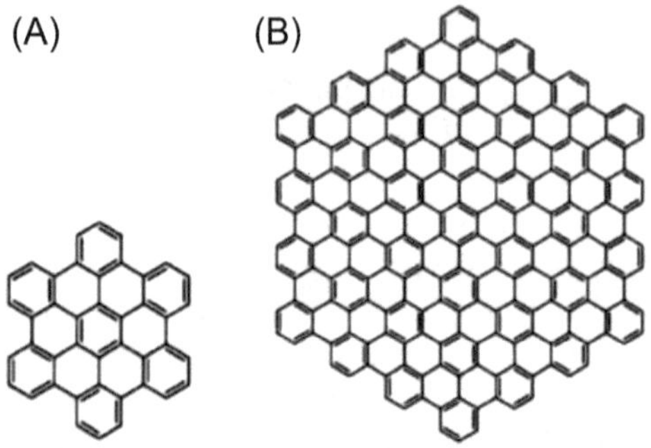

Figure 2.36 Precursor molecules for total organic synthesis of graphene: (A) hexabenzocoronene ($C_{42}H_{18}$) and (B) $C_{222}H_{42}$. Reproduced from ref. 87 with permission from Elsevier, Copyright 2012.

authors found that the formation/decomposition of Fe_xC_y was responsible for the growth of graphene at temperatures as low as 400 °C, forming layers of 0.7 nm in thickness. Here, the thickness of graphene films can be controlled precisely, as the growth is influenced by the chemical reaction between iron and carbon, instead of conventional temperature-dependent precipitation processes.

2.3.2.3 *Total Organic Synthesis*

The total organic synthesis of graphene employs polycyclic aromatic hydrocarbons, which are essentially graphene segments. Precise fabrication of small graphene molecules can be achieved through the application of synthetic organic chemistry with small precursor molecules, such as hexabenzocoronene ($C_{42}H_{18}$) (Figure 2.36(A)), or larger precursors such as $C_{222}H_{42}$ (Figure 2.36(B)).[87] However, it should be noted that synthesis of larger graphene samples is challenging as competing side reactions and insolubility of the products become more significant. The use of organic synthesis techniques is also limited by the high cost, making it unsuitable for industrial applications.

2.4 Synthesis of Graphene Derivatives

2.4.1 Graphene Oxide

The synthesis of graphene oxide (GO) is commonly achieved by top-down methods as bottom-up methods, such as CVD and epitaxial growth, on substrates are time consuming and difficult to scale-up to industrial level. The synthesis of GO could be said to date back to the work on graphite oxidation in 1859 by chemist B. C. Brodie.[88] His method involved treatment of graphite using potassium chlorate ($KClO_3$) in fuming nitric acid (HNO_3). Oxidative treatments were performed up to four times, increasing the oxygen content

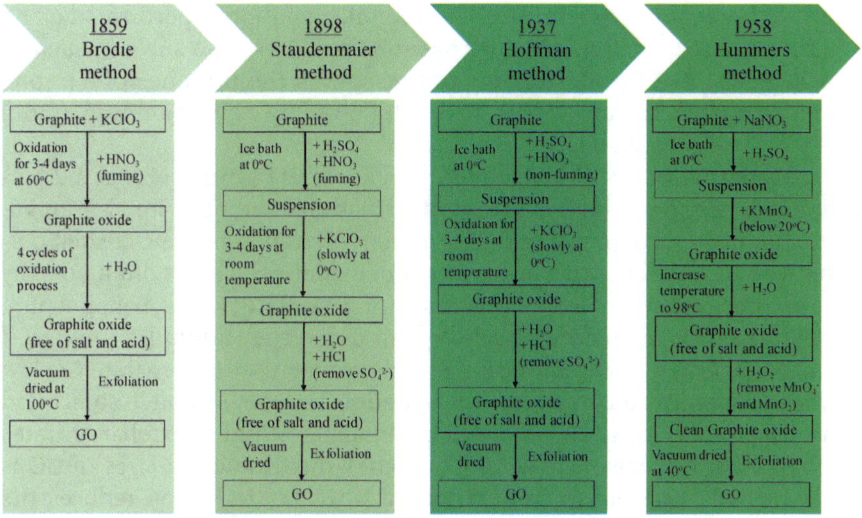

Figure 2.37 Synthesis methods for GO in chronological order and their respective procedures. Reproduced from ref. 38 with permission from Elsevier, Copyright 2018.

to a limit. The graphite oxide produced in this way had a molecular formula of $C_{2.19}H_{0.80}O_{1.00}$ with a C:O ratio of 2.19:1. Exfoliation of this product can then be carried out to produce graphene oxide as shown in Figure 2.37. The later Staudenmaier method was an improvement over the Brodie method with the use of concentrated sulfuric acid to increase the acidity of the mixture.[89] The chlorate was incorporated into the mixture in multiple aliquots over the course of the reaction resulting in a more practical approach with a similar degree of oxidation. It should be noted that the oxidation performed using nitric acid released gaseous nitrogen dioxide (NO_2) and/or nitrogen tetraoxide (N_2O_4), observed with the liberation of yellow fumes. In 1937, Hofmann made a modification in which highly concentrated HNO_3 was substituted with non-fuming HNO_3 and the reaction was performed at 0 °C, hence eliminating the formation of NO_2 and N_2O_4.[90] Nevertheless, the use of $KClO_3$ in the three (Brodie, Staudenmaier and Hoffman) methods resulted in the evolution of toxic chlorine dioxide (ClO_2) gases.

An alternative oxidation method introduced by Hummers and Offeman relied on the use of potassium permanganate ($KMnO_4$) and concentrated H_2SO_4 resulting in a similar extent of graphite oxidation.[91] This method offered a safer route to the oxidation of graphite, since permanganate salt was used as an oxidiser instead of chlorate, and the addition of sodium nitrate ($NaNO_3$) eliminated the use of nitric acid. Subsequently, other researchers made changes to the synthesis route resulting in a procedure now known as the modified Hummers' method.[92] This method increases the amount of $KMnO_4$ and incorporates phosphoric acid (H_3PO_4) to achieve a higher degree of oxidation with no evolution of toxic gases and easy control of reaction temperature.[93] In general, oxidation is performed using a carbon source, such as graphite flakes or powders, in a solvent, such as H_2SO_4, HNO_3, H_3PO_4 or a mixture of these, with the presence of a strong oxidising agent such as $KMnO_4$. This is followed by exfoliation into monolayers or few-layered stacks.

Pre-oxidation of graphite prior to its use in GO synthesis may be performed to achieve a higher extent of oxidation. The additional graphite oxidation procedure prevents the formation of graphite-core/GO-shell particles which contribute to a lower extent of oxidation. This is often done by pre-treating graphite powders in potassium persulfate ($K_2S_2O_8$) or phosphorus pentoxide (P_2O_5) before its subsequent use in the Hummers' method.[94]

Various thermal or mechanical methods can be used to separate GO layers, such as sonication or mechanical agitation in water. Although sonication in water or other polar organic media can achieve higher exfoliation rates as compared to mechanical agitation, a large distribution of sizes of GO is observed. The external force experienced through sonication reduces the dimensions of GO platelets from several microns to hundreds of nanometres. Results of atomic force microscopy (AFM) on ultrasonicated GO platelets in water have shown near-complete exfoliation.[95]

2.4.1.1 Recent Approaches to GO Synthesis

Ultra-large graphene oxide (ULGO) is an important material for the development of advanced devices and high-performance nanocomposites. However, the industrial production of ULGO sheets remains challenging, as size reduction of GO sheets occurs during the oxidation and exfoliation processes in the conventional methods. The introduction of oxygen-rich functional groups into graphite can cause a build-up of elastic strain energy, whereas vigorous agitation and sonication during exfoliation can lead to a decrease in GO sizes. Zhang *et al.* (2018) proposed a stationary oxidation-monolithic crystalline swelling strategy for a scalable synthesis of ULGO.[96] Graphite flakes undergo static oxidation in a mixture of H_2SO_4 and $KMnO_4$ without external agitations, followed by the addition of hydrogen peroxide (H_2O_2). The resulting pristine graphite oxide is then swelled with deionised water and exfoliated into single-layer ULGO sheets through mild manual shaking or mechanical agitation.

The use of electrochemical oxidation for the synthesis of GO has gathered popularity in recent years due its environmental friendliness, low cost and safety. However, this method usually suffers from low yields due to the large amount of unoxidised graphite sheets in the product. To improve this, Yu *et al.* (2022) reported a containment strategy confining the graphite particles in the reacting region during the electrochemical oxidation process.[97] Through their improved strategy, the authors managed to synthesise GO with a C/O ratio of 2.4, whereas a C/O of 2.18 was obtained *via* the Hummers' method. The lower oxidation degree in the GO resulted in a high conductivity of the sample after reduction. The introduction of intermittent ultrasonication during the synthesis of GO has improved exfoliation, as a greater surface area was exposed for oxidation to occur.[98] To increase the yield of GO and eliminate emissions of toxic gases from the use of $NaNO_3$ in the Hummers' method, Zhang *et al.* (2022) investigated the use of boric acid to replace $NaNO_3$.[99] The authors observed that this modification was able to increase the yield by 10% in a shorter amount of time at room temperature. In addition, GO sheets prepared using boric acid possessed fewer defective structures compared to that of the Hummers' method.

2.4.2 Reduced Graphene Oxide

Due to the similarities between pristine graphene and reduced graphene oxide (rGO), the reduction of GO is an important route for obtaining graphene-based materials. This reduction process can be performed through chemical, thermal or electrochemical methods, each yielding different morphologies and properties. The products of these reduction methods resemble pristine graphene to an extent in terms of their properties (electrical, mechanical and thermal) and surface morphology. However, it should be noted that there are differences between pristine graphene and rGO, hence they should be clearly distinguished.

In accomplishing the reduction of GO, several factors are considered, such as the C/O ratio of the product, selectivity in oxygen group removal, recovery of surface defects on GO, choice of reducing agents and the retention or improvement of desired physical and chemical properties from GO.[92]

2.4.2.1 Chemical Reduction

Reagents, such as hydrazine, metal hydrides or hydrohalic acids, have been reported for the reduction of GO. Mohan *et al.* (2015) investigated the effects of three different reducing agents (hydrobromic acid, hydrazine hydrate and hydroiodic acid) on the electrical conductivity of the resulting rGO.[100] The authors found that the hydroiodic acid treatment resulted in a noticeable increase in electrical conductivity (up to 1.03×10^4 S m^{-1}) as the reduction time was increased from 24 to 72 hours. This was attributed to the more aggressive reducing power of hydroiodic acid, yet it did not lead to the removal of carbon from the GO sheets. Similarly, Kumar *et al.* (2015) immersed large-area GO films in hydroiodic acid at 80 °C, for 2 to 3 hours to achieve a higher degree of reduction.[101] Films of rGO produced in this manner showed a conductivity value of 2.43×10^4 S m^{-1}. Notably, the thermal conductivity was also enhanced (1390 W m^{-1} K^{-1}) with superior values compared to those from conventional reduction methods.

In an attempt to reduce the number of defects present in rGO, Kim *et al.* (2018) utilised iron(III) chloride ($FeCl_3$) as a catalyst for the intramolecular cross-dehydrogenative coupling to yield healed rGO (Figure 2.38).[102] Subsequent to the reduction using hydrazine, the rGO was dispersed in dichloromethane and purged with nitrogen gas. $FeCl_3$ in nitromethane was then added dropwise and the reaction was allowed to continue for 3 days, after which

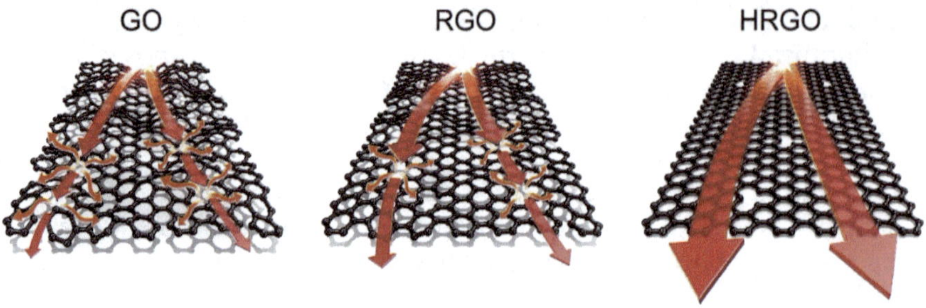

Figure 2.38 Restoration of the graphitic structure in healed reduced graphene oxide (HRGO) for enhancing thermal conductivity (red arrows). Reproduced from ref. 102 with permission from Elsevier, Copyright 2018.

quenching was performed using methanol. The healing of rGO produced a more graphene-like structure which significantly improved the thermal conductivity with values five-fold that of rGO, as there was less resistance to heat flow with the minimisation of chemical defects over the graphitic surface.

2.4.2.2 *Thermal Reduction*

The thermal treatment of GO results in the decomposition of the oxygenated groups of GO into CO_2 and CO gases. This can be achieved through thermal annealing at high temperatures (up to 1050 °C) in the absence of oxygen. Exfoliation of the stacked structure also occurs through the generation of gases as they create enormous pressure within the stacked layers at high temperatures. However, the evolution of CO_2 and CO gases also lead to structural damage of the final product with approximately 30% mass loss during this process. These vacancies and topological defects in the rGO plane produce inevitable effects, particularly on the electronic properties. Nevertheless, reports on the bulk conductivity have shown increased values with the reduction of GO, which indicated effective reduction and recovery of the sp^2 electronic structure. Lin *et al.* (2012) investigated the deposition of a copper film onto GO, followed by thermal annealing at 900 °C, in the presence of hydrogen, to achieve reduction of GO.[103] The authors observed that copper acted as a catalyst for the removal of oxygen-containing functional groups, as well as the recovery of the graphitic structure. Raman spectroscopy measurements revealed increased graphitic domain sizes leading to enhanced electrical conductivity deduced from the low sheet resistance of 13.9 $k\Omega\ sq^{-1}$.

Wang *et al.* (2018) realised that thermal reduction techniques would not introduce impurities, which is vital to the quality of rGO and allowed control over the reduction degree by varying the temperature of the reaction.[104] The authors employed a two-step reduction, which consisted of thermal reduction with a minimum heating rate of 1 K min^{-1} to 1000 K in an argon atmosphere for 1 hour, followed by a Joule heating procedure, which involved the application of direct current through the curved film to achieve a temperature of 3000 K. The resulting product from the two-step reduction exhibited an outstandingly high conductivity of 63×10^4 S m^{-1}, greatly exceeding previously reported values in the literature.

Other methods for preparing rGO involve the irradiation of microwaves onto GO powders[105] and the use of high intensity light for the flash reduction of GO films.[106]

2.4.2.3 Electrochemical Reduction

Compared to the chemical and thermal reduction methods, which require the removal of excess reductants and high temperatures, respectively, electrochemical reduction offers a more convenient, green and energy efficient method to prepare rGO. Electrochemical reduction is often conducted *via* cyclic voltammetry or chronoamperometry with applied potentials ranging from −0.9 to −1.6 V in aqueous or organic media.[107] The required electrochemical potential depends on the pH of the electrolytic medium, where smaller potentials are needed for highly acidic electrolytes. A study conducted by Quezada-Renteria *et al.* (2019) investigated the role of protons on the reduction degree and the formation of defects in electrochemically produced rGO.[108] The presence of protons facilitated the removal of oxygen-containing functional groups and the hydrogenation of carbon atoms in graphitic regions. A higher degree of reduction was observed for basic electrolytes, followed by acidic and organic electrolytes. However, electrochemical reductions conducted in acidic conditions produced the lowest sp^2/sp^3 ratio, which indicated the introduction of defects caused by hydrogenation in the graphitic regions.

Investigations on the electrochemical reduction of GO have reported the removal of epoxy and carbonyl groups, with most carboxyl groups being unaffected by the reduction. Nevertheless, carboxyl groups can be electrochemically reduced at more negative potentials (−2.1 V) and highly acidic conditions.[107] Notably, out of the numerous investigations conducted on electrochemical treatment, only a few studies have reported that defects were introduced to graphene sheets.

2.4.2.4 Recent Approaches to GO Reduction

Zeng *et al.* (2018) investigated a similar copper-catalyzed setup to that used by Lin *et al.* (2012),[103] but at a lower temperature for the controlled reduction of GO.[109] Their approach eliminated the use of very high temperatures, which was deemed unfeasible for the integration of rGO into the fabrication processes of electronics. A thin layer of copper was first deposited onto a polyethylene terephthalate (PET) substrate, followed by dipping into a GO dispersion. Their work offered a scalable method for the fabrication of large area and conductive rGO nanosheets. In addition, the films showed higher electrical conductivity (1.43×10^4 S m^{-1} at room temperature) and lower thermal conductivity as compared to reported films prepared through chemical or thermal means.

2.4.3 Fluorinated Graphene

Fluorographenes can be prepared *via* two main routes: fluorination using fluorinating agents and chemicals, or mechanical exfoliation of pristine graphite fluoride. These approaches will now be outlined.

2.4.3.1 *Fluorination of Graphene*

The fluorination process is initiated by the introduction of graphene into a reactor containing fluorinating agents under the influence of pressure, heat or radiation. Xenon fluoride (XeF_2) can be used as the fluorinating agent at various temperatures in an inert atmosphere. Chronopoulos *et al.* (2017) used a room temperature technique to fluorinate graphene supported on a silicon-on-insulator substrate using XeF_2. This formed fully fluorinated graphene with a dominant stoichiometry of $C_{1.0}F_{1.0}$.[110] The direct use of fluorine gas to fluorinate reduced graphene at various temperatures and pressures has also been reported.[111]

An alternative way to achieve fluorination involves using a plasma to generate fluorine radicals, which adsorb onto the graphene layer and form C–F bonds. Fluorine-containing compounds, such as CF_4, SF_6, and F_2, are commonly utilised as radical-generating precursors.[110] An advantage of this technique is that it allows the controllable fluorination of graphene materials. The selective fluorination of graphene is particularly desirable for its application in nanoelectronics. The patterning of graphene, *i.e.* defining conductive and insulating areas in the same graphene sheet, was demonstrated by Ho *et al.* (2014) using CF_4 plasma treatment, which allowed fluorination at low temperatures (<200 °C) and minimisation of defect formation.[112]

The fluorination of graphene is reversible through thermal annealing in the presence of hydrogen gas or anhydrous hydrazine vapour.[45] The defluorination process has been found to be more effective with the use of anhydrous hydrazine vapour at milder temperatures between 100 °C and 200 °C. In contrast, thermal annealing with hydrogen gas at temperatures between 400 °C and 600 °C induced carbon loss during the defluorination process, evolving products, such as CF_4, C_2F_4 and C_2F_6.

2.4.3.2 *Exfoliation of Graphite Fluoride*

Liquid-phase exfoliation involves suspending bulk graphite fluoride in a solvent, which acts as an intercalating agent to weaken the van der Waals interactions between the fluorographitic layers. Several organic solvents can be used including sulfolane, isopropanol, acetonitrile, ethanol and chloroform, in addition to ionic liquids and surfactants.[110]

Thermal exfoliation of graphite fluoride can be conducted in a nitrogen atmosphere at various temperatures.[113] This approach leads to the formation of partially fluorinated graphene: exfoliation at 400 °C produces 49% fluorinated graphene while higher temperatures lead to a decrease in fluorine content (1000 °C resulting in 3% fluorinated graphene).

2.4.4 Other Halogenated Graphenes

Chlorination and bromination of graphene have been attempted by Gopalakrishnan *et al.* (2012) using ultraviolet irradiation in liquid chlorine and bromine media, respectively.[114] The authors conducted the reaction with few-layer graphene resulting in chlorinated graphene of up to 56 wt% and brominated graphene of up to 25 wt%. Halogenation of graphene using ultraviolet irradiation proved advantageous over the CVD approach, with regard to the extent of halogenation, where only around 15 wt% chlorination was achieved.[115]

Other methods include the use of microwave-sparks in a graphite–liquid halogen mixture,[116] thermal exfoliation/reduction of graphite oxide in a halogen atmosphere,[117] autoclave reactions under elevated temperatures and pressures,[118] as well as the halogenation of hydrogenated graphene.[119]

2.4.5 Hydrogenated Graphene

Hydrogenated graphene has generally been produced by the top-down approach, which includes the reduction of graphite and hydrogenation of sp^2 carbon containing materials, such as graphene and graphite oxide, in a hydrogen plasma or atmosphere. It can also be synthesised by a bottom-up approach such as CVD.

2.4.5.1 Top-down Synthesis of Hydrogenated Graphene

The first successful synthesis of hydrogenated graphene was conducted by annealing graphene crystals in an argon atmosphere for 4 hours at 300 °C, followed by exposure to cold hydrogen plasma at a low pressure (0.1 mbar).[40] Elias *et al.* (2009) found that typically 2 hours of plasma treatment was required for the saturation of measured characteristics.[120] The authors also demonstrated stability of the hydrogenated product at room temperature for many days and found that this conversion was reversible since the original state could be restored by annealing at 450 °C in an argon atmosphere for 24 hours. A similar approach was attempted by Burgess *et al.* (2011) in a plasma-enhanced chemical vapour deposition (PECVD) reactor at a pressure of 2 mbar, resulting in a hydrogenation efficiency of <10%.[121]

A scalable method for the hydrogenation of graphene in which graphite oxide was exfoliated thermally in a hydrogen atmosphere at high temperature (up to 500 °C) and pressure (up to 150 bar) was suggested by Poh *et al.* (2012).[122] The authors found that a hydrogen gas pressure of 100 bar at 500 °C was most efficient resulting in 11.6% hydrogenation. In other work, Antonov *et al.* (2016) synthesised multilayer graphane or hydrographite by exposing graphite to gaseous hydrogen at high pressures above 2 GPa and temperatures ranging from 450 °C to 700 °C.[123] The authors were able to conduct the synthesis at a high temperature of 700 °C, consistently producing hydrogenated samples of $(C_{1.4}H_1)_n$.

Various degrees of graphene hydrogenation were achieved *via* electrolytic hydrogenation of thin graphite flakes by Ilyin *et al.* (2011).[124] The treatment was conducted in distilled water with 10 V and 2 mA at room temperature, resulting in hydrogenated clusters or graphane-like fragments throughout the flakes. Ciammaruchi *et al.* (2019) proposed the water splitting reaction as an energetically efficient, environmentally friendly approach to the hydrogenation of GO.[125] The mild conditions involved in the process, such as low-applied voltage, room temperature and atmospheric pressure, offered benefits over the use of high energy plasmas or harsh chemical conditions in other methods. The authors also found that the hydrogenation of graphene-based materials was more favourable in the presence of epoxy and hydroxyl groups, which were found in GO.

The Birch reduction is conventionally used for the partial hydrogenation of aromatic rings and its application has been extended by several groups for the hydrogenation of graphite[126,127] and reduced graphite oxide.[128] The authors used different electron sources for the first step of the reduction, and proton sources for the second step; sodium in liquid ammonia was used as the source of electrons and alcohol as the source of protons. The greatest hydrogenation efficiency was obtained by Yang *et al.* (2012) using graphite, sodium and *tert*-butyl alcohol (*t*-BuOH) as the starting material, electron source and hydrogen source, respectively.[126] Through elemental analysis, the authors determined the hydrogenated graphene to have the formula $(C_{1.3}H_1)_n$ where the ideal stoichiometric ratio for a fully hydrogenated graphene, *i.e.* graphane is $(C_1H_1)_n$. Such high levels of hydrogenation also induced a large band gap of ~4 eV.

2.4.5.2 Bottom-up Synthesis of Hydrogenated Graphene

In an attempt at producing high-quality large-area graphene, Wang *et al.* (2010) applied a plasma-assisted atom-beam process in high vacuum to initially prepare hydrogenated graphene.[129] The CVD procedure used methane and hydrogen sources, and a copper catalyst. The use of excess hydrogen led to the formation of hydrogenated graphene. This intermediate product was next subjected to thermal dehydrogenation to produce graphene.

Fei *et al.* (2022) developed a new technique to minimise the introduction of impurities due to the complex-solution environments in the Birch reduction and electrochemical methods.[130] This new reaction involved lithium hydride (LiH) as the hydrogen source and CO as the carbon source, successfully yielding a 33% hydrogenation efficiency to produce few-layer hydrogenated graphene as depicted in Figure 2.39.

2.4.6 Graphyne

The synthesis of graphynes and graphdiynes is based on coupling reactions of the monomer hexaethynylbenzene. Various techniques with different reaction conditions have been developed for the synthesis as discussed below.

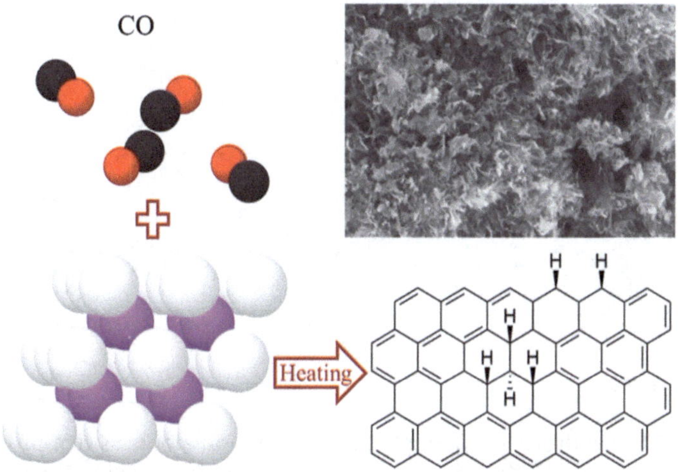

Figure 2.39 Synthesis of hydrogenated graphene (HG) using lithium hydride (LiH) and CO. Reproduced from ref. 130 with permission from Elsevier, Copyright 2022.

2.4.6.1 *Liquid/Solid Systems*

In 2010, Li *et al.* (2010) first synthesised multilayered γ-graphyne-2 or graphdiyne as large area films.[57] The authors demonstrated the production of a 1 μm thick film through a cross-coupling reaction using hexaethynylbenzene, in the presence of pyridine, on the surface of copper for 72 hours at 60 °C under nitrogen atmosphere. Conductivity measurements on the synthesised γ-graphyne-2 films revealed comparable values to that of silicon, suggesting promising properties as a semiconductor. In 2015, the controlled growth of γ-graphyne-2 films on ZnO nanorod arrays, using a reduction and vapour liquid–solid (VLS) growth processes, were investigated achieving thicknesses of 540, 42.6 and 22 nm.[131]

2.4.6.2 *Liquid/Liquid and Gas/Liquid Systems*

To produce thinner γ-graphyne-2 films of higher quality, Matsuoka *et al.* (2017) attempted the synthesis of such films at the interface between two immiscible liquids.[58] The top layer consisting of copper(II) acetate and pyridine served as the cross-coupling catalysts, while the bottom dichloromethane layer contained the precursor hexaethynylbenzene. The reaction was performed under an argon atmosphere at room temperature for 24 hours and resulted in a film thickness of 24 nm. In addition to this liquid/liquid interface synthesis, the authors have also investigated a gas/liquid interfacial synthesis procedure, which allowed the production of thinner films. In this approach, minute amounts of hexaethynylbenzene (20 nmol) in

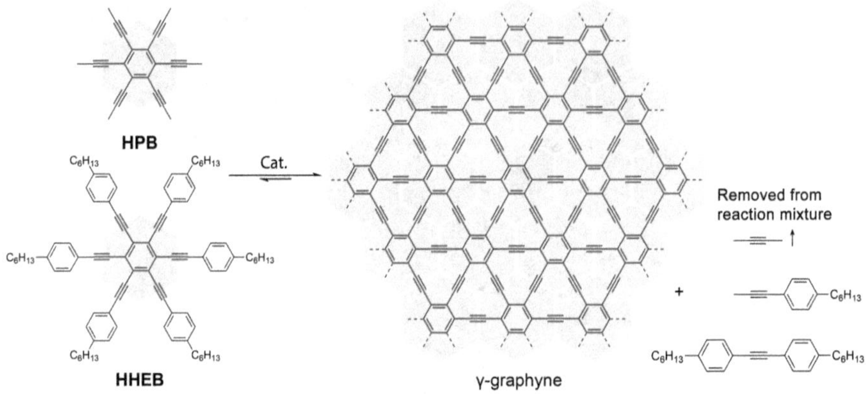

Figure 2.40 Synthesis of γ_1-graphyne from two monomers: hexapropynylbenzene (HPB) and 1,2,3,4,5,6-hexakis[2-(4-hexylphenyl)ethynyl]benzene (HHEB). Reproduced from ref. 132 with permission from Springer Nature, Copyright 2022.

dichloromethane and toluene were added onto the surface of an aqueous copper catalyst solution, such that the organic solvents were rapidly evaporated, and a polymerisation reaction took place at the gas/water interface. Measurements using SEM, TEM and AFM revealed a thickness of 2.97 nm with a narrow distribution of 0.03 nm.

Recently, Hu *et al.* (2022) synthesised γ_1-graphyne by the simultaneous use of two different hexa-alkylnyl-benzenes as the comonomers.[132] Specifically, hexapropynylbenzene and 1,2,3,4,5,6-hexakis[2-(4-hexylphenyl)ethynyl]benzene were dissolved in tetrachloromethane and the reaction was carried out on an activated molybdenum catalyst solution (Figure 2.40), resulting in a product that showed a layered morphology of the graphyne sheet.

2.4.6.3 *Solid/Gas Systems*

A method of synthesising γ-graphyne-2 in air or nitrogen at 120 °C was reported by Zuo *et al.* (2017).[59] The authors made a number of γ-graphyne-2 structures including 1D nanochains, 2D nanoribbons and 3D frameworks using different combinations of heat treatment and atmosphere (Figure 2.41). Specifically, when hexaethynylbenzene was exposed to a gradual heating process in nitrogen, 2D nanoribbons were formed. However, when this treatment was applied in air, an explosive phenomenon occurred resulting in a 6-fold volume increase to a 3D framework. The rapid heating of hexaethynylbenzene in air resulted in a more violent explosion forming ultrafine 1D nanochains entangled like cotton. Notably, this approach provided high production yields of up to 98% in a short reaction time. Nevertheless, the crystallinity of graphynes produced through this method was inferior to the graphynes synthesised from the other systems.

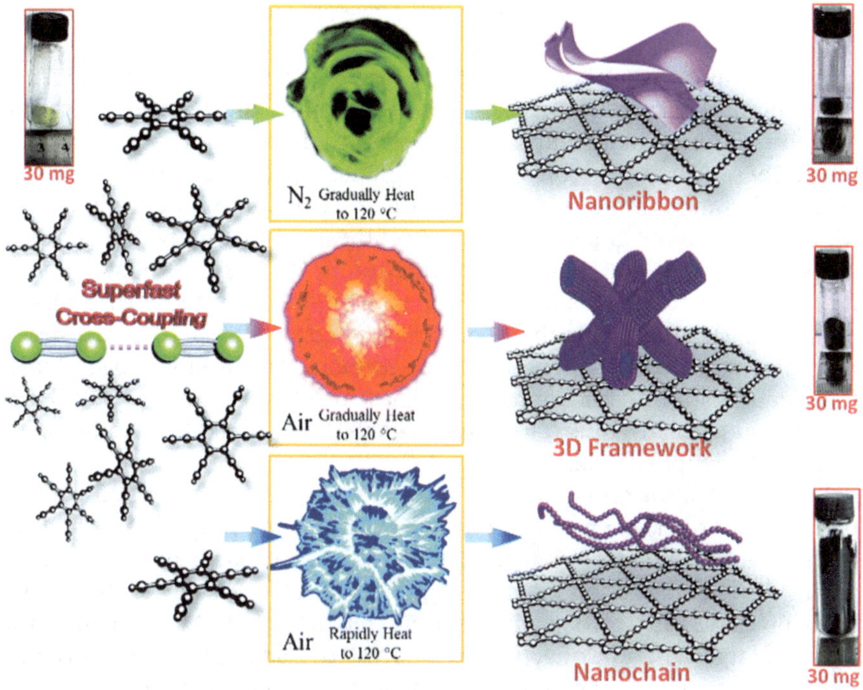

Figure 2.41 1D nanochain, 2D nanoribbon and 3D framework of γ-graphyne-2 obtained *via* different treatments. Reproduced from ref. 59 with permission from the Royal Society of Chemistry.

2.5 Applications of Graphene Materials

Due to their unique physicochemical characteristics, including large specific surface area, superior optical and electron transport properties, extraordinary strength and malleability, graphene materials have been applied in a number of fields, such as environmental remediation, energy storage, lubricant additives and sensor devices. Some of these applications will now be discussed in detail.

2.5.1 Wastewater Treatment

The increase in anthropogenic activities has led to elevated concentrations of hazardous substances, such as synthetic dyes, heavy metals and pharmaceutical residuals in the aquatic environment. These pollutants have been reported to cause bioaccumulation thereby triggering detrimental health impacts and damaging the ecosystems irreversibly. Recently, graphene materials have been widely investigated in adsorption and photocatalysis systems for wastewater treatment.

Adsorption using graphene is an attractive choice for the remediation of polluted water owing to its simplicity, cost effectiveness and high removal efficiency. Among the various graphene forms, GO is the most widely tested for pollutant adsorption. It has a large specific surface area, abundant oxygen-rich functional groups, high mechanical flexibility and chemical stability. These features endow GO with high adsorption potential for various water pollutants. Another unique feature of GO is its ease of modification, which has sparked much interest in synthesising 3D graphene composites for environmental pollution treatment.[38,133,134] This is evident from the extensive literature data shown in Table 2.9 on the adsorption performance of various 2D and 3D graphene materials. As can be seen, the adsorption mechanisms of the graphene materials primarily involved hydrogen bonding, electrostatic and π–π interactions. These resulted from the plentiful oxygenated functional groups, negatively-charged sorption sites and aromatic rings of the graphene materials.

The effects of the aromatic rings of malachite green dye on the adsorption performance of graphene was discussed by Gupta and Khatri (2017).[138] They found that rGO adsorbent had a higher adsorption capacity than GO despite having slower adsorption kinetics (Figure 2.42). This was due to the surface area of rGO (931 $m^2 g^{-1}$) being higher than that of GO (103.7 $m^2 g^{-1}$), thus providing a larger amount of active sites for the dye adsorption. Furthermore, malachite green dye possessed three aromatic rings and a cationic centre (N^+), while the rGO had an abundant π-electron cloud, which facilitated the π–π interactions with the dye during adsorption.[138]

Lai *et al.* (2021) applied the ice-templating method, as a facile and environmentally friendly approach to induce porosity in GO aerogel, for the treatment of cadmium tainted water.[133] Ice-templating involves the use of a solvent as the pore-inducing agent, where ice particles are created upon freezing and removed upon freeze drying to create porous structures within the aerogel. The high content of oxygenated groups within the adsorbent, contributed by GO and xanthan gum, enhanced the adsorption of cadmium up to a maximum 67.3 mg g^{-1}. Notably, the authors were able to use the aerogel in batch adsorption and packed bed adsorption modes. From their study, the column adsorption capacity was found to be 64.05 mg g^{-1}, which was very close to the batch adsorption capacity. This finding illustrates the potential of graphene materials in wastewater treatment.

To improve the adsorption performance of graphene derivatives, additional functionalities can be incorporated into the graphitic network through chemical functionalisation. Magnetic core–shell $MnFe_2O_4/TiO_2$ nanoparticles decorated on rGO were prepared by Chang *et al.* (2021).[144] According to their characterisation results depicted in Figure 2.43, $MnFe_2O_4$ formed the core (black) of the adsorbent while titanium oxide (TiO_2) (lighter colour) formed the shell. When the graphene was integrated with $MnFe_2O_4$ and TiO_2, the adsorbent showed uneven surface morphology. Functionalisation of the rGO also introduced new magnetic properties as evidenced by the measured magnetic strength of 14.71 emu g^{-1}. This allowed simple recovery of the nano-adsorbent using magnetic separation at the end of adsorption.[144]

Table 2.9 Adsorption performance of graphene materials towards various pollutants[a]

Adsorbent	Pollutant	q_m/mg g^{-1}	Isotherm model	Kinetic model	Mechanisms	References
Dye						
Polydopamine and polyethylenimine co-functionalised GO aerogel (PPGA)	Methyl orange	208.8	Langmuir	PSO	Hydrogen bonding between hydrogen atoms of nitrogen containing functional groups in PPGA and oxygen atom of dye	135
Bismuth oxide decorated GO nanocomposite	Rhodamine B	320.0	Temkin	PSO	Hydrogen bonding	136
Sodium dodecyl sulfate exfoliated graphene	Methylene blue	782.3	Freundlich	PSO	Electrostatic interaction	137
rGO	Malachite green	476.2	Langmuir	PSO	π–π interaction between aromatic rings of malachite green and graphitic skeleton, and electrostatic interaction	138
Heavy metal						
Anionic polypeptide poly(γ-glutamic acid) functionalised iron oxide-GO hybrid nanocomposite	Cadmium(II) Copper(II) Nickel(II)	625.0 574.7 384.62	Langmuir	PSO	Coordination bonding, electrostatic interaction	139
Magnetic ionic liquid/chitosan/GO composite	Chromium(VI)	145.4	Langmuir	PSO	Hydrogen bonding, electrostatic interaction	140
GO/titanium dioxide/xanthan gum sponge	Cadmium(II)	67.33	Freundlich	PSO	Electrostatic interaction, complexation and hydrogen bonding	133
Pharmaceutical residue						
3D manganese dioxide/rGO hybrid aerogel	Acetaminophen	252.9	Langmuir	PFO/PSO	Hydrogen bonding and pore entrapment	134
rGO aerogel	Diclofenac	596.7	Freundlich	PFO	Hydrogen bonding, hydrophobic interaction	141
Magnetic adsorbent genipin-crosslinked chitosan/GO-sulfonic acid	Ibuprofen	138.2	Freundlich	PSO	Physical bonding	142
Manganese dioxide/graphene nanocomposite	Tetracycline	1789	Langmuir	PSO	Water solubility, electrostatic interaction, hydrogen bonding, coordination effect, π–π interaction	143

[a]PFO: Pseudo-first-order; PSO: Pseudo-second-order.

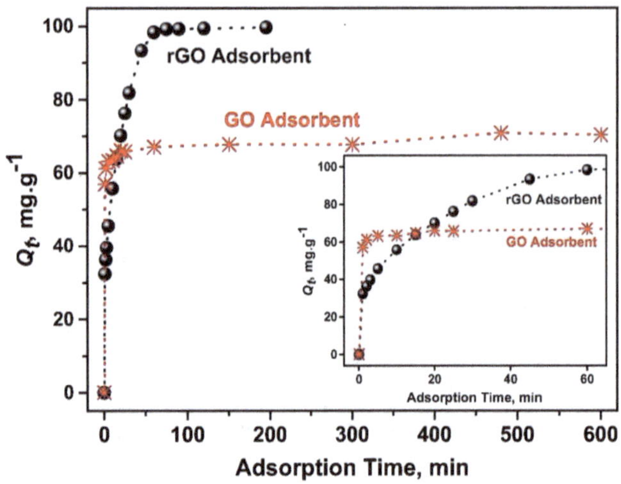

Figure 2.42 Adsorption of malachite green dye onto GO and rGO as a function of time. Reproduced from ref. 138 with permission from Elsevier, Copyright 2017.

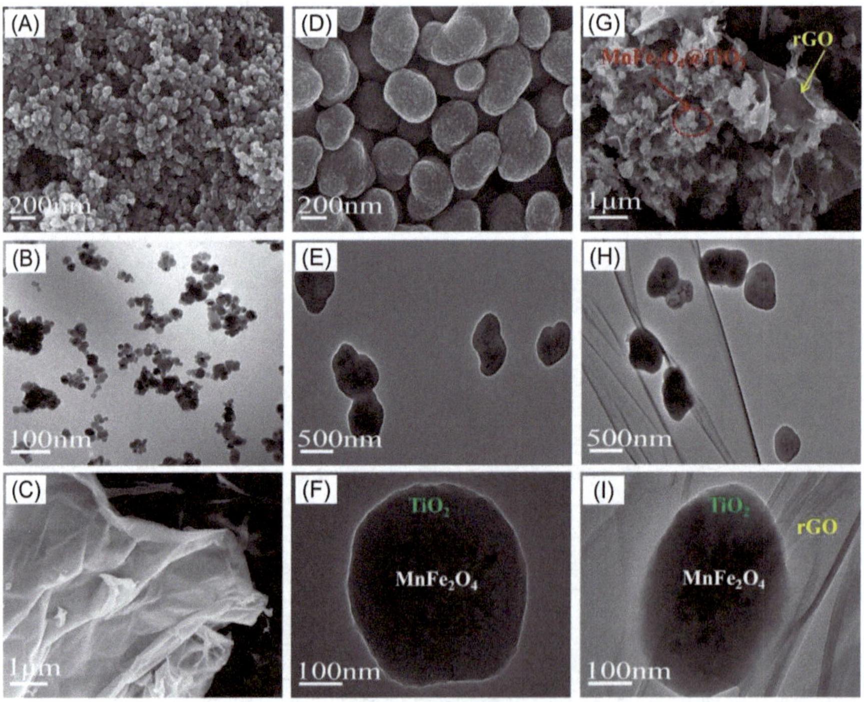

Figure 2.43 Morphology analysis of $MnFe_2O_4$ and its composites (A) Field emission scanning electron micrograph (FESEM), (B) TEM; rGO (C) FESEM; $MnFe_2O_4/TiO_2$ (D) FESEM, (E and F) TEM and $MnFe_2O_4/TiO_2$-rGO (G) FESEM, (H and I) TEM. Reproduced from ref. 144 with permission from Elsevier, Copyright 2021.

Graphene materials can also be applied as catalysts for the degradation of organic pollutants *via* photocatalysis, which is a relatively new technology for treating water contaminants under the influence of light. Conventional semiconductors, such as titanium oxide, zinc oxide and zirconia oxide, have been used as photocatalysts to absorb light to reach an excitation state of greater energy than the bandgap for initiation of electron–hole pairs. The photo-excited holes and electrons are subsequently transported from the catalyst to the active sites. A redox reaction will occur at the active sites, which degrades the adsorbed pollutant on the catalyst surface. However, conventional photocatalysts often suffer from rapid recombination of the hole–electron pairs, which hampers the efficiency of photocatalytic degradation. Therefore, it is desired to develop an effective photocatalyst with strong light harvesting potential, high charge separation and electron transfer efficiency.

The use of graphene materials in photocatalysts can help to overcome the disadvantages of traditional photocatalysts. Properties such as optical transparency, high specific surface area, superior electron mobility and aromatic ring structure, give graphene materials important advantages as prospective catalysts for the photocatalytic removal of aqueous pollutants. Table 2.10 shows the photocatalytic degradation performance of graphene materials towards various pollutants.

Table 2.10 Photocatalysis performance of graphene materials towards various pollutants.

Photocatalyst	Pollutant	Bandgap/eV	Working range	Efficiency/%	Time/min	References
Dye						
ZnO–rGO composites	Methylene blue	—	UV-light	99	120	145
CdS–rGO heterojunction	Rhodamine B	—	Visible light	97.2	60	146
Core/shell rGO/BiOBr particles	Orange II	2.33	Visible light	89	105	147
Graphene/TiO_2-Ag composites	Amaranth	3.04	UV-light	98	120	148
Pharmaceutical residue						
rGO–WO_3 composites	Sulfamethoxazole	2.15	Visible light	98	180	149
Calcined layered double hydroxides/rGO composites	Paracetamol	—	Visible light	95	420	150
BiOBr/MoS_2/GO composites	Tetracycline	1.11	Visible light	98	40	151

As shown in the table, graphene materials, particularly GO and rGO, are highly effective as photocatalysts for environmental pollution control. The active oxygenated sites of both GO and rGO provide good integration with the selected oxides (TiO_2, ZnO, WO_3, *etc.*) to improve the photocatalysis potential. The delocalised π-bonds within the sp^2 carbon lattice and the resultant free electrons enhance the electrical conductivity and photoconversion efficiency of the graphene-modified photocatalyst through the inhibition of charge recombination.[152] Additionally, in graphene-loaded photocatalysts, a positive shift occurs in the Fermi level thus enhancing the electrochemical potential and charge separation efficiency.[153]

The aforementioned properties were observed in a study of the photodegradation of tetracycline using bismuth bromide oxide/molybdenum disulfide/GO composites conducted by Li *et al.* (2021).[151] The band gap of pristine bismuth bromide oxide (BiOBr) was 2.73 eV and the value was further decreased to 2.49 and 1.11 eV for BiOBr incorporated with GO and molybdenum disulfide, respectively.[151] This was due to the improved interfacial interaction between the unpaired π-electrons of carbon atoms and the free electrons on the BiOBr surface.[154] This finding indicated that the photocatalyst had great photocatalytic activity due to the free electron-holes generated with a narrow band gap. This is evident from the highest tetracycline photodegradation (93.4%) by the photocatalyst in comparison to BiOBr (70.7%) and MoS_2/GO (54.8%) as illustrated in Figure 2.44.[151]

Similarly, in the rGO–WO_3 photocatalyst developed by Zhu *et al.* (2017), the addition of rGO to tungsten trioxide (WO_3) triggered the possible electronic transition of π to π^* of rGO and n to π^* between the n-orbital of the oxygen functional groups of rGO and WO_3.[149] This enhanced the surface electric charge of WO_3 and narrowed the band gap. The calculated band gaps for pristine WO_3 and rGO–WO_3 photocatalyst are 2.75 and 2.15 eV, respectively.[149] The photocatalysis performance towards sulfamethoxazole removal was reported to be >98% within 3 hours at visible light irradiation. They also proposed

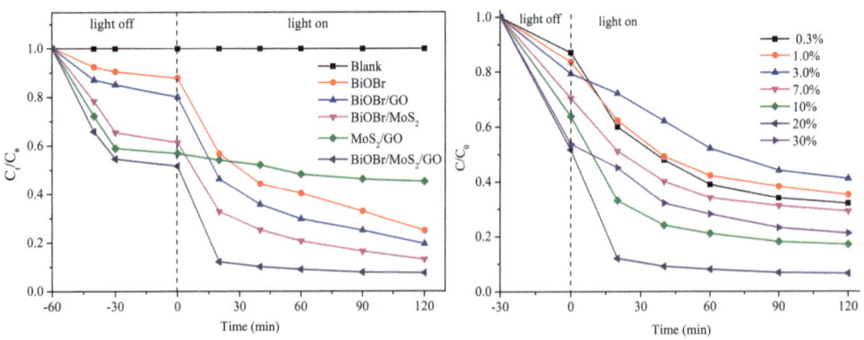

Figure 2.44 Photodegradation profile of tetracycline using different compositions of BiOBr/MoS_2/GO composites. Reproduced from ref. 151 with permission from Elsevier, Copyright 2021.

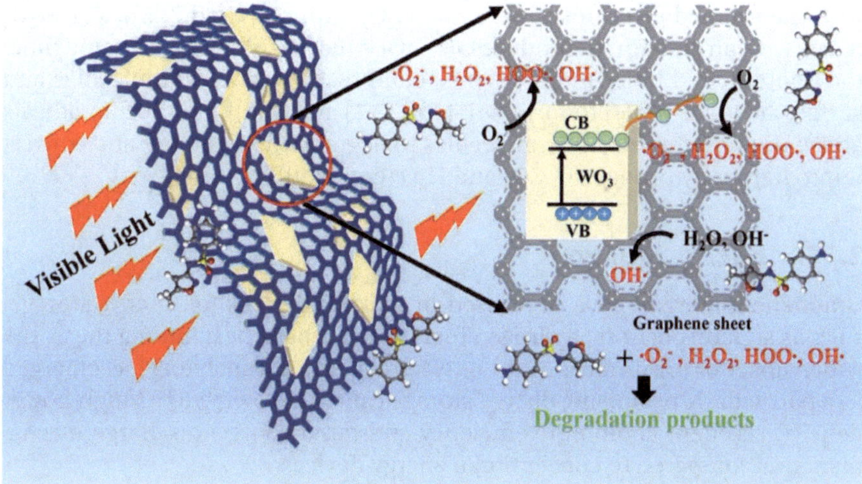

Figure 2.45 Schematic illustration of photodegradation of sulfamethaxazole using RGO–WO_3 composites under visible light irradiation. Reproduced from ref. 149 with permission from Elsevier, Copyright 2017.

the following mechanism for sulfamethoxazole degradation by rGO–WO_3 photocatalyst as follows (Figure 2.45):[149]

$$WO_3 + \text{light} \rightarrow WO_3\left(e^- + h^+\right) \tag{2.1}$$

$$e^- + \text{rGO} \rightarrow \text{rGO}\left(e^-\right) \tag{2.2}$$

$$\text{rGO}\left(e^-\right) + O_2 \rightarrow \text{rGO} + O_2^- \tag{2.3}$$

$$O_2^- + H^+ \rightarrow HOO^\bullet \tag{2.4}$$

$$HOO^\bullet + e^- + H^+ \rightarrow H_2O_2 \tag{2.5}$$

$$H_2O_2 + e^- \rightarrow OH^\bullet + OH^- \tag{2.6}$$

$$H_2O + h^+ \rightarrow OH^\bullet + H^+ \tag{2.7}$$

$$h^+ + OH^- \rightarrow OH^\bullet \tag{2.8}$$

$$OH^\bullet + \text{sulfamethoxazole} \rightarrow \text{by-product} + CO_2 + H_2O \tag{2.9}$$

As already noted, the introduction of rGO narrows the band gap of WO_3, extending its light response range to the visible. Thus, when irradiated by visible light, WO_3 absorbs incident photons, generating electrons and holes (eqn (2.1)). The excited electrons can then transfer from the conduction band of WO_3 to GO (eqn (2.2)). The photogenerated electrons will be scavenged by dissolved oxygen in water, resulting in efficient electron (e^-)–hole (h^+) separation and producing peroxide radicals ($^\bullet O^{2-}$) (eqn (2.3)).[148,149] These peroxide radicals ($^{\bullet}O^{2-}$) then combine with H^+ to form $HOO^{\bullet}$ which in turn combines with the trapped electrons to generate H_2O_2 (eqn (2.4) and (2.5)). The H_2O_2 reacts with an electron to produce $OH^{\bullet} + OH^-$ (eqn (2.6)).[145] At the same time, the photoexcited holes in the valence band react with water molecules and surface hydroxyl (^-OH) to form $OH^{\bullet}$ (eqn (2.7) and (2.8)). Lastly, the radicals attack the sulfamethoxazole molecules to degrade the structure and convert it into lighter by-products + CO_2 and H_2O (eqn (2.9)).[149]

2.5.2 Energy Storage

Graphene materials have been used in making devices for energy storage, such as supercapacitors, hydrogen fuel cells and batteries. Among these, the supercapacitor (or ultracapacitor or electrochemical capacitor) has emerged as a particularly promising energy storage candidate owing to its high-power density, excellent coulombic efficiency and rapid charge–discharge mechanisms, as compared to conventional energy devices.

Electric double layer capacitors (EDLC) are the primary kind of supercapacitor that employs graphene materials. In EDLCs, the electrostatic charge is stored *via* reversible ionic exchange at the electrode/electrolyte interface. The electric charges accumulate on the electrode surface while the electrolyte ions with opposite charges are formed from the electrolyte side to retain charge neutrality.[155] The electrons from the negative electrode migrate towards the positive electrode through an external load during the charging process. At the same time, the cations transfer towards the negative electrode while the anions transfer towards the positive electrode. Carbon materials, such as graphene and graphene derivatives, are candidates for EDLC electrodes due to their good chemical stability, good electrical conductivity and high specific surface area.[156] The performance of graphene materials in EDLC supercapacitor devices is set out in Table 2.11. It can be seen from the table that mixed transition metal oxides are often blended with graphene material to form effective supercapacitors. This could be attributed to synergistic influences between multiple metal compositions and low activation energy requirement for electron transportation, which subsequently result in a higher specific capacitance and electrical conductivity than those of a single transition metal oxide.[157] These properties are enhanced with the incorporation of graphene materials. Graphene materials not only increase the exposure surface for electrical conduction of the electrode, but they also reduce the charge transfer between the metal oxides and current collector, thus improving the supercapacitor performance.[158] As an example of this approach, Kaverlavani *et al.* (2017) successfully designed graphene-wrapped nanoporous copper cobaltate ($CuCo_2O_4$) hollow-sphere electrodes with an ultrahigh specific capacitance of 1813 F g^{-1} at 2 A g^{-1} current density. The electrodes demonstrated a remarkable cycle stability with only 4.8% loss reported after 6000 cycles.[159] One notable finding from this work is that the electrode exhibited spherical structures with rough surfaces in which the $CuCo_2O_4$ nanoparticles formed a highly porous shell, as shown

Table 2.11 EDLC supercapacitor performance using graphene materials.

Electrode	Electrolyte	Specific capacitance/F g^{-1}	Current density/A g^{-1}	Cycling stability	References
Holey graphene	1 M H_2SO_4	250	0.5	80% for 500 cycles	160
MnO_2/rGO composite	1 M Na_2SO_4	759	2	88% after 3500 cycles	161
Graphene-wrapped nanoporous $CuCO_2O_4$ hollow spheres	3 M KOH	1813	2	95.2% after 6000 cycles	159
Fe–Co oxides nanoparticles/ crumpled graphene	5 M KOH	325	0.1	80% after 2000 cycles	162
$NiCo_2O_4$ hexagonal nanoplates/rGO	3 M KOH	947.4	0.5	97.9% after 3000 cycles	163
Spinel $NiCo_2O_4$ nanowires on rGO	6 M KOH	1248	2 mA cm^{-2}	90% after 2000 cycles	164
$ZnCo_2O_4$-rGO composite	2 M KOH	3222	1	94% after 5000 cycles	165

in Figure 2.46. Furthermore, the graphene nanosheet wrapped around the $CuCo_2O_4$ microspheres and bridged them together to form a 3D network. This combination not only increased the specific surface area but also served as a conducting medium to promote charge transfer.[159]

An enhancement of the supercapacitor performance of spinel nickel cobaltate ($NiCo_2O_4$) nanowires using rGO coating was demonstrated by Zhang *et al.* (2018). In this work, capacitive and Faradaic materials were combined.[164] They reported that the mass loading of rGO is an important factor to optimise the supercapacitor performance. The specific capacitance was 1248 F g^{-1} at current density of 2 mA cm^{-2} with an excellent cyclic stability of 90% capacitance retention after 2000 charge–discharge cycles.[164] The study also revealed that the specific capacitance increased, when the rGO loading increased from 0 to 1.1 mg cm^{-2}, before dropping, when the rGO loading further increased from 1.4 to 1.5 mg cm^{-2}.[164] The addition of a rGO porous coating improved the electrical conductivity of the supercapacitor device, increasing the specific capacitance in proportion to the rGO loading.[164] However, an excessive rGO loading covered the mesopores of the supercapacitor device, hampering the transportation of the electrolyte ions across the active materials, and lowered the effective surface area for redox reactions.[164] Hence, it is critical to determine the amount of graphene materials to be

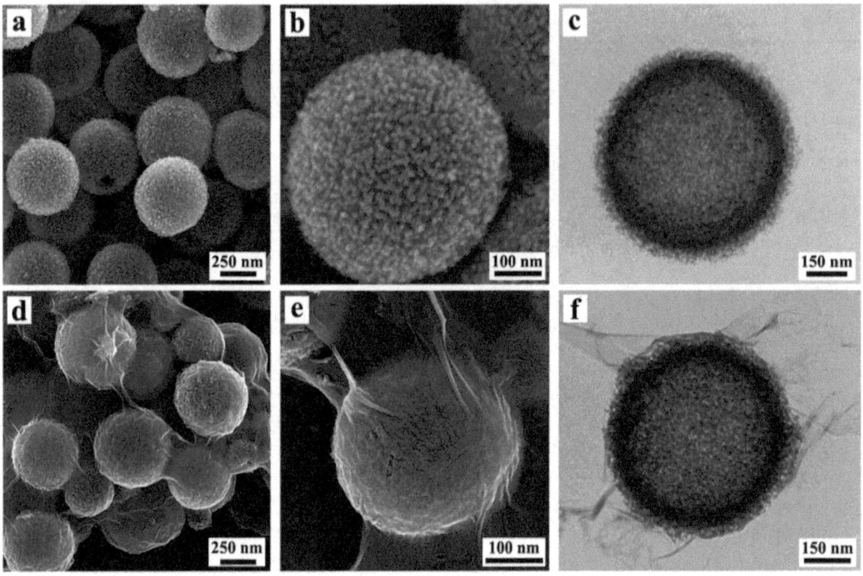

Figure 2.46 $CuCo_2O_4$ hollow spheres and graphene-wrapped $CuCo_2O_4$ hollow spheres. (A and B) FESEM images and (C) TEM image of the $CuCo_2O_4$-HSs. (D and E) FESEM images and (F) TEM image of graphene-wrapped $CuCo_2O_4$ hollow spheres. Reproduced from ref. 159 with permission from the Royal Society of Chemistry.

incorporated into the energy storage device to maximise the performance of the supercapacitor.

Battery development is another example where graphene materials have been heavily applied. A wide variety of battery types are currently in use including Na-ion,[166] nickel–hydrogen,[167] lithium-ion[168] and vanadium-redox flow.[169] Owing to their long cyclic life, portable size, applicable dosage and high power density, lithium-ion batteries have become the most widely used.[170] In general, lithium-ion batteries are composed of three components, namely the anode, electrolyte and cathode, as illustrated in Figure 2.47. Most lithium-ion batteries currently contain graphite anodes, as discussed in Chapter 1.

During the charging process, lithium ions are extracted from the cathode and transported through the electrolyte before being deposited onto the anode, while the discharging process is the reverse of this step. Compared to cathode materials, there is a greater potential range of anode materials for lithium-ion batteries. Carbon-based materials, particularly graphene and its derivatives, have been frequently used in the development of battery anodes due to their high electrical conductivity, facile lithium ion intercalation, low charge–discharge (~10%) volume change and relatively low cost.[172] Theoretically, the lithium storage capacity of graphene is 744 mA h g^{-1}, which is almost double that of traditional graphite anode (372 mA h g^{-1}), making graphene a promising candidate for anode fabrication.[173]

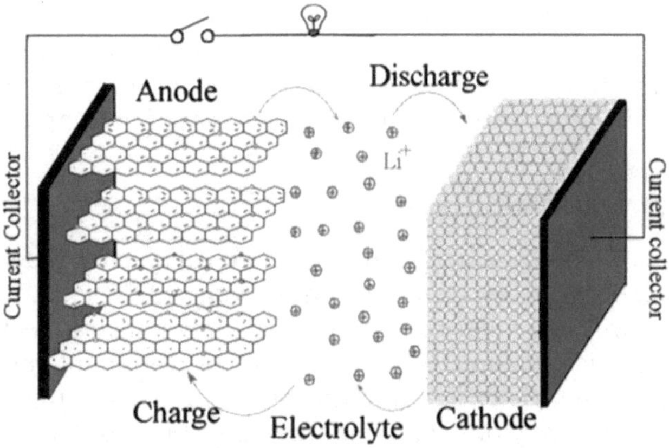

Figure 2.47 Schematic diagram of a rechargeable lithium battery. Reproduced from ref. 171 with permission from the Royal Society of Chemistry.

As with supercapacitors, transition metal oxides, such as titanium dioxide (TiO_2), iron(III) oxide (Fe_3O_4), cobalt(III) oxide (Co_3O_4) and manganese dioxide (MnO_2), are often integrated into graphene battery anodes as they can enhance the theoretical lithium storage capacity and rate performance.[174] As an example, Wu *et al.* (2020) described the successful fabrication of a novel architecture comprising hollow Co_3O_4 nanocrystals anchored onto holey graphene. The authors reported that each of the hollow Co_3O_4 nanospheres, ranging from 15 to 30 nm, were uniformly located in each of the holes on the graphene surface.[175] This unique structure shortened the mass diffusion pathway significantly to enhance the lithium-ion transport. This could be seen by the high discharge capacity of 893 mA h g^{-1} at 1 A g^{-1} current density,[175] as compared to Co_3O_4/graphene anodes (600 mA h g^{-1} at 1 A g^{-1} current density).[175]

In another study, Almodóvar *et al.* (2021) studied the electrochemical performance of hexagonal molybdenum trioxide (h-MoO_3) micro-composites with graphite, graphene and GO for application in lithium-ion batteries. It was reported that h-MoO_3 graphene and h-MoO_3 GO demonstrated excellent performance as a lithium-ion battery anode.[176] The specific capacity of h-MoO_3 graphene and h-MoO_3 GO electrodes were 665 mA h g^{-1} at 2 A g^{-1} current density and 789 mA h g^{-1} at 1 A g^{-1} current density, respectively.[176] This was attributed to the uniformly dispersed graphene and GO layers, which improved the structural stability as well as creating a conductive network to facilitate effective ambipolar diffusion of electrons and lithium ions.[176] This improved the overall electrochemical performance of the lithium-ion battery.

An effective manganese(II,III) oxide (Mn_3O_4)/rGO nanocomposite, synthesised using a facile, low temperature and chemical reduction method, was tested by Weng *et al.* (2020) as a lithium-ion anode. They reported that the Mn_3O_4/rGO nanocomposite possessed an appreciable rate capability of 640 mA h g^{-1} at 1.2 A g^{-1} current density, which could be a result of the enhanced lithium-ion diffusion coefficient (2.4×10^{-10} m^2 s^{-1}), higher than MnO_2 nanoneedle (1.3×10^{-10} m^2 s^{-1}), rGO (1.85×10^{-10} m^2 s^{-1}) and GO (7.32×10^{-10} m^2 s^{-1}).[177] This indicated that the lithium ions were delivered at a faster rate, hence improving the electrochemical performance of the graphene anode.

2.5.3 Lubricant Additives

The occurrence of friction and wear in mechanical systems is a result of the direct contact between two surfaces in any moving machinery. Often, the friction and wear will contribute to operational setbacks, such as energy wastage (~30%), mechanical failures (~70%), surface corrosion and environmental pollution.[178,179] The use of liquid lubricants can minimise the friction and wear, and consequently overcome the mechanical issues. Liquid lubricants reduce friction by preventing the sliding interface at metal–metal surface and generating an elasto-hydrodynamic lubrication film with low shear and high hydrodynamic durability on the rubbing surfaces.[178] However, most traditional lubricants contain elements such as sulfur, phosphorus and trace metals, which are not environmentally friendly. In addition, the lubrication film can be unstable or breakdown under extreme operating conditions. Lubricant additives have the potential to modify the tribological properties of the lubricants to generate a highly protective boundary film to be used in various operations.[180]

Lately, graphene materials are commonly incorporated into lubricants to improve their overall tribological performance, owing to the unique attributes of graphene materials, such as high mechanical strength, easy shear capability, ability to modify surface properties and atomically smooth surfaces.[181] A literature compilation of graphene additives in lubricant applications is listed in Table 2.12. The data in this table show that the incorporation of graphene materials into the base lubricant matrix can significantly improve the tribological performance as evidenced by the low coefficient of friction and high-wear reduction. The lubrication mechanisms of graphene additives can be categorised into four effects, namely tribofilm formation, microstructural transformation, interlayer slipping and synergistic effect.[178]

As a 2D nanosheet of carbon with single atomic thickness, graphene materials could easily slip into the friction interlayers and prevent the immediate rubbing between two surfaces. Their nanosized structure also allows graphene materials to generate a boundary film, which enables them to be adsorbed onto the friction surface *via* physisorption or chemisorption during the lubrication process.[182] For example, Meng *et al.* (2017) successfully synthesised an Au/GO nanocomposite under supercritical CO_2 fluid as a lubricant additive. In their study, the Au/GO nanocomposite was dispersed in PAO 6 oil (0.08 wt%) and the tribological assessment revealed that the coefficient of friction was approximately 0.08, while the wear rate was 8.9×10^{-7} mm^3 N^{-1} m^{-1}.[183] This corresponded to relative reductions of 33.6% and 72.8%, respectively, as compared to the lubricant without additives.[183] During the sliding process, the GO sheets and Au nanoparticles were deposited on the contact pair

Table 2.12 Tribological properties of graphene-modified lubricants.

Additive	Lubricant	Friction test conditions (material, load, speed)	Coefficient of friction	Wear assessment	References
Au/GO nanocomposite	PAO 6 oil	GCr15 steel, 10 N, 0.1 m s^{-1}	~0.08	Rate: 8.889 × 10^{-7} mm^3 N^{-1} m^{-1}	183
SiO_2/graphene	Water	AZ31 Magnesium alloy/ AISI 52100 steel, 3 N, 0.08 m s^{-1}	0.068	Volume: 4.86 mm^3	186
Superhigh-exfoliation rGO	Base oil	AISI 52100 steel/brass (H62), 3 mm s^{-1}	<0.08	Wear volume >60%	185
Mn_3O_4/ graphene	PAO 6 oil	GCr15 steel, 2 N, 2.4 mm s^{-1}	0.07	Wear reduction: 97%	187
Multilayer graphene/ micron graphite	IRIS-200BB	GCr15 steel, 392 N, 1200 rpm	0.075	Wear reduction: 84.16% (depth)	188
Modified graphene using octadecylamine and dicyclohexylcarbodiimide	PAO 6 oil	AISI 52100 steel, 2 N, 5 mm s^{-1}	~0.10	Wear reduction: 90% (depth)	189
Carbon quantum dots/GO	PEG synthetic lube oil	AISI 52100 steel, 196–588 N, 1200 rpm	0.048	Volume: 18.8 × 10^{-3} mm^3	190

surfaces and formed a protective thin coat. These materials smoothen the surfaces, thus minimising the mechanical friction and wear of the steel surface.[183] They also reported that the concentration of Au/GO nanoparticles in oil was an important controlling parameter. At excessive Au/GO concentration, the GO agglomerated between the friction surfaces, thus blocking oil film and becoming coated on the surface, possibly causing dry friction.[183]

The ability of tribofilm to resist friction and wear depends on the original properties of the graphene microstructure and the interlayer spacing.[178] Generally, graphene nanosheets with pre-existing structural defects, such as dislocations and vacancies, would have higher coefficients of friction than ones without major structural defects.[184] This is demonstrated by the work of Zhao *et al.* (2020) who have reported enhanced lubricating properties using superhigh-exfoliation graphene (SRGO). The SRGO was prepared by thermally reducing the GO under the activation of KOH. The synthesised SRGO possessed a high specific surface area of 1665 m^2 g^{-1} and an interlayer spacing of 0.425 nm.[185] The high specific surface area indicated a good dispersion of SRGO in the lubricant, which promoted the deposition of SRGO at the friction interface to produce a protective film, as a result of the unique 2D laminar microstructure of graphene materials.[185] Furthermore, the large interlayer spacing effectively decreased the sliding force between the layers, thus enabling slippage between the adjacent sites of the rubbing layers. The improvement of tribological properties by SRGO was proven by the reduction of coefficient of friction and wear volume of 70% and >60%, respectively.[185]

The tribological performance of lubricants can also be enhanced by graphene additives through the synergistic combination of lubricating effects between graphene and another material. As shown in Table 2.12, most of the graphene additives are composite materials as graphene can combine with other chemicals to form new structures with remarkable tribological properties.[191] In work by Zhao *et al.* (2019), the sandwich-like Mn_3O_4/graphene nanostructure produced a significant reduction in friction coefficient (75%) and wear depth (97%) as compared to pristine PAO 6 oil, even at an ultralow concentration (0.075 wt%) and high operating temperature (125 °C).[187] This was attributed to 'slide-roll' effects from the combination of graphene nanosheets and Mn_3O_4 nanoparticles. The nanoparticles anchored on the graphene surface weakened the interaction with the graphene nanosheet, thus causing the Mn_3O_4/graphene additives to form a stable dispersion in the base oil.[187] Thereafter, the Mn_3O_4/graphene additives can easily be inserted into the contact areas during sliding while the Mn_3O_4 nanoparticles are released from the additives by frictional force and adsorbed onto the contact areas.[187] The synergistic effects are realised as graphene can slide layer by layer during friction, while the Mn_3O_4 nanoparticles are rolling, thus attaining excellent tribological performance.[187]

2.5.4 Sensor Devices

In view of its large surface area and remarkable electrical conductivity, graphene offers great potential in sensing applications. The adsorption of external molecules, whether environmental pollutants or biomolecules, onto graphene sensors, causes a significant change in electrical conductivity, which enables the detection of these molecules through the measurement of electrical conductivity or resistance. Graphene-based sensors have been shown to be highly responsive compared to conventional ones allowing detection at sub-ppm levels of the target analytes.

Graphene, in its pristine form, offers advantages in sensing, such as low electrical noise, high electrical conductivity and high sensitivity, towards the analytes. In the earliest known graphene sensor reported by Schedin *et al.* (2007), various gases, such as NO_2, ammonia, CO and water, were detected at 1 ppm with a detection limit of 1 ppb.[192] Since then, much research has been devoted to graphene-based materials for their potential in sensing applications. Table 2.13 illustrates recent advances in graphene-based sensors for the detection of a diverse range of molecules and ions. In most cases,

Table 2.13 Recent advances in graphene-based chemical sensors.

Graphene-based chemical sensor	Analyte	Limit of detection	Working range	Sensitivity	References
Pristine graphene					
Graphene flakes stabilised with flavin monocleotide sodium salt	Ammonia gas	1.6 ppm	10 to 1000 ppm	2.8% at 10 ppm 18.5% at 1000 ppm	203
Chitosan-decorated graphene	Hydrogen peroxide	5 μM	20 μM to 20 mM 20 mM to 60 mM	6.42 μA mM^{-1} cm^{-2} 2.86 μA mM^{-1} cm^{-2}	204
Graphene on glassy carbon electrode	Ascorbic acid	6.45 μM	9 to 2314 μM	0.06674 μA μM^{-1}	205
	Dopamine	2.00 μM	5 to 710 μM	0.1125 μA μM^{-1}	
	Uric acid	4.82 μM	6 to 1330 μM	0.1029 μA μM^{-1}	
Graphene/nickel hexacyanoferrate	Ascorbic acid	0.25 μM	2 to 200 μM	7.029 μA mM^{-1} cm^{-2}	206
Nitric acid-treated graphene	Ammonia	27 ppb	20 to 100 ppm	41.9% at 100 ppm	193
Boron nitride/porous graphene	Sulfonamide	0.011 μM	0.5 to 362.5 μM	8.6 μA at 50 μM	207
GO					
GO	Ammonia		50 to 200 ppm	2.1 at 50 ppm	208
Nickel oxide-GO				20 at 50 ppm	
Zinc oxide-GO				60 at 50 ppm	
Bismuth oxide-GO				81.23 at 50 ppm	
Cerium oxide-GO	Glyphosate	30 nM	110 to 290 nM		198
Thymine-GO-carbohydrazide	Mercury(II)	1 ppb	5 to 600 ppb		197
	Chromium(VI)	20 ppb			
rGO					
3D cobalt sulfide/rGO	Nitrogen dioxide gas	0.05 ppm	50 ppb to 100 ppm	10.5%	199
Poly(3-hexylthiophene)/rGO	Methanol	1 ppm	1 to 100 ppm	110.6% at 10 ppm	209
	Ethanol			69.9% at 10 ppm	
	Acetone			84.9% at 10 ppm	
	Isopropyl alcohol			38.1% at 10 ppm	
rGO/zinc phthalocyanine	Ammonia	82 ppb	30 to 210 ppm		202
rGO/cobalt phthalocyanine		140 ppb			
rGO		233 ppb			
Iron-doped rGO/tungsten oxide	Acetone	0.05 ppm	1 to 10 ppm	78% at 10 ppm	200
rGO-alanine-polyaniline	Cadmium(II)	0.03 nM	0.08 to 100 nM	0.43 μA nM^{-1} cm^{-2}	201
	Lead(II)	0.045 nM		0.61 μA nM^{-1} cm^{-2}	
	Copper(II)	0.063 nM		0.71 μA nM^{-1} cm^{-2}	

graphene requires modifications to further enhance the chemisorption of molecules onto its surface, as well as improving selectivity toward target analytes. This is typically achieved by doping, chemical functionalisation, or the formation of hybrid composites with other materials. For instance, the simple treatment of graphene with concentrated nitric acid described by Li *et al.* (2022) produced a highly sensitive ammonia sensor with a very low detection limit (27 ppb). The response of the sensor was three times that of untreated graphene.[193]

Graphene oxide possesses several advantages over pristine graphene as an electrochemical probe, particularly after surface modification. One such advantage is that surface chemical functional groups can be tailored to adsorb target molecules that can result in a strong binding affinity towards these targets. Therefore, the use of GO-based electrochemical probes can facilitate the analysis of specific molecules without the need for extensive sample preparation prior to analysis. Various established chromatographic and spectroscopic techniques applied for the detection of molecules, such as amino acids and diclofenac, require complicated sample preparation.[194] Electrochemical methods employing GO provide an attractive alternative because of their higher sensitivity and accuracy, as well as simplicity and lower cost. The abundant chemical functional groups and high surface area of GO also facilitate the capture of target molecules, even at trace levels down to the picoscale (pM). The lowest detected concentration by GO sensors for nitrogen dioxide (NO_2) was 20 ppb.[195] On the other hand, the oxygen containing groups of GO reduce the electron transport function as compared to graphene, thus diminishing its sensing abilities. In addition, the higher adsorption energies of molecules on the surface of GO result in difficulties in recovering GO-based sensors after detection.[196]

Jayaraman *et al.* (2022) functionalised GO with two different groups for the detection of mercury and hexavalent chromium ions.[197] The sensor had a high sensitivity, towards these heavy metal ions, of 1 ppb for mercury and 20 ppb for chromium, illustrating the potential of this sensor for heavy metal detection in various contaminated sources. A graphene oxide–cerium oxide sensor has been investigated for the detection of glyphosate, a commonly used herbicide which is difficult to detect and quantify by conventional methods.[198] The sensor was capable of detecting glyphosate down to 30 nM and discriminating between, as well as quantifying, the presence of potential interferents, such as amino methyl phosphonic acid and *N*-nitroso glyphosate.

Despite the excellent sensitivity of GO-based sensors, there are difficulties in recovering the GO after the detection of molecules or ions, as already noted. The abundance of oxygen functional groups on the surface of GO also limits the selectivity for particular analytes. Precise control over the GO surface would be required in order to selectively enhance the molecular adsorption of specific targets. The restoration of the graphitic structure through the reduction of GO partially recovers the electrical conductivity crucial for sensing applications while maintaining some functional groups. Thus, rGO-based sensors are expected to perform better than those of graphene and GO.

Various rGO-based sensors have achieved ultrasensitive detection of molecules and ions with detection limits as low as 50 ppb for NO_2[199] and acetone,[200] and 0.03 nM for cadmium(II) ions.[201] Yabaş *et al.* (2023) reported the functionalisation of rGO with zinc phthalocyanine and cobalt phthalocyanine for the fabrication of a sensor for volatile organic compounds such as ammonia.[202] The functionalised sensors showed improved detection limits changing from 233 ppb for pristine rGO to 82 ppb for the functionalised counterpart. In addition, the response time was reduced from 250 s to 190 s, further promoting the potential of this sensor in the medical diagnosis of diseases.

The use of graphene-based materials in sensors is not limited to the detection of molecules and ions. The superior thermal conductivity of graphene and its temperature-responsive properties gives it potential as a temperature sensor. One application of temperature sensors is to provide continuous thermal data on human skin. This requires complex circuits to accurately detect the low thermal responses obtained.[210] Graphene and related materials have great potential here, considering their benefits, such as flexibility, structural strength and biocompatibility. All of these characteristics suggest that these materials may be suitable for use in wearable devices. Table 2.14 presents the various types of graphene-based temperature sensors for different

Table 2.14 Graphene-based temperature sensors. Adapted from ref. 211 with permission from Elsevier, Copyright 2022.

Graphene-based temperature sensor	Operating range	Temperature coefficient of resistance	Response time
Graphene-coated polypropylene textile fibres	30–45 °C	−0.17% $°C^{-1}$	NA
Graphene nanowalls-polydimethylsiloxane	35–45 °C	21.4% $°C^{-1}$	~1.6 s
rGO nanofiller in polyhydroxybutyrate	20–70 °C	−0.8% $°C^{-1}$	NA
Alginate nanofibril/GO/polyacrylamide	25–65 °C	−2% $°C^{-1}$	~2.5 s
GO/aluminium/silicon	−40–0 °C 10–60 °C	59.3 kHz $°C^{-1}$ 46.1 kHz $°C^{-1}$	NA
rGO on polyethylene terephthalate	30–100 °C	−0.63% $°C^{-1}$	~1.2 s
rGO on polyimide	35–45 °C	−0.743% $°C^{-1}$	NA
rGO on platinum interdigital electrodes	−33–127 °C	−0.38% $°C^{-1}$	NA
Poly(3,4-ethylenedioxythiophene) polystyrene sulfonate and rGO	30–50 °C	−1.69% $°C^{-1}$	NA
Graphene nanoplatelets/poly(dimethylsiloxane)	25–75 °C	0.8% $°C^{-1}$	~1.32–3.91 s
Laser-induced graphene/poly(dimethylsiloxane)	20–60 °C	−0.1% $°C^{-1}$	NA
Graphene coated textile yarn	30–40 °C	−0.041% $°C^{-1}$	NA
Graphene nanoplatelets/alginate	20–45 °C	−1.55% $°C^{-1}$	~26.3 s
Graphene nanoribbon ink on paper	30–80 °C	−1.27% $°C^{-1}$	~0.5 s

applications. The temperature coefficient of resistance (TCR) is an important parameter in temperature sensors, being a measure of the sensitivity. Most graphene-based sensors have relatively high TCR values (as compared to platinum-based sensors) indicating that they have a higher sensing ability.[211]

2.6 Conclusions

Graphene materials have attracted huge attention as a result of their exceptional physicochemical attributes, such as ultra-high specific surface area, unique electrical properties, superior thermal conductivity and ease of modification. In its pristine form, graphene exhibits a honeycomb network of elemental carbon with unique tunable properties. Defects within the graphene framework can influence its physicochemical properties, enabling modification of properties through configuration, density and distribution of defects. Graphene oxide and rGO are the two most important graphene derivatives. They possess varying oxygenated functional groups providing them with properties different from pristine graphene, such as thermal insulation and higher reactivity. Similarly, the halogenation of graphene attaches fluorine, chlorine, bromine or iodine onto the graphene plane, while hydrogenation of graphene leads to graphane, graphone and hydrogenated graphene, causing disruption of the carbon lattice. These forms of graphene display unique configurations, such as *boat*, *chair* and *twist-boat*, owing to the repulsion of non-carbon atoms. With the numerous graphene variants, it is crucial to understand the structural characteristics and physicochemical properties in order to establish a suitable synthesis pathway prior to utilisation in real-life applications.

The various forms of graphene materials discussed here can be synthesised using top-down and bottom-up methods. The top-down methods often introduce structural defects, but they have higher potential for upscaling. On the other hand, bottom-up methods provide potential routes for fabricating high quality graphene with minimum defects, but their synthesis is complex, which can limit the product yield. Most of the work on graphene synthesis has been conducted at laboratory-scale, with many attempts to upscale the production. When taking into consideration cost, graphene quality and yield, process scalability and safety, chemical vapour deposition, oxidation–reduction and liquid-phase exfoliation have high potential for up-scaling. Nevertheless, issues, such as inconsistent product quality, waste generation and labour-intensive operation, remain the major constraints to realise a sustainable production of graphene at large scale. Therefore, an appropriate graphene synthesis method should be carefully selected according to its application, and where necessary, surface modification and functionalisation can then be implemented to enhance graphene functionality.

Graphene materials have exhibited improved performances in a wide range of applications including wastewater treatment, energy storage and sensors. Frequently, the graphene materials are incorporated with other materials (transition metal oxide, non-metallic compounds and other nanomaterials) or undergo pre-treatment processes to develop new graphene composites to improve their performance. These modified graphene materials provide benefits, such as increased specific surface area, new functional groups, creation of reactive bonds and tunable surface properties. Undoubtedly the growing number of graphene-based materials with specifically tailored properties will have a significant commercial impact in the near future.

Acknowledgements

The authors would like to thank the Ministry of Higher Education (MOHE) Malaysia for their support for this work under the Fundamental Research Grant Scheme (FRGS/1/2020/STG05/UNIM/02/2).

References

1. K. S. Novoselov, A. K. Geim, S. V. Morozov, D. Jiang, Y. Zhang, S. V. Dubonos, I. V. Grigorieva and A. A. Firsov, *Science*, 2004, **306**, 666.
2. X. J. Lee, B. Y. Z. Hiew, K. C. Lai, L. Y. Lee, S. Gan, S. Thangalazhy-Gopakumar and S. Rigby, *J. Taiwan Inst. Chem. Eng.*, 2019, **98**, 163.
3. A. Abbas, Z.-B. Lin, R.-L. Ma, K.-M. Lin and H.-C. Lin, *Int. J. Hydrogen Energy*, 2022, **52**, 1175.
4. F. Arshad, F. Nabi, S. Iqbal and R. H. Khan, *Colloids Surf., B*, 2022, **212**, 112356.
5. A. Morenghi, S. Scaravonati, G. Magnani, M. Sidoli, L. Aversa, R. Verucchi, G. Bertoni, M. Riccò and D. Pontiroli, *Electrochim. Acta*, 2022, **424**, 140626.
6. A. D. Sontakke and M. K. Purkait, *Chem. Eng. J.*, 2021, **420**, 129914.
7. L. Zhao, B. Yu, F. Xue, J. Xie, X. Zhang, R. Wu, R. Wang, Z. Hu, S.-T. Yang and J. Luo, *J. Hazard. Mater.*, 2015, **286**, 449.
8. S. Pilathottathil, J. Kavil and M. S. Thayyil, *J. Indian Chem. Soc.*, 2021, **98**, 100169.
9. Z. Naseem, E. Shamsaei, K. Sagoe-Crentsil and W. Duan, *Cem. Concr. Res.*, 2022, **158**, 106843.
10. A. Bianco, H.-M. Cheng, T. Enoki, Y. Gogotsi, R. H. Hurt, N. Koratkar, T. Kyotani, M. Monthioux, C. R. Park, J. M. D. Tascon and J. Zhang, *Carbon*, 2013, **65**, 1.
11. G. Yildiz, M. Bolton-Warberg and F. Awaja, *Acta Biomater.*, 2021, **131**, 62.
12. G. Yang, L. Li, W. B. Lee and M. C. Ng, *Sci. Technol. Adv. Mater.*, 2018, **19**, 613.
13. C. Shearer, A. Slattery, A. Stapleton, J. Shapter and C. Gibson, *Nanotechnology*, 2016, **27**, 125704.

14. S. M. M. Dubois, Z. Zanolli, X. Declerck and J. C. Charlier, *Eur. Phys. J. B*, 2009, **72**, 1.
15. R. Nandee, M. A. Chowdhury, A. Shahid, N. Hossain and M. Rana, *Results Eng.*, 2022, **15**, 100474.
16. Y. Cao, V. Fatemi, S. Fang, K. Watanabe, T. Taniguchi, E. Kaxiras and P. Jarillo-Herrero, *Nature*, 2018, **556**, 43.
17. A. S. Mayorov, R. V. Gorbachev, S. V. Morozov, L. Britnell, R. Jalil, L. A. Ponomarenko, P. Blake, K. S. Novoselov, K. Watanabe, T. Taniguchi and A. K. Geim, *Nano Lett.*, 2011, **11**, 2396.
18. N. O. Weiss, H. Zhou, L. Liao, Y. Liu, S. Jiang, Y. Huang and X. Duan, *Adv. Mater.*, 2012, **24**, 5782.
19. T. Xu and L. Sun, in *Defects in Advanced Electronic Materials and Novel Low Dimensional Structures*, ed. J. Stehr, I. Buyanova, and W. Chen, Woodhead Publishing, 2018, pp. 137–160.
20. Y. Zhu, S. Murali, W. Cai, X. Li, J. W. Suk, J. R. Potts and R. S. Ruoff, *Adv. Mater.*, 2010, **22**, 3906.
21. S. Deng and V. Berry, *Mater. Today*, 2016, **19**, 197.
22. J. Kotakoski, C. Mangler and J. C. Meyer, *Nat. Commun.*, 2014, **5**, 1.
23. M. Singla and N. Jaggi, *Sens. Actuators, A*, 2021, **332**, 113118.
24. M. B. Burkholder, F. B. A. Rahman, E. H. Chandler, J. R. Regalbuto, B. F. Gupton and J. M. M. Tengco, *Carbon Trends*, 2022, **9**, 100196.
25. R. J. Young, M. Liu, I. A. Kinloch, S. Li, X. Zhao, C. Vallés and D. G. Papageorgiou, *Compos. Sci. Technol.*, 2018, **154**, 110.
26. P. Zhu, F. Tang, S. Wang, W. Cao and Q. Wang, *Mater. Today Commun.*, 2022, **33**, 104280.
27. R. Chegel, *Diamond Relat. Mater.*, 2023, **137**, 110154.
28. S. N. Faisal, E. Haque, N. Noorbehesht, W. Zhang, A. T. Harris, T. L. Church and A. I. Minett, *RSC Adv.*, 2017, **7**, 17950.
29. D. R. Dreyer, S. Park, C. W. Bielawski and R. S. Ruoff, *Chem. Soc. Rev.*, 2010, **39**, 228.
30. U. Hofmann and R. Holst, *Ber. Dtsch. Chem. Ges. (A and B Series)*, 1939, **72**, 754.
31. G. Ruess, *Monatsh. Chem. Verw. Teile Anderer Wiss.*, 1947, **76**, 381.
32. W. Scholz and H. P. Boehm, *Z. Anorg. Allg. Chem.*, 1969, **369**, 327.
33. T. Nakajima, A. Mabuchi and R. Hagiwara, *Carbon*, 1988, **26**, 357.
34. A. Lerf, H. He, M. Forster and J. Klinowski, *J. Phys. Chem. B*, 1998, **102**, 4477.
35. M. Mermoux, Y. Chabre and A. Rousseau, *Carbon*, 1991, **29**, 469.
36. T. Szabó, O. Berkesi, P. Forgó, K. Josepovits, Y. Sanakis, D. Petridis and I. Dékány, *Chem. Mater.*, 2006, **18**, 2740.
37. K. Erickson, R. Erni, Z. Lee, N. Alem, W. Gannett and A. Zettl, *Adv. Mater.*, 2010, **22**, 4467.
38. B. Y. Z. Hiew, L. Y. Lee, X. J. Lee, S. Thangalazhy-Gopakumar, S. Gan, S. S. Lim, G.-T. Pan, T. C.-K. Yang, W. S. Chiu and P. S. Khiew, *Process Saf. Environ. Prot.*, 2018, **116**, 262.

39. C. Gómez-Navarro, J. C. Meyer, R. S. Sundaram, A. Chuvilin, S. Kurasch, M. Burghard, K. Kern and U. Kaiser, *Nano Lett.*, 2010, **10**, 1144.
40. M. Inagaki and F. Kang, *J. Mater. Chem. A*, 2014, **2**, 13193.
41. X. Zhu, Z. Cheng, B. Zhang, Y. Wang, Z. Lu and G. Zhang, *Phys. E*, 2022, **143**, 115379.
42. F. Karlicky and M. Otyepka, *J. Chem. Theory Comput.*, 2013, **9**, 4155.
43. A. N. Rudenko, F. J. Keil, M. I. Katsnelson and A. I. Lichtenstein, *Phys. Rev. B: Condens. Matter Mater. Phys.*, 2010, **82**, 035427.
44. Y. Liang and L. Yang, *MRS Online Proc. Libr.*, 2011, **1370**, 37.
45. F. Karlický, K. Kumara Ramanatha Datta, M. Otyepka and R. Zbořil, *ACS Nano*, 2013, **7**, 6434.
46. H. T. Nguyen, L. K. Huynh and T. N. Truong, *Carbon*, 2017, **121**, 248.
47. M. Pumera and C. H. A. Wong, *Chem. Soc. Rev.*, 2013, **42**, 5987.
48. V. Georgakilas, M. Otyepka, A. B. Bourlinos, V. Chandra, N. Kim, K. C. Kemp, P. Hobza, R. Zboril and K. S. Kim, *Chem. Rev.*, 2012, **112**, 6156.
49. D. K. Samarakoon and X.-Q. Wang, *ACS Nano*, 2009, **3**, 4017.
50. D. K. Samarakoon and X.-Q. Wang, *Phys. Appl. Graphene: Theory*, 2011, 113.
51. Q. Cao, X. Geng, H. Wang, P. Wang, A. Liu, Y. Lan and Q. Peng, *Crystals*, 2018, **8**, 357.
52. L. Feng and W. X. Zhang, *AIP Adv.*, 2012, **2**, 042138.
53. R. Baughman, H. Eckhardt and M. Kertesz, *J. Chem. Phys.*, 1987, **87**, 6687.
54. K. Khan, A. K. Tareen, M. Iqbal, Z. Shi, H. Zhang and Z. Guo, *Nano Today*, 2021, **39**, 101207.
55. Q. Tang, Z. Zhou and Z. Chen, *Nanoscale*, 2013, **5**, 4541.
56. E. Belenkov, M. Brzhezinskaya and V. Mavrinskii, in *Handbook of Graphene Set*, ed. E. Celasco, A. N. Chaika, T. Stauber, M. Zhang, C. Ozkan, C. Ozkan, U. Ozkan, B. Palys, and S. W. Harun, John Wiley & Sons, Inc., Hoboken, USA, and Scrivener Publishing, Beverly, USA, 2019, vol. III, pp. 113–150.
57. G. Li, Y. Li, H. Liu, Y. Guo, Y. Li and D. Zhu, *Chem. Commun.*, 2010, **46**, 3256.
58. R. Matsuoka, R. Sakamoto, K. Hoshiko, S. Sasaki, H. Masunaga, K. Nagashio and H. Nishihara, *J. Am. Chem. Soc.*, 2017, **139**, 3145.
59. Z. Zuo, H. Shang, Y. Chen, J. Li, H. Liu, Y. Li and Y. Li, *Chem. Commun.*, 2017, **53**, 8074.
60. Y. Jiao, A. Du, M. Hankel, Z. Zhu, V. Rudolph and S. C. Smith, *Chem. Commun.*, 2011, **47**, 11843.
61. J. Li, Z. Zhang, Y. Kong, B. Yao, C. Yin, L. Tong, X. Chen, T. Lu and J. Zhang, *Chem*, 2021, **7**, 1284.
62. P. Sang, X. Ma, Q. Wang, W. Wei, F. Wang, J. Wu, X. Zhan, Y. Li and J. Chen, *Appl. Surf. Sci.*, 2021, **536**, 147836.

63. H. Shang, Z. Zuo, L. Li, F. Wang, H. Liu, Y. Li and Y. Li, *Angew. Chem., Int. Ed.*, 2018, **57**, 774.
64. M. Yi and Z. Shen, *J. Mater. Chem. A*, 2015, **3**, 11700.
65. J. Chen, M. Duan and G. Chen, *J. Mater. Chem.*, 2012, **22**, 19625.
66. B. Jayasena and S. Subbiah, *Nanoscale Res. Lett.*, 2011, **6**, 95.
67. N. Kumar, R. Salehiyan, V. Chauke, O. Joseph Botlhoko, K. Setshedi, M. Scriba, M. Masukume and S. Sinha Ray, *FlatChem*, 2021, **27**, 100224.
68. Y. Xu, H. Cao, Y. Xue, B. Li and W. Cai, *Nanomaterials*, 2018, **8**, 942.
69. S. Liang, Z. Shen, M. Yi, L. Liu, X. Zhang, C. Cai and S. Ma, *J. Nanosci. Nanotechnol.*, 2015, **15**, 2686.
70. P. G. Karagiannidis, S. A. Hodge, L. Lombardi, F. Tomarchio, N. Decorde, S. Milana, I. Goykhman, Y. Su, S. V. Mesite, D. N. Johnstone, R. K. Leary, P. A. Midgley, N. M. Pugno, F. Torrisi and A. C. Ferrari, *ACS Nano*, 2017, **11**, 2742.
71. P. Yu, S. E. Lowe, G. P. Simon and Y. L. Zhong, *Curr. Opin. Colloid Interface Sci.*, 2015, **20**, 329.
72. F. Liu, C. Wang, X. Sui, M. A. Riaz, M. Xu, L. Wei and Y. Chen, *Carbon Energy*, 2019, **1**, 173.
73. D. B. Shinde, J. Brenker, C. D. Easton, R. F. Tabor, A. Neild and M. Majumder, *Langmuir*, 2016, **32**, 3552.
74. N. Li, Z. Wang, K. Zhao, Z. Shi, Z. Gu and S. Xu, *Carbon*, 2010, **48**, 255.
75. E. S. Agudosi, E. C. Abdullah, A. Numan, N. M. Mubarak, M. Khalid and N. Omar, *Crit. Rev. Solid State Mater. Sci.*, 2020, **45**, 339.
76. Z. G. Wang, Y. F. Chen, P. J. Li, X. Hao, Y. Fu, K. Chen, L. X. Huang and D. Liu, *Vacuum*, 2012, **86**, 895.
77. M. Portail, A. Michon, S. Vézian, D. Lefebvre, S. Chenot, E. Roudon, M. Zielinski, T. Chassagne, A. Tiberj, J. Camassel and Y. Cordier, *J. Cryst. Growth*, 2012, **349**, 27.
78. A. Srivastava, C. Galande, L. Ci, L. Song, C. Rai, D. Jariwala, K. F. Kelly and P. M. Ajayan, *Chem. Mater.*, 2010, **22**, 3457.
79. D. M. Sedlovets, M. A. Knyazev, A. V. Zotov, A. P. Naumov and V. I. Korepanov, *J. Mater. Res. Technol.*, 2021, **14**, 1339.
80. P. R. Kidambi, D. D. Mariappan, N. T. Dee, A. Vyatskikh, S. Zhang, R. Karnik and A. J. Hart, *ACS Appl. Mater. Interfaces*, 2018, **10**, 10369.
81. T. Kobayashi, M. Bando, N. Kimura, K. Shimizu, K. Kadono, N. Umezu, K. Miyahara, S. Hayazaki, S. Nagai, Y. Mizuguchi, Y. Murakami and D. Hobara, *Appl. Phys. Lett.*, 2013, **102**, 023112.
82. T. Hesjedal, *Appl. Phys. Lett.*, 2011, **98**, 133106.
83. N. Mishra, J. Boeckl, N. Motta and F. Iacopi, *Phys. Status Solidi A*, 2016, **213**, 2277.
84. F. Presel, H. Tetlow, L. Bignardi, P. Lacovig, C. A. Tache, S. Lizzit, L. Kantorovich and A. Baraldi, *Nanoscale*, 2018, **10**, 7396.
85. R. Zheng, Z. Xu, A. Khanaki, H. Tian, Z. Zuo, J.-G. Zheng and J. Liu, *Thin Solid Films*, 2017, **627**, 39.

86. U. Wurstbauer, T. Schiros, C. Jaye, A. S. Plaut, R. He, A. Rigosi, C. Gutiérrez, D. Fischer, L. N. Pfeiffer, A. N. Pasupathy, A. Pinczuk and J. M. Garcia, *Carbon*, 2012, **50**, 4822.
87. H. Grennberg and U. Jansson, *Science and Technology of Atomic, Molecular, Condensed Matter & Biological Systems*, Elsevier, 2012, vol. 2, pp. 105–127.
88. B. C. Brodie, *Philos. Trans. R. Soc. London*, 1859, **149**, 249.
89. L. Staudenmaier, *Ber. Dtsch. Chem. Ges.*, 1898, **31**, 1481.
90. U. Hofmann and E. König, *Z. Anorg. Allg. Chem.*, 1937, **234**, 311.
91. W. S. Hummers and R. E. Offeman, *J. Am. Chem. Soc.*, 1958, **80**, 1339.
92. A. T. Smith, A. M. LaChance, S. Zeng, B. Liu and L. Sun, *Nano Mater. Sci.*, 2019, **1**, 31.
93. D. C. Marcano, D. V. Kosynkin, J. M. Berlin, A. Sinitskii, Z. Sun, A. Slesarev, L. B. Alemany, W. Lu and J. M. Tour, *ACS Nano*, 2010, **4**, 4806.
94. N. I. Kovtyukhova, P. J. Ollivier, B. R. Martin, T. E. Mallouk, S. A. Chizhik, E. V. Buzaneva and A. D. Gorchinskiy, *Chem. Mater.*, 1999, **11**, 771.
95. C. Mellado, T. Figueroa, R. Baez, M. Meléndrez and K. Fernández, *Mater. Today Chem.*, 2019, **13**, 1.
96. J. Zhang, Q. Liu, Y. Ruan, S. Lin, K. Wang and H. Lu, *Chem. Mater.*, 2018, **30**, 1888.
97. Q. Yu, L. Wei, X. Yang, C. Wang, J. Chen, H. Du, W. Shen, F. Kang and Z.-H. Huang, *Appl. Surf. Sci.*, 2022, **598**, 153788.
98. R. Jayawardena, D. S. Eldridge and F. Malherbe, *Colloids Surf., A*, 2022, **650**, 129587.
99. Q. Zhang, Y. Yang, H. Fan, L. Feng, G. Wen and L.-C. Qin, *Colloids Surf., A*, 2022, **652**, 129802.
100. V. B. Mohan, R. Brown, K. Jayaraman and D. Bhattacharyya, *Mater. Sci. Eng., B*, 2015, **193**, 49.
101. P. Kumar, F. Shahzad, S. Yu, S. M. Hong, Y.-H. Kim and C. M. Koo, *Carbon*, 2015, **94**, 494.
102. C. B. Kim, J. Lee, J. Cho and M. Goh, *Carbon*, 2018, **139**, 386.
103. Y.-C. Lin, K.-K. Liu, C.-Y. Wu, C.-W. Chu, J. T.-W. Wang, C.-T. Liang and L.-J. Li, *Phys. Chem. Chem. Phys.*, 2012, **14**, 3083.
104. Y. Wang, Y. Chen, S. D. Lacey, L. Xu, H. Xie, T. Li, V. A. Danner and L. Hu, *Mater. Today*, 2018, **21**, 186.
105. X. Xiang, Y. Zhu, M. Yin, S. Xia and C. Guo, *J. Mater. Sci.*, 2022, **57**, 3280.
106. D. Yang and C. Bock, *J. Power Sources*, 2017, **337**, 73.
107. A. Ambrosi and M. Pumera, *Chem. – Eur. J.*, 2013, **19**, 4748.
108. J. A. Quezada-Renteria, C. O. Ania, L. F. Chazaro-Ruiz and J. R. Rangel-Mendez, *Carbon*, 2019, **149**, 722.
109. W. Zeng, X.-M. Tao, S. Lin, C. Lee, D. Shi, K.-h. Lam, B. Huang, Q. Wang and Y. Zhao, *Nano Energy*, 2018, **54**, 163.
110. D. D. Chronopoulos, A. Bakandritsos, M. Pykal, R. Zbořil and M. Otyepka, *Appl. Mater. Today*, 2017, **9**, 60.

111. V. Mazánek, O. Jankovský, J. Luxa, D. Sedmidubský, Z. Janoušek, F. Šembera, M. Mikulics and Z. Sofer, *Nanoscale*, 2015, **7**, 13646.
112. K.-I. Ho, J.-H. Liao, C.-H. Huang, C.-L. Hsu, W. Zhang, A.-Y. Lu, L.-J. Li, C.-S. Lai and C.-Y. Su, *Small*, 2014, **10**, 989.
113. O. Jankovský, V. Mazánek, K. Klímová, D. Sedmidubský, J. Kosina, M. Pumera and Z. Sofer, *Chem. - Eur. J.*, 2016, **22**, 17696.
114. K. Gopalakrishnan, K. S. Subrahmanyam, P. Kumar, A. Govindaraj and C. N. R. Rao, *RSC Adv.*, 2012, **2**, 1605.
115. B. Li, L. Zhou, D. Wu, H. Peng, K. Yan, Y. Zhou and Z. Liu, *ACS Nano*, 2011, **5**, 5957.
116. J. Zheng, H.-T. Liu, B. Wu, C.-A. Di, Y.-L. Guo, T. Wu, G. Yu, Y.-Q. Liu and D.-B. Zhu, *Sci. Rep.*, 2012, **2**, 662.
117. H. L. Poh, P. Šimek, Z. Sofer and M. Pumera, *Chem. - Eur. J.*, 2013, **19**, 2655.
118. O. Jankovský, P. Šimek, K. Klimová, D. Sedmidubský, S. Matějková, M. Pumera and Z. Sofer, *Nanoscale*, 2014, **6**, 6065.
119. Y. Yang, X. Jiang, Y. Li, W. Jin and X. Huang, *Compos. Commun.*, 2021, **25**, 100771.
120. D. C. Elias, R. R. Nair, T. M. G. Mohiuddin, S. V. Morozov, P. Blake, M. P. Halsall, A. C. Ferrari, D. W. Boukhvalov, M. I. Katsnelson, A. K. Geim and K. S. Novoselov, *Science*, 2009, **323**, 610.
121. J. S. Burgess, B. R. Matis, J. T. Robinson, F. A. Bulat, F. Keith Perkins, B. H. Houston and J. W. Baldwin, *Carbon*, 2011, **49**, 4420.
122. H. L. Poh, F. Šaněk, Z. Sofer and M. Pumera, *Nanoscale*, 2012, **4**, 7006.
123. V. E. Antonov, I. O. Bashkin, A. V. Bazhenov, B. M. Bulychev, V. K. Fedotov, T. N. Fursova, A. I. Kolesnikov, V. I. Kulakov, R. V. Lukashev, D. V. Matveev, M. K. Sakharov and Y. M. Shulga, *Carbon*, 2016, **100**, 465.
124. A. M. Ilyin, N. R. Guseinov, I. A. Tsyganov and R. R. Nemkaeva, *Phys. E*, 2011, **43**, 1262.
125. L. Ciammaruchi, L. Bellucci, G. C. Castillo, G. M.-D. Sánchez, Q. Liu, V. Tozzini and J. Martorell, *Carbon*, 2019, **153**, 234.
126. Z. Yang, Y. Sun, L. B. Alemany, T. N. Narayanan and W. E. Billups, *J. Am. Chem. Soc.*, 2012, **134**, 18689.
127. R. A. Schäfer, J. M. Englert, P. Wehrfritz, W. Bauer, F. Hauke, T. Seyller and A. Hirsch, *Angew. Chem., Int. Ed.*, 2013, **52**, 754.
128. K. S. Subrahmanyam, P. Kumar, U. Maitra, A. Govindaraj, K. P. S. S. Hembram, V. Waghmare Umesh and C. N. R. Rao, *Proc. Natl. Acad. Sci. U. S. A.*, 2011, **108**, 2674.
129. Y. Wang, X. Xu, J. Lu, M. Lin, Q. Bao, B. Özyilmaz and K. P. Loh, *ACS Nano*, 2010, **4**, 6146.
130. Y. Fei, T. Tong, J. Bao and Y. H. Hu, *J. Phys. Chem. Solids*, 2022, **167**, 110772.
131. X. Qian, H. Liu, C. Huang, S. Chen, L. Zhang, Y. Li, J. Wang and Y. Li, *Sci. Rep.*, 2015, **5**, 7756.
132. Y. Hu, C. Wu, Q. Pan, Y. Jin, R. Lyu, V. Martinez, S. Huang, J. Wu, L. J. Wayment, N. A. Clark, M. B. Raschke, Y. Zhao and W. Zhang, *Nat. Synth.*, 2022, **1**, 449.

133. K. C. Lai, B. Y. Z. Hiew, W. T. Tee, S. Thangalazhy-Gopakumar, S. Gan and L. Y. Lee, *J. Environ. Chem. Eng.*, 2021, **9**, 106057.
134. B. Y. Z. Hiew, W. T. Tee, N. Y. L. Loh, K. C. Lai, S. Hanson, S. Gan, S. Thangalazhy-Gopakumar and L. Y. Lee, *J. Environ. Sci.*, 2022, **118**, 194.
135. J. Xu, P. Du, W. Bi, G. Yao, S. Li and H. Liu, *Chem. Eng. Res. Des.*, 2020, **154**, 192.
136. T. R. Das, S. Patra, R. Madhuri and P. K. Sharma, *J. Colloid Interface Sci.*, 2018, **509**, 82.
137. A. Y. W. Sham and S. M. Notley, *J. Environ. Chem. Eng.*, 2018, **6**, 495.
138. K. Gupta and O. P. Khatri, *J. Colloid Interface Sci.*, 2017, **501**, 11.
139. L. Wang, D. Hu, X. Kong, J. Liu, X. Li, K. Zhou, H. Zhao and C. Zhou, *Chem. Eng. J.*, 2018, **346**, 38.
140. L. Li, C. Luo, X. Li, H. Duan and X. Wang, *Int. J. Biol. Macromol.*, 2014, **66**, 172.
141. B. Y. Z. Hiew, L. Y. Lee, K. C. Lai, S. Gan, S. Thangalazhy-Gopakumar, G.-T. Pan and T. C.-K. Yang, *Environ. Res.*, 2019, **168**, 241.
142. Y. Liu, R. Liu, M. Li, F. Yu and C. He, *Carbohydr. Polym.*, 2019, **220**, 141.
143. Z. Song, Y.-L. Ma and C.-E. Li, *Sci. Total Environ.*, 2019, **651**, 580.
144. L. Chang, Y. Pu, P. Jing, Y. Cui, G. Zhang, S. Xu, B. Cao, J. Guo, F. Chen and C. Qiao, *Appl. Surf. Sci.*, 2021, **541**, 148400.
145. B. Xue and Y. Zou, *J. Colloid Interface Sci.*, 2018, **529**, 306.
146. X.-N. Wei, C.-L. Ou, X.-X. Guan, Z.-K. Peng and X.-C. Zheng, *Appl. Surf. Sci.*, 2019, **469**, 666.
147. L. Allagui, B. Chouchene, T. Gries, G. Medjahdi, E. Girot, X. Framboisier, A. B. h. Amara, L. Balan and R. Schneider, *Appl. Surf. Sci.*, 2019, **490**, 580.
148. M.-C. Roşu, C. Socaci, V. Floare-Avram, G. Borodi, F. Pogăcean, M. Coroş, L. Măgeruşan and S. Pruneanu, *Mater. Chem. Phys.*, 2016, **179**, 232.
149. W. Zhu, F. Sun, R. Goei and Y. Zhou, *Appl. Catal., B*, 2017, **207**, 93.
150. J. Zhu, Z. Zhu, H. Zhang, H. Lu, W. Zhang, Y. Qiu, L. Zhu and S. Küppers, *Appl. Catal., B*, 2018, **225**, 550.
151. Y. Li, Z. Lai, Z. Huang, H. Wang, C. Zhao, G. Ruan and F. Du, *Appl. Surf. Sci.*, 2021, **550**, 149342.
152. Y.-h. Si, Y. Xia, S.-k. Shang, X.-b. Xiong, X.-r. Zeng, J. Zhou and Y.-y. Li, *Nanomaterials*, 2018, **8**, 526.
153. A. V. Karim and A. Selvaraj, *Process Saf. Environ. Prot.*, 2021, **146**, 136.
154. W. Tie, Z. Du, H. Yue, S. Sarathi Bhattacharyya, Z. Zheng, W. He and S. Hee Lee, *J. Colloid Interface Sci.*, 2020, **579**, 862.
155. V. D. Nithya, *J. Energy Storage*, 2021, **44**, 103380.
156. R. Kumar, S. Sahoo, E. Joanni and R. K. Singh, *J. Energy Chem.*, 2022, **74**, 252.
157. W. H. Low, P. S. Khiew, S. S. Lim, C. W. Siong and E. R. Ezeigwe, *J. Alloys Compd.*, 2019, **775**, 1324.
158. R. Lakra, R. Kumar, P. K. Sahoo, D. Thatoi and A. Soam, *Inorg. Chem. Commun.*, 2021, **133**, 108929.

159. S. K. Kaverlavani, S. Moosavifard and A. Bakouei, *J. Mater. Chem. A*, 2017, **5**, 14301.
160. S. K. Jhajharia and K. Selvaraj, *Nanoscale*, 2015, **7**, 19705.
161. S. Jadhav, R. S. Kalubarme, C. Terashima, B. B. Kale, V. Godbole, A. Fujishima and S. W. Gosavi, *Electrochim. Acta*, 2019, **299**, 34.
162. C. Lee, S. K. Kim, J.-H. Choi, H. Chang and H. D. Jang, *J. Alloys Compd.*, 2018, **735**, 2030.
163. L. Ma, X. Shen, H. Zhou, Z. Ji, K. Chen and G. Zhu, *Chem. Eng. J.*, 2015, **262**, 980.
164. C. Zhang, C. Lei, C. Cen, S. Tang, M. Deng, Y. Li and Y. Du, *Electrochim. Acta*, 2018, **260**, 814.
165. Z. Gao, L. Zhang, J. Chang, Z. Wang, D. Wu, F. Xu, Y. Guo and K. Jiang, *Appl. Surf. Sci.*, 2018, **442**, 138.
166. Y. Li, X. Wang, Z. Wang and L. Chen, *ES Energy Environ.*, 2019, **3**, 55.
167. W. Chen, Y. Jin, J. Zhao, N. Liu and Y. Cui, *Proc. Natl. Acad. Sci. U. S. A.*, 2018, **115**, 11694.
168. G. Zubi, R. Dufo-López, M. Carvalho and G. Pasaoglu, *Renewable Sustainable Energy Rev.*, 2018, **89**, 292.
169. A. Fetyan, G. A. El-Nagar, I. Derr, P. Kubella, H. Dau and C. Roth, *Electrochim. Acta*, 2018, **268**, 59.
170. C. Shang, X. Li, R. Wei, X. Liu, S. Xu and J. Zhang, *J. Non-Cryst. Solids*, 2023, **618**, 122547.
171. M. Liang and L. Zhi, *J. Mater. Chem.*, 2009, **19**, 5871.
172. J. P. Mensing, T. Lomas and A. Tuantranont, *Sustainable Mater. Technol.*, 2020, **25**, e00190.
173. H. Ye, G. Zheng, X. Yang, D. Zhang, Y. Zhang, S. Yan, L. You, S. Hou and Z. Huang, *J. Electroanal. Chem.*, 2021, **898**, 115652.
174. Y.-Q. Jing, J. Qu, X.-Q. Jia, X.-Z. Zhai, W. Chang, M.-J. Zeng, X. Li and Z.-Z. Yu, *Chem. Eng. J.*, 2021, **408**, 127266.
175. D. Wu, C. Wang, H. Wu, S. Wang, F. Wang, Z. Chen, T. Zhao, Z. Zhang, L. Y. Zhang and C. M. Li, *Carbon*, 2020, **163**, 137.
176. P. Almodóvar, M. L. López, J. Ramírez-Castellanos, S. Nappini, E. Magnano, J. M. González-Calbet and C. Díaz-Guerra, *Electrochim. Acta*, 2021, **365**, 137355.
177. S.-C. Weng, S. Brahma, P.-C. Huang, Y.-C. Huang, Y.-H. Lee, C.-C. Chang and J.-L. Huang, *Appl. Surf. Sci.*, 2020, **505**, 144629.
178. J. Zhao, T. Gao, Y. Li, Y. He and Y. Shi, *Mater. Today Commun.*, 2021, **29**, 102755.
179. T. He, N. Chen, J. Fang, G. Cai, J. Wang, B. Chen and Q. Liang, *J. Mol. Liq.*, 2022, **357**, 119090.
180. J. Zhao, D. Wang, F. Zhang, Y. Liu, B. Chen, Z. L. Wang, J. Pan, R. Larsson and Y. Shi, *ACS Nano*, 2021, **15**, 11869.
181. R. Li, X. Yang, D. Hou, Y. Wang and J. Zhang, *Mater. Lett.*, 2020, **271**, 127748.
182. Z. Li, C. Xu, G. Xiao, J. Zhang, Z. Chen and M. Yi, *Materials*, 2018, **11**, 2110.

183. Y. Meng, F. Su and Y. Chen, *ACS Appl. Mater. Interfaces*, 2017, **9**, 39549.
184. Y. J. Shin, R. Stromberg, R. Nay, H. Huang, A. T. S. Wee, H. Yang and C. S. Bhatia, *Carbon*, 2011, **49**, 4070.
185. J. Zhao, Y. Huang, Y. Li, T. Gao, Z. Dou, J. Mao, H. Wang, Y. He, S. Li and J. Luo, *Appl. Surf. Sci.*, 2020, **513**, 145608.
186. H. Xie, S. Dang, B. Jiang, L. Xiang, S. Zhou, H. Sheng, T. Yang and F. Pan, *Appl. Surf. Sci.*, 2019, **475**, 847.
187. J. Zhao, Y. Li, Y. He and J. Luo, *ACS Appl. Mater. Interfaces*, 2019, **11**, 36931.
188. Z. Sun, C.-m. Xu, Y.-x. Peng, Y.-y. Shi and Y.-w. Zhang, *Wear*, 2020, **454–455**, 203333.
189. P. Wu, X. Chen, C. Zhang, J. Zhang, J. Luo and J. Zhang, *Friction*, 2021, **9**, 143.
190. W. Shang, T. Cai, Y. Zhang, D. Liu and S. Liu, *Tribol. Int.*, 2018, **118**, 373.
191. S. R. Chandrabhan, V. Jayan, S. S. Parihar and S. Ramaprabhu, *Beilstein J. Nanotechnol.*, 2017, **8**, 1476.
192. F. Schedin, A. K. Geim, S. V. Morozov, E. W. Hill, P. Blake, M. I. Katsnelson and K. S. Novoselov, *Nat. Mater.*, 2007, **6**, 652.
193. Q. Li, W. Chen, W. Liu, M. Sun, M. Xu, H. Peng, H. Wu, S. Song, T. Li and X. Tang, *Appl. Surf. Sci.*, 2022, **586**, 152689.
194. D. Sharma, S. Kanchi, M. I. Sabela and K. Bisetty, *Arabian J. Chem.*, 2016, **9**, 238.
195. M. Donarelli, S. Prezioso, F. Perrozzi, L. Giancaterini, C. Cantalini, E. Treossi, V. Palermo, S. Santucci and L. Ottaviano, *2D Mater.*, 2015, **2**, 035018.
196. R. Ghosh, M. Aslam and H. Kalita, *Mater. Today Commun.*, 2022, **30**, 103182.
197. N. Jayaraman, Y. Palani, R. R. Jonnalagadda and E. Shanmugam, *Sens. Actuators, B*, 2022, **367**, 132165.
198. Í. A. Costa, M. A. Gross, E. D. O. Alves, F. J. Fonseca and L. G. Paterno, *J. Electroanal. Chem.*, 2022, 116719.
199. Z. Zhang, Y. Wang, S. Ying, C. Yang, J. Zhao, N. Hu and C. Peng, *J. Alloys Compd.*, 2022, **926**, 166868.
200. S. Sen, S. Maity and S. Kundu, *Sens. Actuators, B*, 2022, **361**, 131706.
201. M. Akhtar, A. Tahir, S. Zulfiqar, F. Hanif, M. F. Warsi, P. O. Agboola and I. Shakir, *Synth. Met.*, 2020, **265**, 116410.
202. E. Yabaş, E. Biçer and A. Altındal, *J. Mol. Struct.*, 2023, **1271**, 134076.
203. S. Huang, L. A. Panes-Ruiz, A. Croy, M. Löffler, V. Khavrus, V. Bezugly and G. Cuniberti, *Carbon*, 2021, **173**, 262.
204. Q. Zhu, B. Liang, Y. Cai, Q. Cao, T. Tu, B. Huang, L. Fang and X. Ye, *Talanta*, 2018, **190**, 70.
205. S. Qi, B. Zhao, H. Tang and X. Jiang, *Electrochim. Acta*, 2015, **161**, 395.
206. R. Banavath, A. Abhinav, R. Srivastava and P. Bhargava, *Electrochim. Acta*, 2022, **419**, 140335.
207. Y. Zeng, Q. Li, W. Wang, Y. Wen, K. Ji, X. Liu, P. He, B. Campos Janegitz and K. Tang, *Microchem. J.*, 2022, **182**, 107898.

208. B. G. Ghule, N. M. Shinde, S. D. Raut, S. F. Shaikh, A. M. Al-Enizi, K. H. Kim and R. S. Mane, *J. Colloid Interface Sci.*, 2021, **589**, 401.
209. H. J. Cheon, S. Y. Shin, V. V. Tran, B. Park, H. Yoon and M. Chang, *Chem. Eng. J.*, 2021, **425**, 131424.
210. J. Yang, D. Wei, L. Tang, X. Song, W. Luo, J. Chu, T. Gao, H. Shi and C. Du, *RSC Adv.*, 2015, **5**, 25609.
211. A. Nag, R. B. V. B. Simorangkir, D. R. Gawade, S. Nuthalapati, J. L. Buckley, B. O'Flynn, M. E. Altinsoy and S. C. Mukhopadhyay, *Mater. Des.*, 2022, **221**, 110971.

CHAPTER 3

Carbon Black: Manufacturing Processes and Fundamental Properties

CHARLES R. HERD* AND LEWIS B TUNNICLIFFE

Birla Carbon, 1800 West Oak Commons Court, Marietta, Georgia 30062, USA
*Email: Charles.herd@adityabirla.com

3.1 Introduction

Carbon black is one of the most ubiquitous fine particles manufactured in the world today for a wide array of uses that brings tremendous value to the applications in which it is used. If it was not for carbon black, simple activities, such as riding a bicycle or driving an automobile – either an internal combustion engine or even an electric powered vehicle – to work, school or medical facilities, as well as basic transportation of food and goods and services, would not be possible. Additionally, the transmission of electricity over the power grid through medium and high voltage cables to power our homes, hospitals, workplaces and even our automobiles in the future would also not be possible. These are just a few examples of the positive social value of carbon black, and there are many more, that provide for sustainable mobility and growing, learning and thriving populations and economies that collectively benefit us all. Thus, it is important to understand carbon black from a fundamental standpoint and how it works in the various applications in which it is used. It is hoped this

Inorganic Materials Series No. 15
Elemental Carbon
Edited by Peter J. F. Harris

Published by the Royal Society of Chemistry, www.rsc.org

chapter supplies the reader with this critical information, perhaps further growing our understanding of it for future generations as materials and our everyday products evolve to meet our everchanging needs. The readers are also referred to several treatises[1–4] on carbon black, previously published by notable scientists in the field of carbon black and elastomer reinforcement.

Carbon black is a product that has been in use for what is believed to be thousands of years as it is generally thought to have been used in inks for writing in ancient times in greater Asia. At that time, probably animal fats or oils were burned in small pots and soot was captured on the inside of a lid. This process essentially represented the earliest lampblack produced by man. In general, the primary use of carbon black from that period forward was as a colourant for inks and coatings and other applications. It was not until the advent of the automobile and mass production techniques in the early 1900s that carbon black was discovered as an effective reinforcing agent in rubber compounds for tyres. According to an article in *Jalopnik Reviews*,[5] carbon black was initially trialled as a colourant in tyres around the time of World War I to replace zinc oxide, which was in high demand for use in ammunition casings. Significant benefits to the rubber compounds were realised, and this discovery subsequently required mass production of carbon black, which eventually enabled the first commercial shipment of carbon black for tyre applications in 1911 to the Diamond Rubber Company, owned by B. F. Goodrich. This first order was for one-million-pounds of channel black manufactured by Binney & Smith, the precursor to Columbian Chemicals Company and now Birla Carbon, one of the largest carbon black manufacturers in the world today.

The benefit of carbon black in terms of its contribution as a sustainable product for use in tyres, was not appreciated in the early 1900s as the concept of sustainability was not in the forefront of society as it is today. In fact, it should be recognised that by the addition of carbon black to rubber compounds in tyres in the early 1900s, their durability and lifetime were dramatically improved, which meant over the course of the following 100 years, significantly fewer tyres were required to be produced. This means that a significant amount of rubber, carbon black, steel, fabrics and various rubber chemicals were never deployed, saving significant energy and resources. These early technological advancements in tyre and carbon black technology, had a direct impact in moving mobility forward and subsequently in the modernisation of the global economies and increased standard of living, education and healthcare.

3.1.1 Classification of Carbon Black

As carbon black manufacturing grew and matured over the years and its use became more widespread in tyres and rubber goods, it became beneficial to develop a common basis for the classification of carbon black. Hence the ASTM (American Society for Testing and Materials) organisation, Committee

ASTM D24.41, developed a detailed classification system, based upon various colloidal and physical tests, to standardise the range of carbon black products and provide a common basis for certification of carbon black quality as sold into the marketplace[6] today.

In the earlier years, broad classifications for carbon black were introduced and included general terms, such as intermediate super abrasion furnace (ISAF), high abrasion furnace (HAF), and semi-reinforcing (SRF) carbon black, as originally defined by the ASTM Committee D24.14 in and around 1966. Practically speaking these descriptive terms, such as ISAF or SRF, are no longer used as the number of carbon black products has proliferated over the years. According to the ASTM D24.41 logbook for carbon black products, the first product given an official ASTM designation occurred in 1966. Subsequently, several products, such as N110, N299, N330, and N550, were given official ASTM designations and were the start of the comprehensive nomenclature system devised for carbon blacks. Since that time, ASTM has been the global leader in carbon black nomenclature and today the ASTM procedure D1765-19[7] defines the names and colloidal targets for all the commercial carbon blacks used in the tyre and rubber goods industries. In this ASTM standard there are currently 44 different carbon black products listed with various colloidal targets for surface area, structure, tint, and bulk density.

The basic nomenclature used for carbon blacks as described by ASTM (Table 3.1) uses a combination of lettering and numbers to distinguish the various carbon black products. The letter designation can be an N or an S and these represent normal (N) cure or slow (S) cure products. In this sense 'cure' relates to the effect of the carbon black on the kinetics of crosslinking of a rubber compound. Additionally, the first number, which can range from 1 to 9, is representative of the carbon black's surface area range and is divided as shown in Table 3.1. Note that the letter designation and the first number are the only two meaningful terms in the carbon black designations as the second or third number in a product name, such as N110 or N550, are arbitrary.

It should be noted that there are many more 'non-ASTM' carbon blacks manufactured and in use today for tyres, specialty rubber goods, inks, coatings,

Table 3.1 Current ASTM nomenclature and surface area and particle size range according to ASTM D1765-19[7]

N or S number range	Average nitrogen surface area/$m^2\ g^{-1}$	Example products
0 to 99	>150	—
100–199	121–150	N110, N115, N121, N134
200–299	100–120	N299, N231, N220, N234
300–399	70–99	N326, N330, N347, N339
400–499	50–69	—
500–599	40–49	N539, N550, N582
600–699	33–39	N642, N650, N660, N683
700–799	21–32	N754, N772, N774, N787
800–899	11–20	—
900–999	0–10	N990

and plastics. Some of these products remain confidential and exclusive while others have been introduced to the broader markets. Eventually, if such carbon blacks for tyres or rubber goods become widely used then they will be submitted to ASTM by an individual supplier for inclusion into the standard range of carbon black products; however, the carbon blacks for inks, coatings and plastics remain outside of ASTM governance and each individual supplier uses their own branding scheme for this class of products.

3.2 Manufacturing Processes for Carbon Black

There are five major processes for manufacturing carbon blacks: the furnace process, channel process, lampblack process, acetylene, and thermal black processes (Table 3.2).[8] The furnace process primarily addresses the needs of the tyre and rubber goods markets, but also inks, coatings, and plastics, and hence this process alone supplies well over 98% of all carbon black requirements in the world today. The other processes mentioned above are considered specialty processes for niche applications that include inks, coatings,

Table 3.2 Breakdown of carbon black volumes (thousands of metric tons) by process and rubber *vs.* non-rubber and specialty products[8]

Process type/ Application	Volumes (1000s of MT)	Primary applications	Product features and benefits
Furnace rubber capacity (furnace, tread and carcass)	18 196	Tyres and rubber goods	Good reinforcement, abrasion resistance, durability
Specialty carbon blacks for non-rubber applications	1123		Varies depending upon application (inks, coatings, plastics), but primarily the emphasis is on purity, colour development (shades of jetness or darkness), conductivity and UV stability
Furnace	951		
Thermal	77	Rubber and plastics	Low reinforcement rubber applications, smooth surface finish easy dispersing
Acetylene	56	Batteries, wire and cable, conductive polymers, rubber curing bladders	High purity, thermal and electrical conductivity
Lampblack and gas blacks	24	Inks and coatings	Easy dispersing, blue undertone
Superconductive	8	Batteries, conductive polymers	High purity, lower specific gravity, high conductivity at low loadings (rapid percolation threshold)

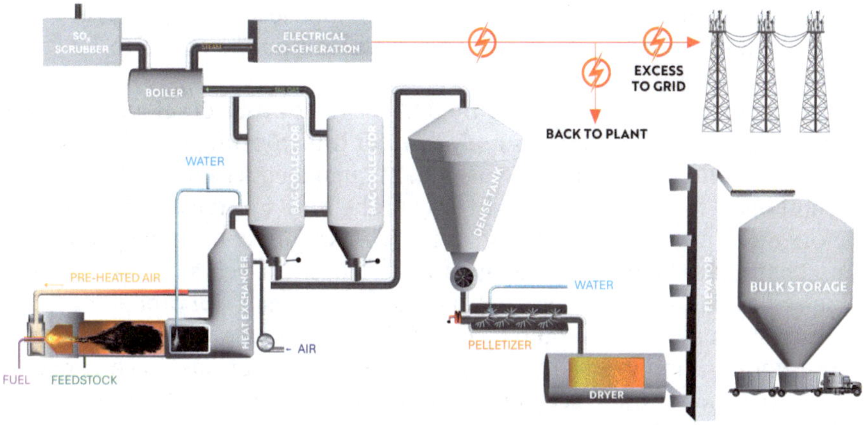

Figure 3.1 Layout of a typical furnace carbon black plant. Reproduced with permission from Birla Carbon USA, Inc.

and plastics among many other non-rubber products, such as batteries, that might use acetylene carbon blacks and even certain furnace carbon blacks.

A typical carbon black plant will have four basic sections: (1) a reactor section where the carbon black is formed, (2) a bag collector section for separating the carbon black from the tail gas, (3) a densification and beading section, and (4) a silo storage and general packaging area (Figure 3.1). The average furnace carbon black factory will produce approximately 100 000 MT of carbon black per year, but some plants can be smaller and some larger, even reaching capacities of up 300 000 MT plus per year of production. Carbon black is typically shipped *via* hopper truck or railcar and even 'big bags' *via* truck that may hold one MT of product. Bulk liners may also be used and hold up to 20 MT of carbon black but require special unloading capability. Most carbon black plants are energy positive, that is, steam and electricity can be produced from the excess heat generated in the production of carbon black, and enough electricity can be produced to run the entire plant and still have remaining capacity to sell to the electrical grid or to sell steam to neighbouring plants and communities.

3.2.1 Furnace Carbon Black Process

The furnace process is generally broken down into two reactor types, those that make tread or hard products and those that make carcass or soft products. These terms are generally derived from the fact that the carbon blacks made on tread reactors are smaller in particle size and are used more in the tread portion of tyres for their excellent abrasion resistance, durability and dry traction characteristics. These tread compounds tend to be harder with low compliance. On the other hand, those carbon blacks made on carcass reactors are larger in particle size and are used in compounds in the non-tread or carcass portion of the tyre for their excellent processing and heat build-up characteristics. These carcass compounds tend to be softer with a high degree of compliance in most cases.

The tread furnace reactor (Figure 3.2)[9] works by first developing a flame in the combustion section to supply heat for transforming liquid feedstocks (oils) into solid carbon black. This is then quenched with water, to stop the carbon black formation and begin to cool the now so-called 'smoke stream' before entering the heat exchanger and downstream collection and packaging. There have been many patents filed[10–13] over the years on various tread reactor designs and each company has its own approach for producing tread-grade carbon blacks. It should be noted that the patents referenced are only representative and are from relatively current time periods. Each company has its own set of trade secrets in addition to their patented processes, but all are similar in the basic approach. Indeed, the patent literature for carbon black production processes is voluminous and extends back well over 175 years; many companies and all different regions of the world have contributed to the technological growth of carbon black and carbon black production processes.

In terms of the tread furnace process, it is best characterised as a fast-flow reactor operating at the maximum temperature range allowable by the insulating refractory and metallurgy, such that maximum productivity and minimum CO_2 emissions can be realised. As can be seen, the carbon black tread process works by first generating a flame in the combustion zone, which then heats the reactor refractory and supplies the necessary energy to pyrolyze and crack the preferred feedstock. Ultimately this pyrolysed and cracked feedstock begins to form the carbon black product, whose primary unit is the carbon black aggregate, nanometres in size, as an individual entity. The flame is generated using air and an appropriate hydrocarbon fuel, usually natural gas or oil (as shown in Figure 3.2), and utilises pre-heated air for

Figure 3.2 Tread Furnace carbon black Reactor. Reproduced from ref. 9 with permission from ORION Engineered Carbons GmBH.

improved flame temperature and overall yield. Typical flame temperatures are in the range of 1400–1800 °C with a reactor pressure in the range of 5–10 psig.

As the air and combustion products, mostly nitrogen, carbon dioxide and water vapour, travel downstream, the reactor narrows and the hot gases pass through a choke section, where the pressure is reduced to approximately ambient pressure (−2 to +2 psig) and the velocity is increased to approximately 0.7 mach. At this point, conversion oil or carbon black feedstock is atomised *via* a fine spray tip and added in the choke section, where the high velocity of the gas stream helps to further break up conversion oil droplets for improved rates of pyrolysis and cracking of the conversion oil into its primary components, which then form carbon black.

Once the carbon black is formed, the reaction is stopped or quenched with water sprays located anywhere from approximately 1–10 metres downstream and the temperature is dropped below 1000 °C to prevent damage to downstream equipment, such as heat exchangers and bag collectors that separate the carbon black from the remaining tail gas. The entire carbon black formation reaction takes only milliseconds and typical reaction times between oil injection and quench sprays range from as little as 2 milliseconds (ms) up to as much as 100 ms.

The second type of furnace reactor mentioned previously was the carcass reactor, where again many companies use their own designs and various trade secrets in their own processes. While the basics of carbon black formation for the larger-particle-sized carcass products remain similar to the tread products, the carcass reactors operate under different temperature, flow and pressure regimes, with lower flame temperatures and slower flow rates for development of the larger-particle-sized products. For example, in the carcass reactor the typical reaction temperatures are in the range of 1000–1200 °C and the velocity of the gases is much lower with reaction/formation times of the order of 1000–3000 ms or 1–3 seconds (s), which is an order of magnitude longer than the tread reactor.

3.2.2 Lampblack Process

The lampblack process is one of the oldest processes known for the production of carbon black, and history indicates it may have been in use for thousands of years in Greater Asia in the preparation of inks for writing. The basic process to produce lampblack involves burning or incomplete combustion of hydrocarbon feedstocks and is described in early patents going back as far as the 1840s.[14–16] In ancient times, the liquid hydrocarbon feedstock was burned in a ceramic pot and carbon black collected on the inside of a lid. Lampblacks today are manufactured on a continuous, industrial scale (Figure 3.3); nonetheless, the basic process remains the same in that a hydrocarbon feedstock or oil is burned in a shallow basin and the resulting lampblack is collected and packaged.

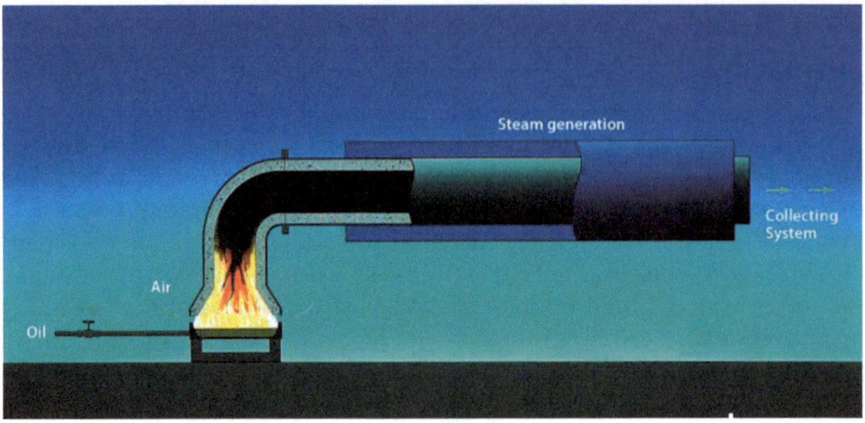

Figure 3.3 Lampblack process. Reproduced from ref. 9 with permission from ORION Engineered Carbons GmBH.

The resulting lampblack-type carbon black is unique, and its products are mainly used in inks, coatings, tinting pastes, cement, sealants, and even some limited use is found in rubber.[17] These products are defined by their generally lower surface area, higher structure, and broader aggregate size distribution and demonstrate a high degree of aggregate breakdown when mixed into high viscosity media, due to an open aggregate architecture and fragile structure.[17] These products would fall in the N700 to N800 surface area and particle-size range. The unique process of making lampblacks, in a slower velocity almost 'lazy' flame, gives them their unique aggregate architecture along with some degree of flexibility in imparting surface oxidation and a lower pH, that is useful in stabilising its dispersion in its various applications. The large particle size and higher aggregate structure mean these products disperse well in their various applications.

3.2.3 Channel and Gas-black Processes

The channel process[18,19] was historically developed in the United States due to the availability of low cost and abundant natural gas. Some of the first channel plants used to make carbon blacks were made of numerous channel houses where a high number of small natural gas flames, sometimes oil-enriched,[18,19] were allowed to impinge on water-cooled iron channels, where the carbon black collects. The carbon black was then scraped off the iron channel and collected for post-processing and shipment to customers. The yield for this process was very poor and in the range of 5%, hence today these channel houses would not be cost effective nor environmentally compliant and this process has long since been abandoned.

The channel process subsequently gave way to the gas-black process,[20,21] developed in Europe by Degussa in the 1930s and 1940s, where the vapours from a highly aromatic oil were used as the feedstock, which greatly improved

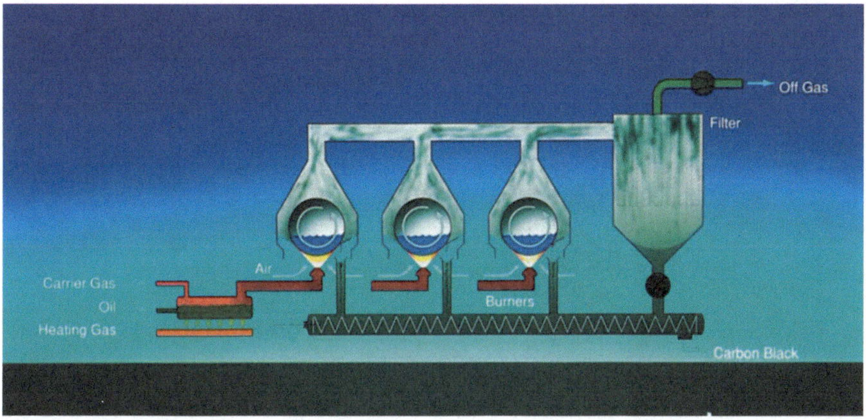

Figure 3.4 Gas-black process. Reproduced from ref. 9 with permission from ORION Engineered Carbons GmBH.

yields, cost effectiveness, and environmental control. The modern channel/gas-black process (Figure 3.4) has even better environmental control and higher yield.

In this process again a flame is allowed to impinge on a cooler, rotating metallic surface and the channel black is continuously removed from the roller with a scraping blade for collection and post-processing.

The gas-black carbon blacks make up a very small portion of the overall carbon black production, somewhere in the range of 0.3–0.5% by weight, but they possess unique surface and colloidal characteristics that make them valuable in their primary applications of inks and high-end, jet-black automotive topcoats.

3.2.4 Acetylene Carbon Black Process

The acetylene carbon black process[22,23] is named as such since the acetylene gas serves as the fuel and feedstock. This process (Figure 3.5) is characterised by numerous smaller 'torts' or burners, where acetylene undergoes partial combustion at even higher temperatures than in the furnace process.

Because the feedstock used is composed of acetylene, the acetylene carbon black is very pure and contains no sulfur. The acetylene black process is also thermodynamically favoured, yielding a spontaneous and on-going reaction once started (without the need to inject air for combustion). This reaction also occurs at temperatures higher than the furnace process, resulting in a product with a more organised microstructure, that is mildly graphitised. Hence the very clean surface, nil sulfur, high carbon content and partially graphitised microstructure, make this carbon black ideally suited for conductive applications in various types of batteries.

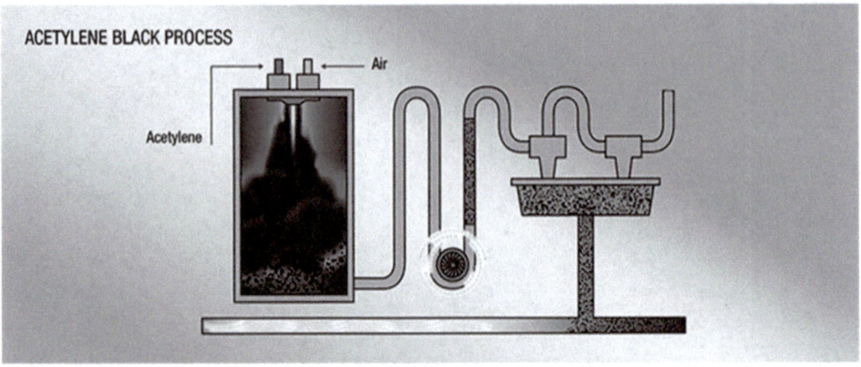

Figure 3.5 Acetylene black process. Reproduced from ref. 9 with permission from ORION Engineered Carbons GmBH.

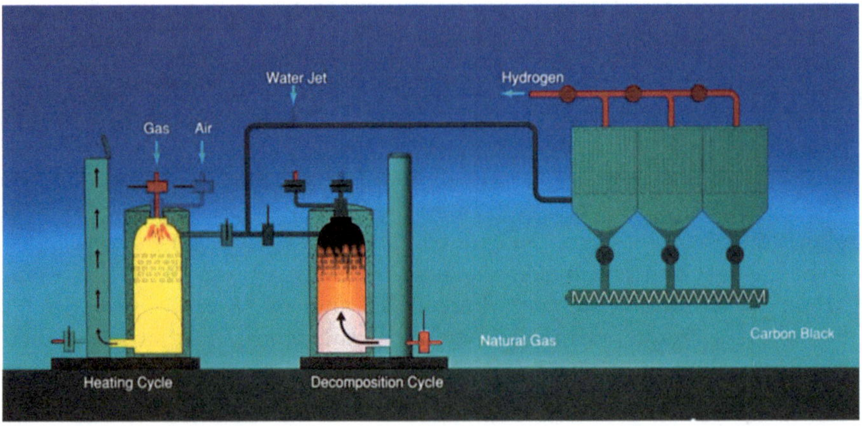

Figure 3.6 Thermal black process. Reproduced from ref. 9 with permission from ORION Engineered Carbons GmBH.

3.2.5 Thermal Carbon Black Process

The thermal process[24] is different to all other carbon black production processes as it is formed by complete thermal cracking of the gaseous or liquid feedstock and not a partial combustion process as for the furnace, channel, and gas processes. In fact, the thermal process is also characterised by much longer reaction times, of the order of seconds, in a thermally heated refractory-lined chamber with little flow or turbulence (Figure 3.6). Typically, in modern day production, the thermal black process is divided into two reactors, one in a 'make cycle' and the other in a sweep/heat cycle. Due to the long reaction time of this type of carbon black in an essentially static reactor, it is economically beneficial to utilise the two reactors for semi-continuous production in what is essentially a batch process.

Thermal carbon blacks are unique and are characterised by large particle sizes of the order of 500 nm with surface areas of 8–12 $m^2\ g^{-1}$ and with very low structure. In fact, these products have the lowest surface area and structure of all carbon blacks produced today. They are primarily used for rubber compounds where low reinforcement, low conductivity, low hysteresis and good compression set are required, and plastics for wire and cable jacketing, as well as other highly specialised applications, such as castable refractory, carburised metal, and insulation of ceramic furnaces.

3.2.6 New Technologies for Carbon Black and Carbon-based Nanomaterial Manufacturing

The furnace process, which accounts for >98% of all carbon blacks produced, has been in existence for well over 75 years. Newer processes with potentially improved sustainability footprints are currently being devised, and in some cases, constructed to commercial scale. These processes may be characterised as those based on recycled rubber and tyres, those with improved carbon conversion yields and those that are considered to work by CO_2 capture and conversion (to various carbonaceous materials).

The processes that use recycled rubber or tyres generally implement pyrolysis of the end-of-life tyres to separate them into three components: oils, carbonaceous materials and steel.[25–29] In fact, this technology is beginning to take hold in the marketplace as it underpins the movement toward circularity. In practice this circular approach works reasonably well and can reduce the CO_2 footprint of the tyre, but some degradation in performance of the recovered raw materials does occur relative to the original raw materials used to build the original tyre. In fact, the carbonaceous material recovered from this process does not perform the same as carbon black, and hence it is referred to as a sustainable carbonaceous material or a sustainable carbon product.

Other processes rely on efficient conversion of a carbonaceous feedstock into carbon black, and one process that claims to have practically a 100% conversion from fuel to solid carbon black is the plasma process.[30,31] This process typically starts with natural gas or methane and cracks the methane, which then forms solid carbonaceous materials, with certain processes forming primarily carbon black. Some of the first attempts to utilise this process were made by Kvaerner,[30] but ultimately the process was not scaled-up to commercial-level production. However, at the time of writing, another company, Monolith,[31] has built a large-scale factory that is being operated on a commercial basis. One by-product of this process is the production of hydrogen, and currently this by-product is being used to make industrial-scale ammonia for other industrial uses. This potentially means this process is almost free of CO_2 generation, except for the CO_2 produced when generating electricity to run the plasma torches. All of these new carbon black processes will require full Life Cycle Analyses (LCAs) to fully validate any advantages in CO_2 reduction.

Other processes that work by capture and conversion of CO_2 into solid carbonaceous materials,[32,33] or by novel introduction of feedstock emulsions[34] include new processes from various inventors and start-up companies. In this instance, these are patented nascent technologies, with scale-up requiring further investment and, ultimately, optimisation to produce commercial-scale quantities of carbonaceous materials with homogeneous morphologies of controlled particle size and structure. Perhaps these new initiatives in production of nanoscale carbonaceous materials represent the future of sustainable processes for the production of reinforcing fine particles, but much work remains to be accomplished.

Regardless of the manufacturing process, whether with mature technology, such as furnace, tread or carcass carbon black reactors, or with newer processes, in early stages of development and industrialisation – all carbon blacks possess a range of similar fundamental properties that determine their performance envelope in the many applications and vehicle systems in which they are used, whether it is in rubber, plastics, inks and coatings, or various specialty applications. As such it is important to be able to characterise and understand the fundamental and physical properties of the carbon black, and to understand their distributional nature in greater detail with the use of sophisticated characterisation techniques. The next sections of this chapter are devoted to summarising the colloidal, physical, and fundamental properties of carbon black; these are important for design and development of new carbon blacks with improved performance to meet the ever-increasing challenges and demands in materials science and the products we use today – and more importantly, for those products that we will use tomorrow.

3.3 Colloidal and Physical Properties of Carbon Black

Carbon black is often specified within industry using colloidal tests of surface area, structure, and tinting strength. Other important properties, often considered as physical or analytical properties, may include pellet properties, residue content, and concentrations of non-carbon components, such as sulfur, ash and metals, as well as hydrogen and oxygen content. These various colloidal and physical properties generally make up the product specifications for carbon blacks as sold to customers of the carbon black industry. As such, it is important to understand the meaning of each of the colloidal and physical tests, as related to their impact on the customers' application, for a better appreciation of the level of engineering and process control required to meet the customers' demands and applications that use carbon black.

As one example, the impact of each of these colloidal and physical properties on a rubber compound are considered, as set out in Table 3.3. Limits of these properties are not shown, only target values, as these ranges remain confidential between supplier and customer. Note that the specification

Table 3.3 Model specification for carbon black showing various colloidal and physical tests to describe and specify carbon black (Birla Carbon USA, Inc.)

Property	ASTM procedure	Typical target value for an N234 (ASTM D1765[a])	Impact on rubber compound
Iodine number/mg g^{-1}	D1510	120[a]	*Surface area*: Primarily impacts abrasion resistance, tensile strength, tear resistance and heat build-up
Nitrogen Surface Area (NSA)/m^2 g^{-1}	D6556	119[a]	*Surface area*: Primarily impacts abrasion resistance, tensile strength, tear resistance and heat build-up. ΔNSA–STSA can be used as a measure of porosity
Statistical Thickness Surface Area (STSA)/m^2 g^{-1}	D6556	112[a]	*Surface area*: Primarily impacts abrasion resistance, tensile strength, tear resistance and heat build-up
Tint strength/%ITRB	D3265	123[a]	Correlated to mean particle size/surface area, secondarily impacted by structure and aggregate size distribution (ASD) (lower tint with wider ASD)
Oil Absorption Number (OAN)/cc per 100 g	D2414	125[a]	*Structure*: Primarily impacts hardness, Mooney viscosity, modulus, tear strength
Crushed Oil Absorption Number (COAN)/cc per 100 g	D3493	102[a]	*Structure*: Primarily impacts hardness, Mooney viscosity, modulus, tear strength
Transmission/%	D1618	>70	Residual hydrocarbon, usually set at a minimum to ensure good carbon black elastomer interaction
pH	D1512	6–10	Acid–base character of surface, can impact cure rate and crosslink density in classical accelerator, sulfur-cured systems
Pellet hardness	D5230		Measure of hardness of pellets or densified form of carbon black produced for ease of shipping, handling, and dispersion
Mean pellet hardness/grams	D5230	20–40	Measure of hardness of pellets or densified form of carbon black produced for ease of shipping, handling, and dispersion
Maximum pellet hardness (average of five hardest, M5H)/grams	D5230	40–80	M5H is usually less than twice the average or mean pellet hardness; very hard beads have a very negative impact on dispersion
Minimum pellet hardness/grams	D5230	10	Impacts 'fines' or powder generation and will negatively impact flow and dispersibility
Pour density/kg m^{-3} (lb ft^{-3})	D1513	320 (20)[a]	Measure of bulk density as packaged, impacted directly by structure
Heat loss/%	D1509	<2%	Measurement of moisture content, can impact carbon black loading accuracy and mixing and compound quality
Ash/%	D1506	<0.75%	Measure of metals content from inputs to the reactor and beading; may impact pH, possibly cure rate in certain circumstances
Water washed residue	D1514		Residue in carbon black can impact ultimate tensile strength and fatigue life
35 mesh (500 μm)/ppm	D1514	<5	
325 mesh (45 μm)/ppm	D1514	<500	

[a]Values taken from ASTM procedure.

parameters are determined according to ASTM procedures. This provides a common suite of tests for specification of the range of carbon blacks in ASTM D1765, for customers across the globe.

3.3.1 Surface Area

Surface area, which is a direct manifestation of the particle size distribution of carbon black, is a key property of carbon black that influences many important properties of the applications in which it is used. In rubber, the surface area determines the abrasion resistance, heat build-up or hysteresis, tensile strength, and tear strength. In inks and coatings, it is the primary factor that determines the jetness, or degree of black colour development, of the product and in plastics it drives jetness, conductivity, and UV resistance.

The surface area of carbon black is specified in two different ways (Table 3.3): surface adsorption of iodine (I_2), or iodine adsorption number,[35,36] and nitrogen adsorption, according to Brunauer, Emmett and Teller (BET) adsorption theory,[37,38] used to measure the surface area of a wide range of fine particles. The nitrogen adsorption data can be further manipulated to determine the non-porous or external surface area of carbon black and this parameter is called the statistical thickness surface area (STSA).[39] The amount of porosity in a carbon black, which is equal to the nitrogen surface area (NSA)–STSA, can be important in the performance of certain applications of a carbon black, but these are few and mostly relate to specific conductive applications.

From the carbon black manufacturing standpoint, the iodine number is primarily used as a quality assurance tool since the test is rapid and reproducible, but the surface chemistry and transmission (or residual hydrocarbon), can influence its measurement. Hence new, faster surface area tests for carbon black that are not influenced by the surface chemistry and surface contaminants are desirable. As one example, a rapid NSA test would be well suited for a carbon black quality assurance tool, as it is mostly free of interference from the carbon black surface chemistry.

3.3.2 Structure Measurement of Carbon Black

The structure of carbon black directly impacts rubber compound properties, such as viscosity and processability, as well as its hardness, modulus, and conductivity (at lower surface areas).

Measurement of the carbon black structure using ASTM-type colloidal test procedures is accomplished in three different ways:

- First by measurement of the oil absorption capacity of nascent powder or pellets, which is known as the oil absorption number (OAN).[40,41]
- Secondly by the crushed oil absorption number (COAN),[42] where the carbon black is crushed four times in a small press chamber at 24 000 psi to mimic breakdown in a rubber compound (which is actually much less than the COAN), prior to measurement of oil absorption number.

- Thirdly, *via* dynamic void volume analysis (DVVA),[43–47] where the carbon black is progressively compressed as a function of pressure, and the volume at each pressure is successively measured.

The OAN test simply measures the amount of oil, in this case, a safe and sustainable sunflower oil, that is absorbed by the carbon black. The structure of the carbon black manifests itself through aggregate branching and the resulting void volume between the branches absorbs more or less oil, as the degree of branching and structure is higher or lower, respectively. The end point of the OAN and COAN test occurs when the oil has been absorbed by intra- and inter-aggregate void volume and forms a stiff paste, at which time the mixer torque increases and plateaus to a pre-set torque cut-off point. Note that it is imperative that the right oil is used in this test as it must effectively wet the carbon black surface. ASTM has conducted many studies around this variable and recommend specific oils for the OAN and COAN test.[41,42]

The same oil absorption occurs in the COAN test, but now the carbon black structure has been broken down by 10–25%. This test is used as a more 'absolute' measure of carbon black structure, since in the OAN test, some apparent structure is formed through the secondary networking of the carbon black, which is a soft secondary structure that can only be moderately reduced through additional mechanical work. However, both structure measurements are indicative of the structure of carbon black, and both are used in QA testing during the manufacture of carbon black, and they show good correlations with physical properties of rubber compounds, such as viscosity, modulus and processibility.

In consideration of the various surface area and structure measurements described in this section, the term 'colloidal balance' should be introduced. In D1765 each ASTM product is listed with their most important colloidal properties, namely surface area and structure (iodine number, NSA, STSA, OAN and COAN), and it is the balance, or colloidal balance, of each of these five parameters that defines these ASTM carbon black products. Therefore, it is very critical to produce each carbon black product with the correct colloidal balance to obtain the desired end-use properties for a given application.

3.3.3 Tinting Strength

The tint strength of carbon black[48] is measured according to ASTM D3265-19b[49] and is performed by mixing a very small amount of carbon black on a Hoover Muller, which is a mixer composed of two parallel glass plates. One plate rotates at a slow rpm to facilitate mixing and dispersion of the carbon black into the tint paste, which is composed of carbon black, epoxidised soybean oil (ESO) and zinc oxide. The tint strength is measured (by a photo-electric reflectance meter) as the reflectance of the tint paste, as drawn down on a smooth, glass surface. Essentially this test measures the ability of the carbon black to sufficiently coat and cover the zinc oxide. Hence smaller particle-sized carbon blacks with higher surface area are more effective in tinting the ESO/zinc oxide tint paste and result in higher tint strength and less reflectance. Tint strength is correlated to the mean particle size of the carbon black and is secondarily impacted by the carbon black structure (lower structure increases tint strength due to more efficient packing and coverage). Tint strength is also secondarily impacted by the distributional width of the carbon black with broader distribution also lowering tint strength.

3.3.4 Transmission

Transmission is a measure of the residual hydrocarbon-based stain left on the carbon black surface following formation in the carbon black reactor. This test is performed according to ASTM D1618-18[50] and uses a solvent to extract the stain and a photometer to measure the extract absorbance at the UV wavelength of 445 nm. General carbon black transmission levels are greater than 70%, with higher values preferred, to ensure a cleaner surface and good carbon-black–elastomer interaction. However, very high transmission values at a level of >98% may indicate a carbon black surface with a very high degree of heat history and such a product may demonstrate mildly diminished carbon-black–elastomer interaction. Such products exist in the ASTM list of carbon black products, for example N115, N110, and N220, and have been found useful for improved tear strength in demanding off-road applications, where the polymer may saltate or slip over the surface due to diminished carbon-black–elastomer interaction, dissipating energy as heat and improving the tear strength of the rubber compound. In addition, these products may show slightly diminished abrasion resistance, also due to the reduced carbon-black–elastomer interaction.

3.3.5 pH

The pH of carbon black is determined according to ASTM D1512-15b[51] and is an indicator of the acid/basic character of the surface. This property is determined by boiling 20 grams of carbon black in hot water for a set time and then measuring the suspension pH with a pH probe. This surface property is important as it has an impact on the cure rate of rubber compounds; an acidic pH has the potential to retard cure and a basic pH has the potential to accelerate the cure rate. Many accelerator components of the sulfur cure package of a rubber compound are amine-based chemicals and, hence, acidic carbon blacks may scavenge these accelerators and cause a slower, less-accelerated cure rate. A basic surface would have the opposite effect. Most carbon blacks have a relatively neutral pH in the range of 6–8. The surface pH may be impacted by the amount of sulfur in the original feedstock used to produce the carbon black, as well as the ash and volatile content; higher ash drives a more basic pH and a higher oxygen-based volatile content drives a more acidic pH.

3.3.6 Pellet Properties

Carbon black, as a powder, has a very low bulk density and is therefore pelletised for densification and subsequent ease in handling, conveying, shipping and mixing (improved dispersion). As described previously, the pelletisation of carbon black is achieved by adding water to the powder in a pin mixer, along with a small amount of bead binder for improved pellet integrity and to minimise pellet breakdown, into what is termed fines (powder) generation, during conveying and shipping. Fines can have a very detrimental impact on conveying, mixing and dispersion and need to be minimised through optimised pellet properties. Carbon black pellets can range in size from tenths of a mm up to several mm with the average size normally in the 35 mesh range of 0.5 mm; the pellet size distribution is determined according to ASTM D1511-12.[52] Typical pellet hardness for carbon black falls in the range of 20–40 grams and is measured using an individual pellet-hardness tester (IPHT) according to ASTM D5230-19.[53] In this test, nominally the 14 mesh, pellet-size-fraction is used as the basis for establishing the carbon black pellet hardness for QA purposes.

3.3.7 Heat Loss

Heat loss, as measured *via* ASTM D1509-18[54] is primarily a measure of the adsorbed moisture on the surface of the carbon black. The equilibrium moisture content of carbon black is driven by surface area and ash content, with higher surface area and higher ash content, increasing the adsorbed moisture and heat loss. Higher moisture content of carbon black is normally detrimental with regards to compounding in various vehicle systems, whether it is rubber, plastics, or coatings, as it may reduce the actual loading of carbon black, slow down incorporation and dispersion, and cause porosity in final compounds.

3.3.8 Ash

The ash content of carbon black as determined by ASTM D1506-15[55] is determined by combusting the carbon black in a muffle furnace with hot air and measuring the weight of remaining inorganic residue. The ash content of carbon black results from various inputs used in the production process. Low ash content is preferred to reduce moisture pick up and ensure the highest level of reinforcing carbon black.

3.3.9 Residue

The residue content of carbon black is measured according to ASTM D1514-15[56] and is performed by placing a 100 gram sample of carbon black on a specified wire mesh, usually 35 mesh (500 μm) and 325 mesh (44 μm), and subjecting it to a high pressure water spray to wash through the smaller-particle-sized carbon black and collect the remaining larger-particle-sized residue. The residue levels in carbon black are very low and have reduced dramatically throughout the history of carbon black development as processes improved and demands for rubber compound durability increased. Typical water-washed-residue levels for 35 mesh are <10 ppm (0.001%) and for 325 mesh are <500 ppm (<0.05%) for tyre compound applications. In certain instances, very clean or very low residue carbon blacks (325 mesh <20 ppm, <0.002%) are required when certain rubber parts are visible to the naked eye, as in automobile sealing systems, and a smooth finish is required. Beyond simply appearance, very low or nil water-washed-residue may be required in certain critical applications, such as conductive strand shields in medium and high voltage cables. Carbon black residue may be composed of coke, refractory, or metal.

The N234 specification and the colloidal and physical properties described above represent measurement of the 'bulk' properties of carbon black and are representative of the collection of carbon black aggregates as a whole. To truly understand carbon black however, one must measure its fundamental properties and their distributional characteristics; these are reviewed in the following sections.

3.4 Fundamental Properties of Carbon Black

Carbon black is one of the most engineered and studied fine particles in commercial production today. As such, an in-depth understanding of its fundamental properties is required, first to understand better how this material functions in the many applications in which it is used, and secondly to allow one to design and develop next generation products.

Generally speaking, the fundamental properties of carbon black, rather arbitrarily defined from years of research and application, include the following: particle size, aggregate morphology (size and structure), surface activity and porosity. In this instance the carbon black microstructure is also being included as a fundamental property, since this property gives tremendous insight into carbon black formation, conductivity and ultimately impacts the surface activity and the interaction of the carbon black surface with its surroundings.

3.4.1 Carbon Black Formation and Microstructure Development

The first fundamental property to consider, the microstructure of carbon black, has been thoroughly studied[3] and refers to the layers of hexagon nets of carbon atoms; or as might be considered today, as graphene layers or lamellas of carbon, layered in an onion-skin orientation (paracrystalline microstructure) that are the building blocks of the primary particles of carbon black. These primary particles are nearly spherical in shape and collide

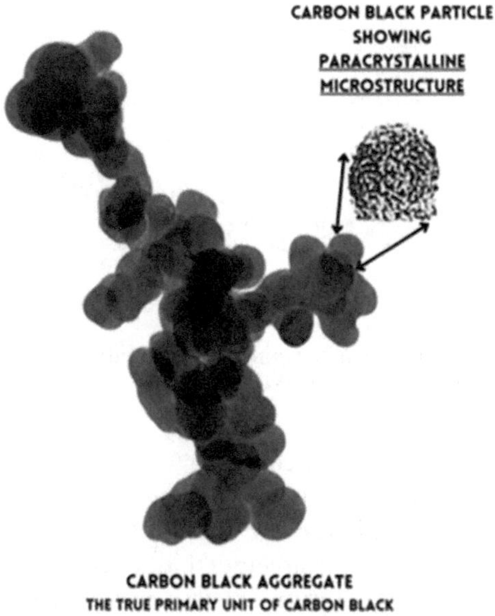

Figure 3.7 Transmission electron microscope (TEM) micrograph of a single carbon black aggregate, the true primary unit of carbon black, also showing the nearly spherical type particles that comprise the aggregate. Reproduced with permission from Birla Carbon USA, Inc.

and adhere during formation to make the final and true unit of carbon black, which is the carbon black aggregate (Figure 3.7). The spherical nature of the particles may be due to similar phenomena as for water droplets, where the primary forces for the outer lamella are inward, creating a surface tension and a spherical shape that minimises the surface area to volume ratio.

As previously described, carbon black formation in a furnace reactor is thought to occur as follows: first the feedstock is sprayed into the reactor *via* specially designed spray nozzles at relatively high pressures to facilitate atomisation of the oil. Next the oil is pyrolysed into the gas phase and decomposes into simpler molecules. These molecules then begin to form polycyclic aromatic hydrocarbon or PAH species, that are thought to condense and begin to form particle nuclei. These nuclei then continue to grow, forming larger, nearly spherical particles due to successive deposition of the PAH-type molecules or graphene-type layers. These particles, which are still somewhat liquid in character, then collide and stick and begin formation of the true primary unit of carbon black, the carbon black aggregate (Figure 3.8).

At some point, about 1–10 metres downstream from the oil injection, water is added to quench or essentially stop the carbon black formation and establish the mean particle size, specific surface area, as well as aggregate size and structure of a given carbon black product. Structure is controlled *via* the addition of alkali metals, primarily potassium-based ionic compounds,[57] where the potassium ion appears to impart a positive charge to the nascent

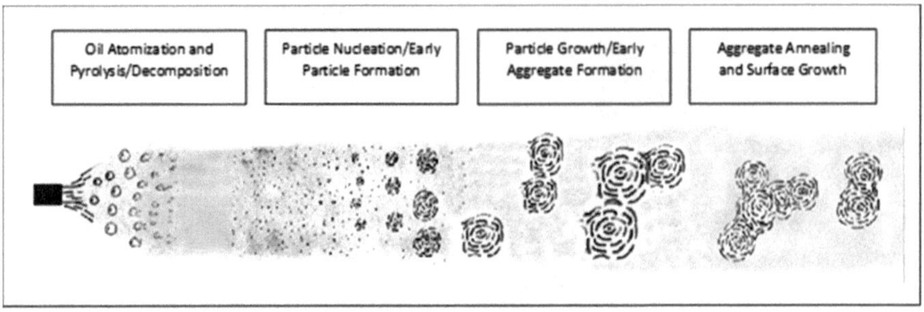

Figure 3.8 Pictorial of the basic carbon black formation mechanism. Reproduced with permission from Birla Carbon USA, Inc.

particles and prevents them from colliding and sticking to one another. The lower structure carbon blacks are made using higher potassium flow rates and higher structure carbon blacks are produced using lower potassium flow rates. The potassium-based salts are added into the carbon black reactor *via* a separate spray in aqueous solution or with the feedstock itself.

The microstructure of carbon black is closely tied to the formation mechanisms of carbon black in the very earliest stages, at the point of atomisation, pyrolysis, decomposition, and condensation of the molecular species in the feedstock itself. The actual formation mechanisms of carbon black have been considered analogous to those postulated for soot formation in flames,[58–61] but certain aspects of the formation remain uncertain, especially the early stages, and the above description is more qualitative than quantitative, yet most likely reasonably reflective of the basic steps as described. It should be noted however that soot and carbon black are not one and the same and differ rather significantly in their purity,[62,63] where soot generally has higher ash and extractable organic components along with a more varied morphology ranging from little to all aciniform carbon. Soot is also considered an unwanted by-product in industrial processes and various engine exhausts and can be randomly and freely admitted to the atmosphere as a pollutant. Meanwhile carbon black is generally purer and it is produced and captured under well-controlled production processes as an engineered, reinforcing fine-particle and ultimately sold in secure packaging for use in its many applications, which also have very controlled production conditions.

The classical formation mechanism of carbon black or soot is assumed to follow a path in which a carbonaceous fuel or feedstock is first atomised, followed by volatilisation and decomposition of the fuel or feedstock into smaller organic species such as C2 intermediates,[64,65] that then start to form the larger molecules of hexagon nets with sp^2 hybridisation. This process is shown in detail for the pathway labelled FCC, for fluidised catalytic cracker feedstock (Figure 3.9).[66]

Newer soot or carbon black formation processes have been devised and do not require the full breakdown and decomposition of the hydrocarbon fuel or feedstock.[62,66] The first step of the breakdown into smaller molecular

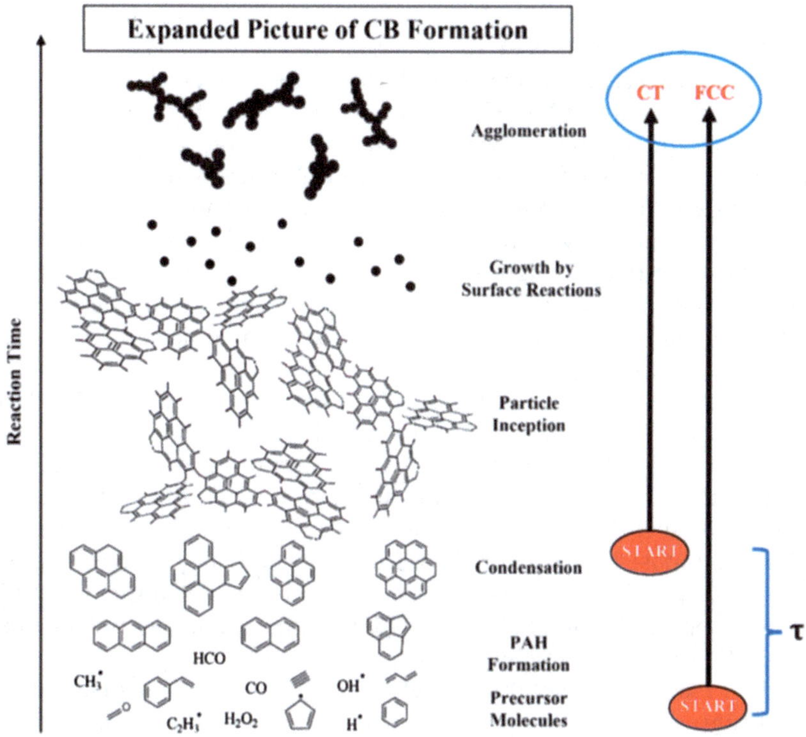

Figure 3.9 Carbon black formation mechanism showing kinetic delay gap for FCC *versus* CT feedstock. Reproduced from ref. 66 with permission from MDPI.

weight intermediates is not required and the aromatic molecules in the feedstocks can begin direct transformation into the lamella graphene-type layers for condensation into spherical particles and finally colliding to form the carbon black aggregate. This formation mechanism is more rapid and bypasses the kinetic delay during the required breakdown and decomposition step that may be necessary for aliphatic species in the fuel or feedstock. This process is shown for the pathway labelled CT for coal tar feedstock (Figure 3.9).[66]

The differences in the feedstock and the resulting microstructural and particle formation mechanisms do have practical implications on the microstructure itself and this will be discussed in more detail in Section 3.4.4. However, first it is necessary to review the basics of the carbon black microstructure and how it is characterised.

3.4.2 Carbon Black Microstructure

Characterisation of the carbon black microstructure has been challenging over the years in terms of identifying even its most basic characteristics, especially before the advent of the transmission electron microscope (TEM), and even then, not until higher resolution transmission electron microscopes were available. In fact, the first characterisation of the carbon black microstructure was performed on the bulk powder by X-ray-diffraction (XRD) in 1934 and forward,[67,68] at which time it was determined that indeed the carbon black microstructure demonstrated some degree of crystallinity. Subsequently, the first models of the carbon black microstructure were developed that described randomly oriented domains of graphitic microstructure; however, over time newer and more accurate models were developed from the 1940s onward, once high resolution TEMs were available and different imaging modes, such as dark-field,[69,70] diffracted beam,[71] and phase-contrast imaging,[72–75] along with other sophisticated spectroscopic methods such as Raman spectroscopy.

Today, carbon black microstructure, as postulated from the TEM and XRD studies referenced in this section, is primarily believed to be considered paracrystalline in nature, as first proposed by Ergun,[76] and composed of continuous lamellas oriented in an onion-skin orientation, with some degree of short-range order (Figure 3.10). As can be seen, the internal microstructure is clearly composed of the layered nature of the graphene-type lamellas, with a finite dimension in the x–y direction. It can also be noted that the microstructure appears to contain features of unorganised lamellas and organised or crystalline zones of lamella layering.

Carbon black is a graphitisable, elemental, carbonaceous material, and upon heat treatment at temperatures above 1000 °C and up to 2500 °C, the internal lamella become mobile, and this results in changes of the carbon black microstructure with longer range order and a high degree of crystallinity. By graphitisable, however, we mean that due to the lamella mobility upon heat treatment under an inert atmosphere, the carbon black takes

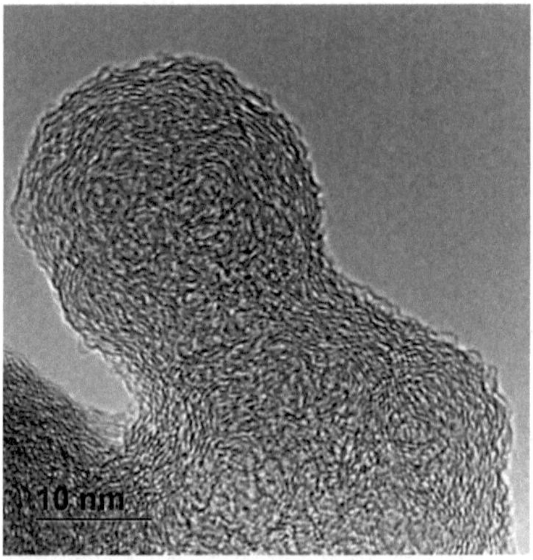

Figure 3.10 High-resolution phase-contrast image of a single carbon black particle. Reproduced with permission from Birla Carbon USA, Inc.

on crystallographic characteristics of graphite, but it does not form graphite itself. This change can be observed in the TEM (Figure 3.11), where a significantly higher degree of lamella order and long-range crystallinity is observable.

In general, the internal microstructure determines and imparts the surface activity characteristics of carbon black. For example, the standard carbon black has a naturally organophilic surface, mostly hydrophobic in nature with numerous surface-layer-graphene edge sites containing various functional groups, including but not limited to hydrogen and various oxygen-containing functional groups. Such surface characteristics have a very natural and strong interaction with hydrocarbon-based materials, such as elastomers, especially those that are unsaturated and contain aromatic structures, that impart the very special reinforcement characteristics of carbon black to rubber,[77] which are utilised so widely in the tyre and rubber goods industries.

Conversely, highly graphitised carbon blacks as shown in Figure 3.11 have a significantly diminished surface activity that reduces its reinforcement potential in rubber.[77] Hence such carbon blacks are generally not favourable for use in rubber due to the poor carbon-black–elastomer interaction that can cause early crack growth and reduced fatigue life for rubber articles used in dynamic applications, such as tyres or antivibration components.

However, such highly graphitised carbon blacks may impart higher levels of thermal conductivity to the composite, rubber or plastic, due to their crystalline nature and ability to transfer thermal phonons through crystalline lattice vibrations.[78,79] Graphitised carbon blacks are also very useful for highly conductive applications in plastics, such as wire and cable and anti-static packaging, as well as other very important applications such as batteries.

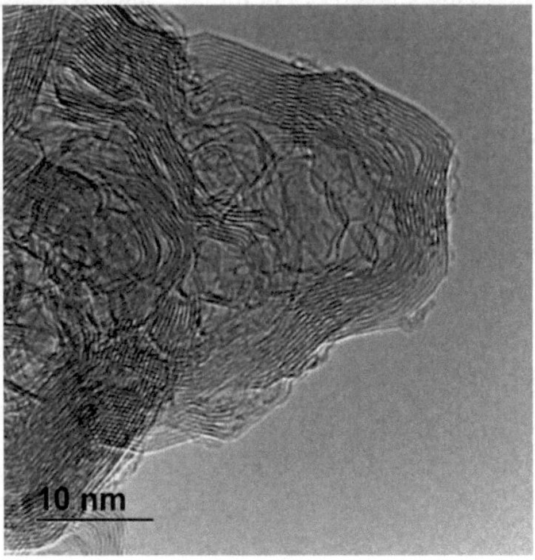

Figure 3.11 High-resolution phase-contrast image of a graphitised (heat treated at 2500 °C) single, carbon-black particle. Reproduced with permission from Birla Carbon USA, Inc.

Model crystallites of carbon atoms for graphite and for carbon black (Figures 3.12 and 3.13) can be used to highlight the basic difference in microstructure between the two materials. Note for graphite (Figure 3.12), the hexagon planes of carbon atoms that are held together by van der Waals forces, form a crystallite pattern in an ABAB stacking arrangement (only ABA shown), with every other plane shifted to a minimum energy configuration. The structure of graphite is discussed in more detail in Chapter 1.

For carbon black the AB stacking is less ordered, where the lamella are shifted and even perhaps rotated relative to other lamella in this form of turbostratic carbon (Figure 3.13), so the lamella are more geometrically randomised relative to one another. This results in a larger interlayer d_{002}

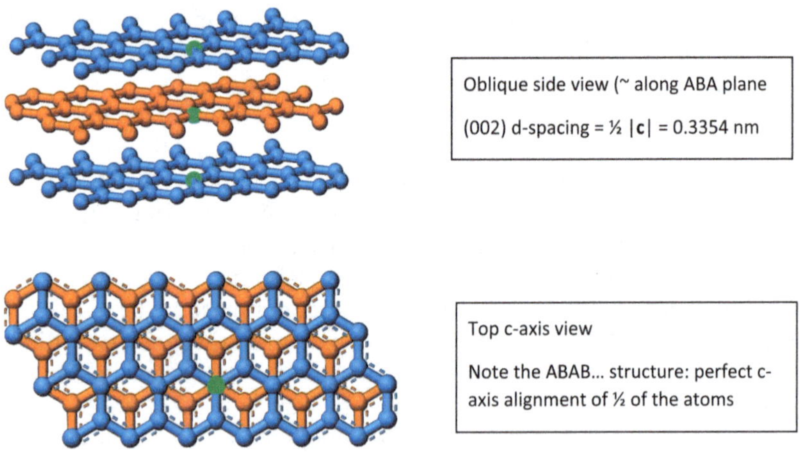

Figure 3.12 Carbon plane stacking in graphite showing crystalline orientation in ABAB stacking unit. Reproduced with permission from Birla Carbon USA, Inc.

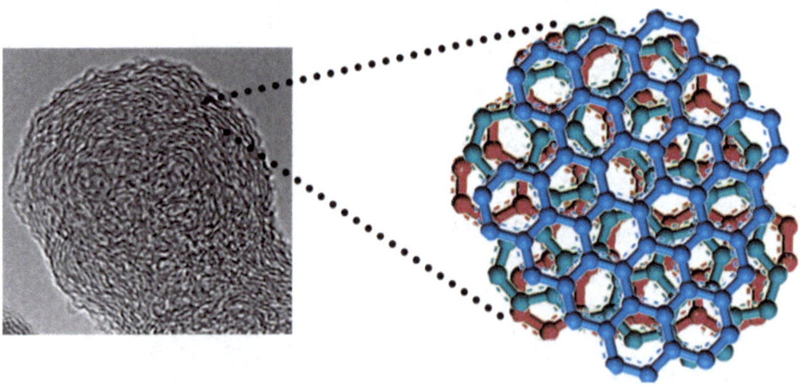

Figure 3.13 Carbon plane stacking in carbon black showing the random orientation of the lamella. Reproduced with permission from Birla Carbon USA, Inc.

spacing in carbon black (~0.36 nm) than graphite (0.3354 nm) due to atom–atom repulsion. This difference in d_{002} has been measured and verified by techniques described in the next section.

3.4.3 Characterisation of the Crystallite Domain Size in the Carbon Black Microstructure

In terms of measuring the average crystallite parameters for carbon black, techniques, such as electron diffraction, X-ray diffraction, and Raman spectroscopy, have been utilised. The gross crystalline parameters (*e.g.* the d_{002} average spacing between lamella; the L_c, or average stacking height of lamella domains; and the average x–y dimension of the lamella, known as L_a) are important in characterising the carbon black microstructure to define the degree of crystallinity, or lack thereof, for insight into formation mechanisms, heat history, and surface characteristics.

3.4.3.1 *Electron Diffraction* via *TEM*

The electron diffraction pattern for a furnace N234 carbon black (Figure 3.14) is composed of three main diffraction planes including the main diffraction plane, d_{002}, that is the strongest diffraction plane and is parallel to the electron beam, and the (010) and (011) diffractions that are weaker and are perpendicular to the electron beam. These diffraction bands are broad in energy distribution and faint due to the paracrystalline nature of the carbon black microstructure. However, for a graphitised N234 carbon black (Figure 3.15) the same d_{002}, d_{010} and d_{011} electron diffraction features are more intense and sharper as the microstructure is more developed and crystalline in nature.

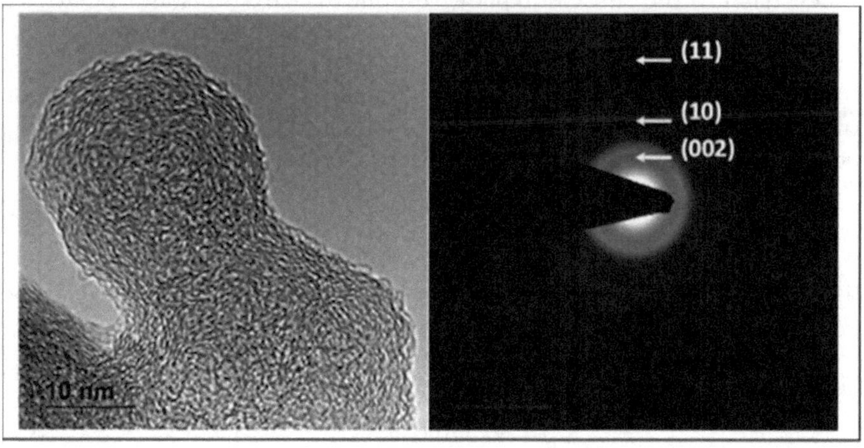

Figure 3.14 TEM image and electron diffraction pattern for an N234 carbon black (single particle microstructure). Reproduced with permission from Birla Carbon USA, Inc.

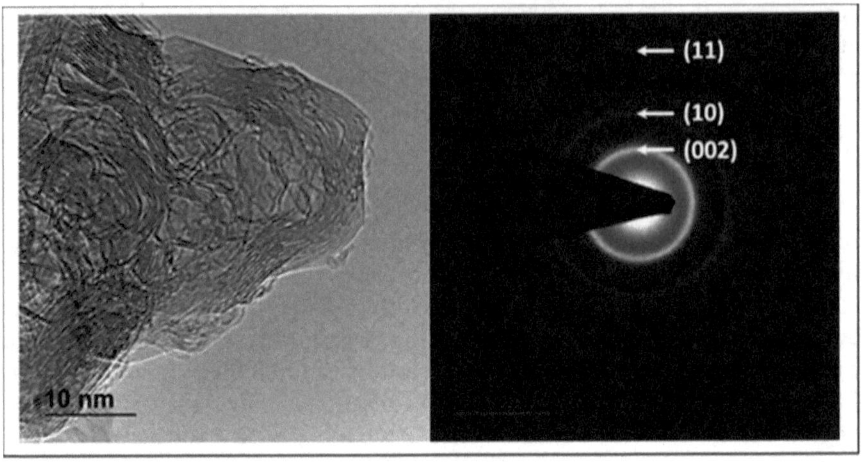

Figure 3.15 TEM image and electron diffraction pattern for a graphitised N234 carbon black (single particle microstructure). Sample was heat treated under nitrogen at 2000 °C. Reproduced with permission from Birla Carbon USA, Inc.

3.4.3.2 *X-ray Diffraction*

The second technique to be employed to evaluate the carbon black microstructure is X-ray diffraction, and this technique utilises principles of diffraction theory originally developed by Bragg[80] for characterisation of crystalline materials by X-rays. The use of X-rays, which happen to be of similar wavelength to atomic distances in crystals, makes diffraction and elucidation of crystalline structures possible. The atomic arrangement and unit cell define the diffraction planes unique to a given crystalline material. This theory was applied to X-ray diffraction by Scherrer[81] to determine crystallite domain size.

The Bragg equation[80] was developed such that the various diffraction planes in crystalline materials could be measured and defined. The Bragg equation for determining the various diffraction planes, d, is as follows:

$$d = n\lambda / 2\sin\theta \tag{3.1}$$

where d = distance between diffraction planes, d_{hkl}; n = an integer indicating diffraction order, usually 1; λ = X-ray wavelength to generate diffraction plot (usually constant); θ = peak position in diffraction pattern.

The Scherrer equation[81] was developed to allow the determination of the sizes of the crystallites and is as below:

$$\tau = \mathrm{K}\lambda / \beta\cos\theta \tag{3.2}$$

where τ = the mean size of the ordered crystalline domains; K = shape factor, normally taken as 0.9; λ = X-ray wavelength; β = line broadening at half maximum intensity (FWHM) in radians.

X-ray diffraction functions by holding the specimen in a stationary position, while the X-ray source and the X-ray detector are scanned around the specimen to measure diffracted intensity as a function of angle 2θ. The typical X-ray source is copper K-α radiation (λ = 1.5406 Å). This X-ray source has been found to be the most convenient due to good signal to noise ratio, easier to cool and allows diffraction from a range of d_{hkl} for a wide range of inorganic materials in the typical range of angles (5–90°) most typically used in XRD. The result is an XRD pattern, with peaks at various angles depending upon the crystalline structure of the material. Subsequent deconvolution and analysis of the XRD pattern then provides fundamental crystallographic data for a crystalline material.

The X-ray diffraction patterns for carbon black, as produced and after heat treatment at 1500 °C and 2000 °C, and for graphite, show narrowing of the peak widths (Figure 3.16) as the microstructure becomes more ordered due to heat treatment (under nitrogen). They more closely approach the peak widths of crystalline graphite, demonstrating the capability of XRD to reveal microstructural information of carbon black.

Note the changes in the main peaks (002), (010) and (011) as the degree of crystallinity changes from paracrystalline for the non-heat-treated (or as-produced) carbon black, to the heat-treated carbon blacks, to crystalline graphite itself. Very quickly at a treatment temperature of 1500 °C for the carbon black, the (002), (010) and (011) peaks become much sharper as the paracrystalline microstructure changes toward a higher degree of crystallinity. The causes of the broadening of the X-ray diffraction peaks have been well studied and debated in the literature.[82–84] In this instance, for the non-heat-treated carbon black, the broadening is due to the sensitivity of the interaction of the X-ray beam and the small crystallite sizes in the carbon black, as

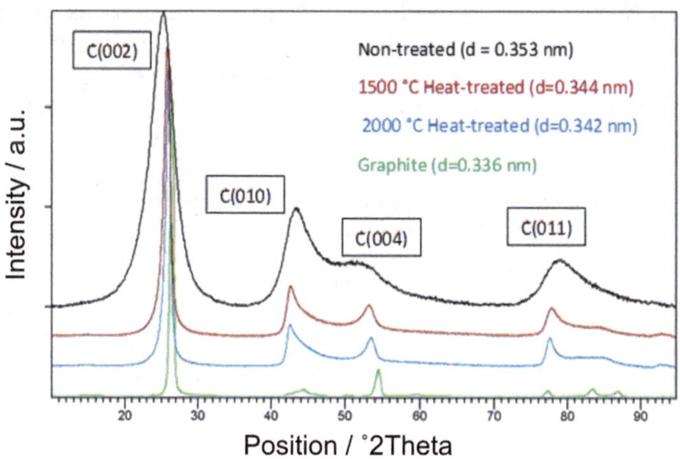

Figure 3.16 X-ray diffraction pattern for graphite and carbon black as produced and heat treated or partially graphitised. Reproduced with permission from Birla Carbon USA, Inc.

well as the characteristics of the turbostratic microstructure (rotation, translation of the graphene layers and differences in interlayer spacing as a result of this rotation and translation, among other reasons). In the end, however, these results demonstrate the ability of X-ray diffraction to clearly detect changes in the carbon black microstructure, which allows quantification of the basic crystallite dimensions for carbon blacks and other carbonaceous materials, such as values for the average d_{002} interlayer spacing of lamella, and crystallite dimensions, L_c or average stacking height of the crystallites and L_a for the x–y average size.

Several excellent studies of carbon blacks *via* X-ray diffraction[85–91] have been reported in the literature. Recent data, generated at Birla Carbon (Table 3.4), shows XRD data for a wide range of carbon blacks that gives insight into the basic crystallographic properties as a function of mean particle size and surface area. For products over 70 m^2 g^{-1} in surface area, which are made on a tread reactor, the reaction temperature is high (~1800 °C) and the reaction time is very short and of the order of milliseconds; while for products with a surface area <50 m^2 g^{-1}, which are made on a carcass reactor, the reaction temperature is lower (1200 °C) and the reaction time is orders of magnitude longer and in the range of 1 second. Finally for the thermal carbon black, with a surface area in the range of 8–12 m^2 g^{-1}, the reaction temperature is lower yet at approximately 1100 °C and the reaction time is orders of magnitude longer and of the order of 2–3 minutes.

The XRD data (Table 3.4) show that as the surface area decreases and particle size increases, the d_{002} spacing becomes progressively smaller (Figure 3.17), approaching a d_{002} for that of N234 heat treated between 1000 °C and 1500 °C. Conversely the L_c shows a gradual and slight increase with lower surface area and larger mean particle size, while the L_a values or x–y crystallite dimension shows a general decrease in size with decreasing surface area and increasing mean particle size. One can surmise that the change in d_{002} and L_c reflect a longer heat history and as a result, the microstructure contains slightly more developed crystallites, potentially also aided by the larger radius of curvature of the lower surface area and larger mean particle size. The changes in L_a are not quite as clear and seemingly show differences only at the extremes of the surface

Table 3.4 XRD data for a wide range of carbon black products of varied surface area and heat treatment (under nitrogen – N234 only), Birla Carbon USA, Inc.

Carbon black	NSA/m^2 g^{-1}	d_{002} spacing/nm	L_c/nm	L_a/nm
N234	119	0.362	1.4	2.2
N234 HT 1000 °C	123	0.360	1.4	2.5
N234 HT 1500 °C	110	0.353	2.1	3.9
N234 HT 2000 °C	104	0.347	3.6	5.6
N234 HT 2500 °C	100	0.345	4.5	6.2
N339	85	0.359	1.5	2.0
N330	74	0.360	1.5	1.9
EXP. CB	60	0.358	1.6	2.0
N550	37	0.357	1.8	2.0
N772	30	0.355	1.8	1.7
N990	11	0.354	1.9	2.5

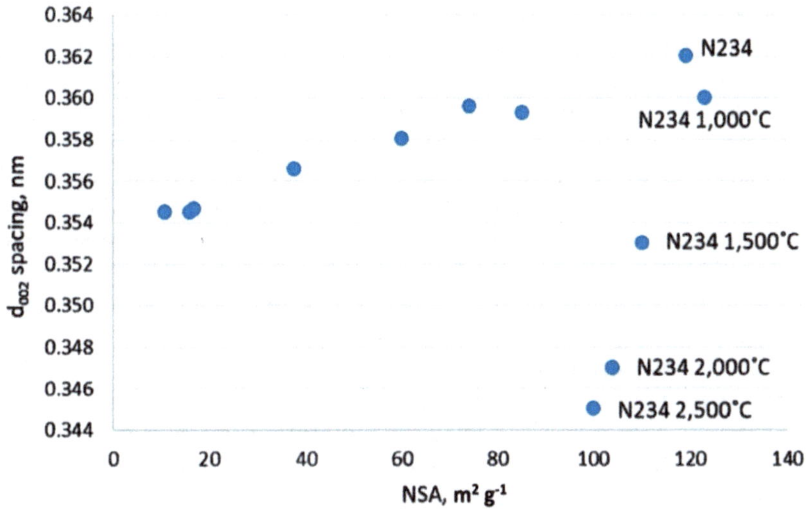

Figure 3.17 The d_{002} spacings from XRD for a wide range of carbon blacks of varied surface area. Reproduced with permission from Birla Carbon USA, Inc.

areas (Table 3.4), where perhaps the trend in the L_a is one of becoming progressively smaller with decreasing surface area and increasing mean particle size. The N990 thermal carbon black product however reverses the trend and shows a larger L_a than the furnace tread or carcass grade carbon blacks. One interpretation of the general lack of variability in L_a may be that the carbon black lamella that make up the microstructure, are in fact curved to some degree (and most likely contain some amount of pentagonal rings to accommodate the curvature of the lamella). Also XRD will only measure the crystallite dimensions of the planes perpendicular to the XRD beam and thus the XRD technique may undermeasure the L_a or x–y dimension of the lamella. Additionally, the XRD technique measures the entire microstructure and some average L_a may be obtained with smaller L_a near the interior and larger at the exterior of the particles.

The XRD data for N234 heat-treated at 1000 °C, 1500 °C, 2000 °C and 2500 °C under nitrogen (Table 3.4 and Figure 3.17) shows the d_{002} average interlayer spacing decreases with temperature, closer to that of graphite, and the L_c and L_a dimensions both increase, as might be expected as the carbon black microstructure becomes more graphite-like. These results demonstrate the insight that can be gained from the use of techniques, such as XRD, to characterise the microstructure of carbon-based materials, such as carbon black, graphite, and graphene.

3.4.3.3 Raman Spectroscopy

Raman spectroscopy is a widely applicable tool for characterisation of a broad array of solid, liquid or gaseous materials useful for characterising their composition and crystalline structure, with each molecule and material having a unique Raman spectrum. The technique relies upon Raman scattering (inelastic) upon irradiation of a material from a laser beam of specified wavelength. The effect was discovered by Chandrasekhara Raman in 1928.[92] Raman scattering (at longer wavelengths than the irradiating laser beam) probes molecular bonding and crystalline lattice vibrations, where it becomes useful for characterising the microstructure of various carbons,[93–96] ranging from coal, to carbon black, to diamond and graphite. Analysis and interpretation of the Raman spectra has proven to be an area of much research with many different techniques and approaches developed over the years for clearer understanding and characterisation of the microstructure of carbonaceous materials.[93–96]

Several Raman studies related to characterisation of the carbon black microstructure have been performed over the years[97–101] by a range of researchers. A typical Raman spectrum for carbon black shows two main peaks (Figure 3.18) which have been used for understanding the microstructural properties, namely the L_a and amount of disordered carbon.

The two peaks at 1575 cm^{-1} and 1345 cm^{-1} represent the graphitic G peak and the so-called disordered or D peak, respectively. The G peak intensity, I_G, for graphite, is narrow and intense, but for carbon blacks, it can become lower in intensity and broader, much the same as for XRD. This peak represents an E_{2g} vibrational mode of a single layer of hexagon nets of carbon atoms within a layer of graphite. For the D peak, the intensity, I_D, for graphite is smaller and narrower, compared to that for carbon black, where the I_D becomes more intense and broader.

The ratio of these two peaks (I_D/I_G) has been used as a measure of the general degree of disorder within a carbon black, with a higher value representing more disorder, and a lower value, less disorder or a higher degree of

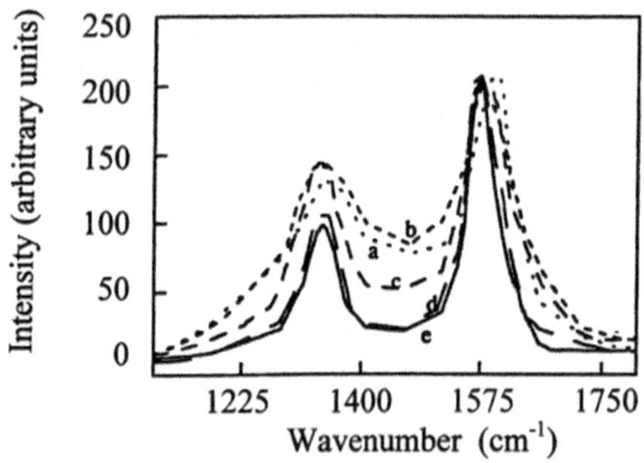

Figure 3.18 Raman spectra of carbon black as-manufactured and at four different levels of heat treatment (laser (Ar^+ ion) irradiation; fluence 800 W cm^{-2}, 2400 W cm^{-2}, 4800 W cm^{-2}, 7200 W cm^{-2}). Reproduced from ref. 99 with permission from American Chemical Society, Copyright 2000.

crystallinity as might be observed upon graphitisation of the carbon black *via* heat treatment or laser annealing.[97,99] Additionally, these peak intensities can be used to calculate the lateral crystallite dimension, L_a, for a carbonaceous material using the following empirically derived equation:[101]

$$L_a(nm) = 4.35/(I_D/I_G) \quad (3.3)$$

In a study by Gruber, Zerda, and Gerspacher,[97] where they evaluated a wide range of carbon blacks using XRD and Raman spectroscopy in the as-manufactured state and after heat treatment at four different temperatures, it was found that the L_a values for the two different techniques were well-aligned and similar in magnitude to each other. These results appear to confirm the validity of these two techniques as applicable for characterising the microstructure of carbon blacks and they give an indication of the relative amounts of disorganised *versus* graphitic components in these materials, along with crystallite dimensions.

3.4.4 Carbon Black Microstructure Development as a Function of Feedstock

In a recent study into the impact of the carbon black feedstock on the carbon black microstructure and its formation mechanisms,[66] a unique and different method not based on diffraction or Raman, was used to characterise the internal carbon black microstructure. In this case, the actual lamellas are imaged at high resolution in the TEM and then directly measured using image analysis to determine their length and tortuosity. This method was developed by Vander Wal and has been applied to numerous materials including soot and carbon black.[102,103]

In a study conducted on a single carbon black grade, produced from two different feedstocks,[66] the lamella imaging (Figure 3.19) and measurement showed that the carbon black made with coal tar (CT) feedstock (CT-CB) contained lamellas that were shorter on average and more randomly oriented in the interior, with multiple nucleation centres and a disordered core. On the other hand, the carbon black made with fluidised catalyst cracker feedstock (FCC-CB) had lamellas that were longer on average with a higher degree of curvature or tortuosity. Also, within a particle, there was a single nucleation centre.

Additional measurement of the lamellas and use of a laser-induced annealing technique to differentiate the mobility of internal lamellas[104] showed that the interior lamellas of the CT-CB were smaller and more mobile upon graphitisation, leading to a hollow internal morphology upon laser-induced annealing or laser-induced derivatisation. By comparison the FCC-CB yielded longer ribbons and a less hollow core (Figure 3.20).

These results were related back to the molecular characteristics of the feedstocks, where coal tar feedstocks have a higher concentration of aromatic molecules and perhaps more complex aromatics than FCC, which

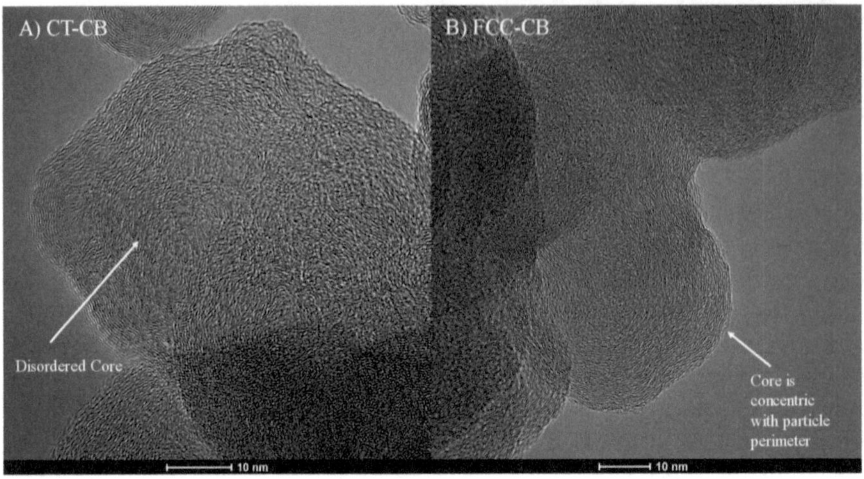

Figure 3.19 High resolution TEM (HRTEM) images of single particles showing the internal microstructure as a result of two different feedstocks: (A) CT-CB is coal tar-carbon black and (B) FCC-CB is fluidised catalyst cracker-carbon black. Reproduced from ref. 66, with permission from MDPI.

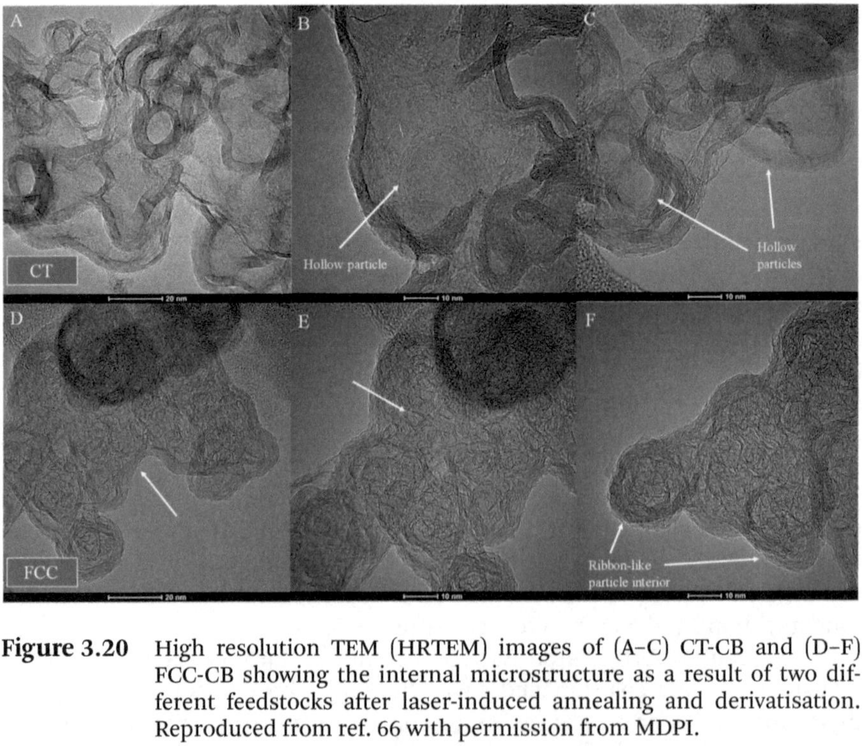

Figure 3.20 High resolution TEM (HRTEM) images of (A–C) CT-CB and (D–F) FCC-CB showing the internal microstructure as a result of two different feedstocks after laser-induced annealing and derivatisation. Reproduced from ref. 66 with permission from MDPI.

contain more saturates, paraffins and aliphatics. The end result is that the coal tar feedstock is a feedstock in which resonance-stabilised hydrocarbon-radical chain-reaction mechanisms[105,106] may dominate and allow covalently bound clusters of PAHs to condense and form particle nuclei at a faster rate than for FCC feedstock. This would lead to multiple nuclei within a given particle and a faster rate of overall carbon black formation. On the other hand, the FCC molecular composition, that contains a higher hydrogen content, may terminate radicals *via* hydrogen abstraction, thus blocking hydrocarbon-radical chain-reaction mechanisms as described above for coal tar. Additionally, it is postulated that aromatic radicals could also be capped by hydrogen-containing alkyl radicals and hydrogen atoms further blocking clustering of aromatic radicals,[107–109] and further delaying the particle inception and growth and eventually carbon black aggregate formation. This delay also means fewer particle nuclei in a more dilute state and hence forming particles tend to have one nucleus, as per the HRTEM imaging.

These results again show that using analytical tools to evaluate the carbon black microstructure can lead to significant information pertaining to formation mechanisms, oxidation rates, their impact on graphitisation and perhaps surface structure and surface activity, thus it is an important fundamental property of carbon black. The challenge has always been in obtaining statistically meaningful and quantitative data on the microstructure itself.

3.4.5 Surface Structure of Carbon Black

The surface structure of carbon black is related to its microstructure in that the graphene type layers (lamellas) that condense to form the particle at the endpoint of the particle/aggregate formation, reside at the surface, imparting what might be termed the surface structure. The surface structure is composed of lamellas, similar as for the internal microstructure, and one might assume that the lamellas, overlapping and layered, are present in discrete zones similar in size to the L_a of the lamellas that compose the particle microstructure. In an overlapping tile structure, edge sites of the lamellas are exposed, and their terminal chemistry and functional groups are impacted by the formation environment and downstream exposure at various points in the process, either as a manifestation of the normal production process or *via* a directed, post-treatment method. These edge sites determine the chemistry of the surface itself, a topic taken up in a later section, while the surface structure in this context is considered more in terms of a morphological or geometrical construct. Nonetheless this surface structure plays a significant role in impacting the interaction of the carbon black with its surroundings and various vehicle systems in which it is compounded. Understanding this surface structure can be important in understanding how elastomers and other polymeric molecules may sterically and chemically interact (functional group spatial arrangement) with the surface.

For many years the surface structure could only be qualitatively evaluated *via* TEM, and it was not until the advent of scanning probe microscopy, in the mid to late 1980s and 1990s, that the surface could be evaluated directly at the atomic to angstrom scale. In this instance a probe tip actually scans over the carbon black surface and the conductive, repulsive and attractive forces between the final atoms of the scanning probe and the carbon black surface were used to obtain nano- and angstrom-scale imaging and characterisation. Additionally, gas adsorption methods have also been developed for evaluating surface roughness and porosity.

The first scanning probe microscope was developed by Binning and Rohrer[110] of the IBM Research Laboratories in Zurich, Switzerland in 1982. This technique, which they termed the scanning tunnelling microscope (STM) works on conductive materials by placing a voltage across the scanning probe tip and the sample and measuring the current flow, *via* electron tunnelling, either by holding the scanning probe at constant height (of the order of 1 nm) or by holding the probe at constant current, which may be used to study the surface topography. The x–y scan size is also very small, of the order of 1 nm^2 to 500 nm^2. The tunnelling current is controlled by the local electron density at each atom that is determined by its overall environment and adjacent atomic nuclei.

For example, when scanning graphite, six carbon atoms in each hexagon ring are not observed, rather only three atoms are observed and these three atoms lie directly above other carbon atoms, while the three carbon atoms not observed in the STM image, represent the atoms located in the centre of the hexagon rings of carbon atoms in the layer below the surface layer[111] (Figure 3.21).

In terms of STM images of carbon black, with a locally turbostratic microstructure, where the lamellas are somewhat rotated relative to one another with larger interlayer spacings, the images of local electron density are not

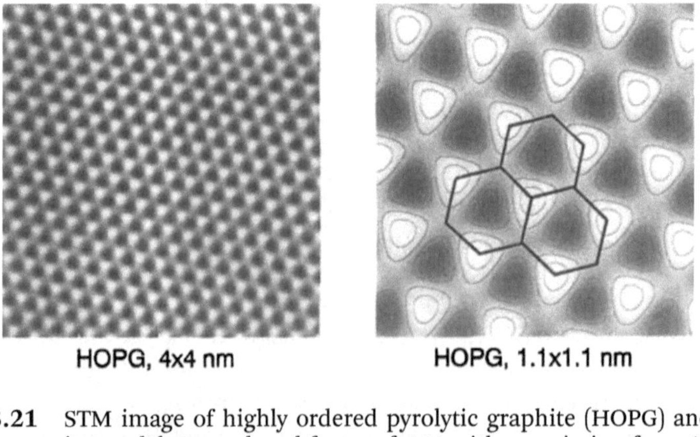

Figure 3.21 STM image of highly ordered pyrolytic graphite (HOPG) and graphite model. Reproduced from ref. 111 with permission from American Chemical Society, Copyright 1994.

quite as easy to predict, but it is clear from various STM studies of carbon black, that there are nanometre-sized zones of carbon atoms on the surface of carbon black with atomic order.

In 1992, Donnet and co-authors were the first to publish STM images of carbon black[112] and the same authors also conducted a thorough STM study of carbon fibres and plasma-treated carbon blacks.[113,114] Wang, Wolff and Freund[111] performed a comprehensive STM study of a wide range of carbon blacks of varied particle size and heat treatment. An STM image (Figure 3.22) for an N550 carbon black[111] shows these surface zones of atomic order, while the STM image (Figure 3.22) of a graphitised N550 carbon black[104] now shows that the surface layers or zones are indeed larger and with a much larger range of atomic graphitic order present on the surface as might be predicted from the XRD data as discussed previously. These results appear to indicate that STM is indeed generating atomic order images that are in alignment with other techniques concerning the characterisation of the carbon black microstructure and lamellas.

Additionally, in an STM evaluation of a series of carbon blacks of varied particle size (Figure 3.23) by Wang *et al.*,[111] the size of the surface layers or zones with atomic order grew larger as the particle size became larger, and is a different result than the average L_a determined from XRD as described previously, where the L_a, x–y dimensions, were found to become slightly smaller with increasing particle size (decreasing surface area). The apparent discrepancy may be at least partially explained by the fact that the STM is a truly surface technique, while the XRD is an average of the entire microstructure.

Additionally, a larger L_a at the surface may be rationalised from the various carbon black formation environments and the radius of curvature of the larger particle size carbon blacks that allows accommodation of larger lamellas, although this fact alone does not mean the lamellas have to be larger.

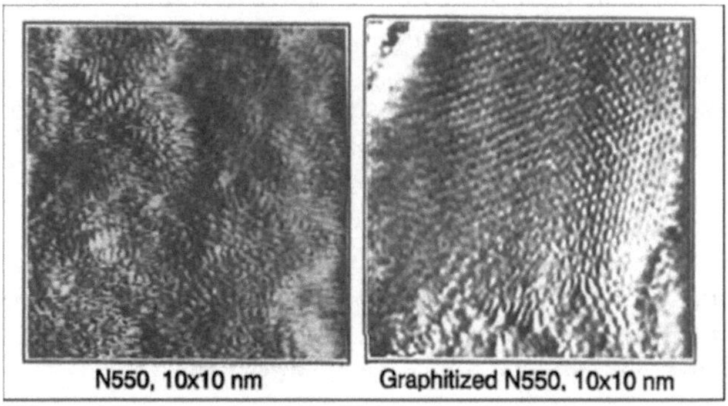

Figure 3.22 STM image of the surface of N550 carbon black in as-manufactured state and after graphitisation at 2700 °C under nitrogen. Reproduced from ref. 111 with permission from American Chemical Society, Copyright 1994.

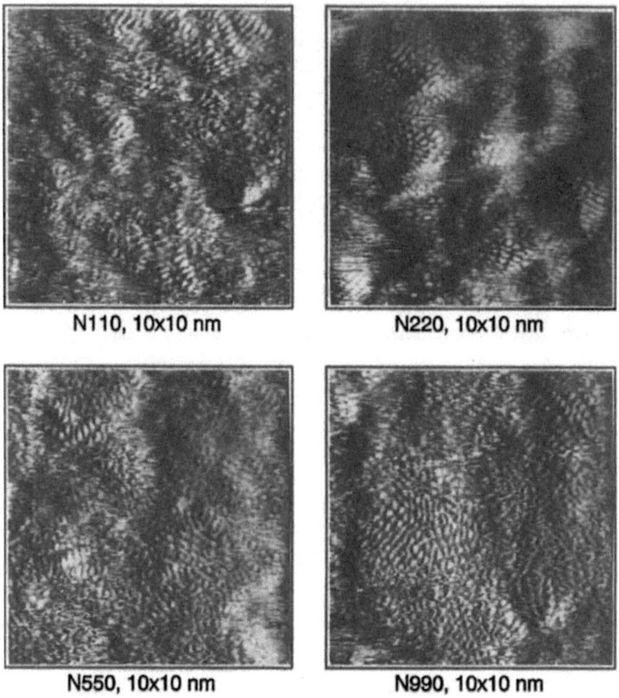

Figure 3.23 STM image of the surface of ASTM grades of carbon black of varied particle size. Reproduced from ref. 111 with permission from American Chemical Society, Copyright 1994.

In summary Wang, Wolff and Freund,[111] in a very thorough study of the STM imaging of various carbon blacks, of both ASTM products and graphitised versions of the same products, found the following:

1. The carbon black surface contains zones or domains of organised and unorganised carbon, with organised carbon dominating the surface coverage.
2. The size of the surface domains grows larger with larger particle size, which is *not* overall consistent with XRD measurements.
3. The surface zones of organised carbon grow larger with graphitisation, consistent with XRD measurements.
4. The carbon blacks of smaller particle size with smaller surface zones or domains of organised carbon and more unorganised carbon, display higher surface energies (as measured by inverse gas phase chromatography, IGC, which will be discussed later).

Overall, STM imaging of the carbon black surface proved useful in confirming ideas about lamella orientation and size and revealed very interesting information around the surface organisation of carbon black at the atomic and angstrom scale.

The STM uses electron tunnelling and current measurement as a way to image the surface structure of carbon black, which does not actually measure a signal based solely on topography, but rather the local electron density and tunnelling current, even if used in the constant current mode as a means to measure topography. Hence another scanning probe technique, atomic force microscopy (AFM), has been employed by other researchers at a later date in an attempt to directly measure the surface structure or surface topography.

W. Niedermeier *et al.*[115] were among the first to report results on the surface topography of the carbon black surface using AFM. They utilised the repulsive force portion of the signal, which is a short-range force of only a few ångströms, due to a localised overlapping force between the tip and several atoms in the localised area, and it is claimed to be able to resolve angstrom level surface structures. An image of the surface roughness of atoms of the mica substrate, used to mount the carbon black for AFM analysis, was shown as evidence of the atomic resolution and surface topography mapping.

The authors reported that the carbon black surface is comprised of 1–3 nm tilted domains, confirming the basic microstructural models devised from TEM high resolution imaging. Again, this surface roughness is highlighted as playing a significant role in the reinforcement potential of carbon black in elastomeric compounds. Several other papers on similar topics were published by others.[116–118] In fact, subsequent to these studies, a patent was filed utilising the concept of extended surface roughness to further increase the carbon black elastomer interaction and as a way to reduce networking and hysteresis in rubber even with a narrow distribution of carbon black.[119] Additionally, this combination of improved elastomer interaction with the normal aggregate size distribution (ASD), was reported to give improved treadwear, overcoming the traditional balance of lower hysteresis *via* broad ASD in conjunction with suppressed treadwear.

Finally, Klüppel[120] developed a model of the carbon black surface from the surface adsorption of ethene, which describes the surface in terms of its potential surface structures (graphitic planes, dislocations, amorphous carbon and pores) as related to their surface energy. This work predicts that at least four different surface structures of varied surface energy may be present on the carbon black surface.

These surface structures include class I, graphitic basal planes with low energy; class II, amorphous carbon also with low energy; class III, crystallite edges with relatively high surface energy; and class IV, slit-shaped cavities, with the highest surface energy. Overall, in the evaluation of a wide range of carbon blacks of varied particle size, the results showed that as particle size increased, the high energy sites decreased. In fact, the N550 and N990 showed essentially nil high-energy sites, and was more similar to graphite.

Beyond the carbon black microstructure, surface structure, and resulting surface energy, the carbon black morphology is one of the richest areas of research and development for new carbon blacks and forms the topic for the next section.

3.5 Carbon Black Morphology

The morphology of carbon black is perhaps one of the most significant fundamental properties that determines the overall performance of carbon black in a given system, whether it is used in a tyre, plastics, inks or coatings application. Of all fine particles that are currently manufactured on an industrial scale, carbon black is one of the most engineered in existence, where extremely fine control can be exerted over the carbon black morphology, generating numerous products with highly tailored properties to deliver the maximum value for an individual application. This aspect of carbon black is sometimes underappreciated by the general rubber, plastics, and inks and coatings industries due to the product maturity of carbon black; however, many carbon blacks with new performance signatures are still being developed today, regardless of the product maturity.

As described previously, the true, fundamental unit of carbon black, the carbon black aggregate (Figure 3.7), is composed of mostly spherical particles that are formed early in the carbon black formation process. Thus, when referring to the size of carbon black, these two fundamental properties, particle size distribution and aggregate size distribution, must be differentiated. Most fine particles are commonly composed of one general particle morphology and hence particle size refers to the size of the individual unit itself, but in carbon black technology, particle size and aggregate size have two distinctly different meanings and it is important to distinguish between them and discuss them as separate entities.

In addition to the particle and aggregate size distributional properties, the number and geometrical arrangement of the particles within an aggregate determine the structure or degree of branching within the aggregate, where aggregate mass distributions with more spherical or ellipsoidal shapes are considered lower structure and those with a higher number of branches are considered higher structure. The aggregate structure is also very important in determining final compound performance and physical properties through entrapment or occlusion of polymer, or other solid or liquid materials, within the voids of the aggregate branches, and indeed this aggregate structure is considered the third morphological distributional property of carbon black. Each of these morphological distributional properties of carbon black, particle size, aggregate size and aggregate structure distribution, are discussed below.

3.5.1 Carbon Black Particle Size Distribution

As already described, the carbon black particle is formed from lamellas that are oriented in an onion-skin fashion, forming a practically spherical carbonaceous particle, that then collides in the reactor with other particles, ultimately forming the carbon black aggregate. While the carbon black aggregate is the true, fundamental unit of carbon black, the particle size is the first morphological feature to be controlled in the carbon black reactor and it drives the final surface area of the carbon black, hence its importance in influencing carbon black technology and the applications in which it is used. Additionally, it is critical to have an understanding of the particle size statistical parameters describing its distributional nature. However, it should be noted that all aspects of carbon black, that is, its morphology, surface activity and porosity demonstrate a rather broad distributional nature, which is both a strength and a weakness of carbon black. In the end, these distributional characteristics make carbon black a very versatile fine particle and provide more advantages and benefits than disadvantages for the end user.

First and foremost, the particle size has a direct impact on the specific surface area (or surface area per unit weight) of the carbon black, which plays a significant role in the applications in which it is used, driving performance properties, ranging from abrasion resistance, to tensile strength and heat build-up in a rubber compound, to UV resistance, conductivity, and masstone (blackness or jetness) and undertone (blue or red hues) colour development in plastics, inks and coatings.

The particles or spherical entities that make up the carbon black aggregate are visually identifiable, yet at the same time there can be coalescence during formation due to their liquid nature early in the formation process, as well as due to additional deposition of lamella on the surface during the final aggregate formation steps. This makes the particle boundaries less precise, nonetheless it is important to be able to measure the particle size distributional properties of carbon black, which have proven to be challenging due to the coalescence described above and the sheer number of particles that must be measured within and between numerous aggregates.

Early measurements of the carbon black particle size were not achievable until the invention of the transmission electron microscope by Knoll and Ruska[121] in 1931. Until then only light microscope imaging of carbon black was possible, which made it impossible to gain a fundamental understanding of the true morphology of carbon black due to its nanoscale size. In an interesting side note, Ruska, along with the inventors of the scanning tunnelling microscope (as already described), Gerd Binning and Heinrich Rohrer, were awarded the Nobel Prize in 1986 for their pioneering work in development of the electron and scanning probe microscopes, respectively. Additionally, in 2018, Harris[122] published a comprehensive summary on the development and use of electron microscopy in analysis of carbonaceous materials that further highlights the progression of TEM analysis as applied to carbonaceous materials. In this review, it was noted that carbon black was indeed the first carbonaceous material analysed by TEM during the time period from 1944 to 1946,[123,124] however, internal publications of Columbian Carbon Company in 1940[125] and 1942,[126] followed by external publication of the same in 1942, by Wiegand and Ladd[127,128] of Columbian Carbon Company, were reported earlier.

The main finding from early TEM investigations was that, for the first time, one could measure the carbon black particles manually from TEM micrographs and count a sufficient number to begin to draw conclusions about the mean particle size, its distributional nature and carbon black surface area. Wiegand and Ladd[127,128] were the first to take advantage of the electron microscope in the early 1940s to manually measure the carbon black particle size from TEM micrographs and calculate its specific surface area based on the TEM-generated particle size distribution. For the first time, it allowed researchers the ability to understand the impact of carbon black particle size and surface area on in-rubber properties. This technique was of course very time consuming until the advent of semi-automatic measuring devices[129,130] and subsequently even faster methods utilising computers and image analysis on TEM photographic images.[131,132] Even with the relatively high level of automation at that time, this technique was a very time-consuming task and not practical until cameras were coupled to transmission electron microscopes and computer-aided or automated image analysis (AIA) was available.[133,134] In this instance more particles and more aggregates could be measured in a shorter time-period, making statistical evaluations more meaningful and making it possible to begin to characterise a wider range of carbon black products to provide deeper insight into carbon black morphology.

In measuring mean particle size, Hess and McDonald[135] pioneered a TEM/AIA method that allowed an automated method for measuring the particle size distributional properties of carbon black that was adopted as a standard ASTM method (ASTM D3849).[136] The method is based upon a mean-chord model that was solely developed for measuring the particle size distributional properties of carbon black.[135] This technique utilises a cellulose acetate butyrate (CAB) chip dispersion that breaks down the aggregate structure and creates aggregates more spherical in nature that makes the area (A) to perimeter (P) ratio, A/P, representative of the particle size and forms the basis of the methodology. Measuring particle size of carbon black from other dispersion forms, such as dry black or as-produced, or from a rubber compound, will distort the model algorithm and not be as reliable.

The equation for calculation of the particle size, d_p, from a single aggregate *via* the mean-chord model[135] and ASTM D3849-14a, is as follows:

$$d_p = \alpha(\pi A / P) \quad (3.4)$$

where A is the aggregate two-dimensional projected area; P is the perimeter of the aggregate two-dimensional projected area; $\pi A/P$ is the mean-chord of the two-dimensional projection of the carbon black aggregate.

The coefficient, α, is <1 and is the multiplier of the mean-chord since the mean-chord is representative of several particle diameters. Note that α is an empirical factor derived from the measurement of particles within an aggregate across a broad particle/aggregate size range.[135]

The number of particles in a given aggregate are required for calculation of the mean and weight mean particle size and it is determined by dividing the volume of the particle into the volume of the aggregate, as follows:[136]

$$\text{number of particles per aggregate} = V_a / V_p \quad (3.5)$$

where V_a = volume of an individual aggregate, (8/3)(A2/P); V_p = volume of an individual particle, $(\pi d_p^3/6)$.

Measurement of up to 2000 aggregates normally supplies a sufficient number of particles for statistically meaningful particle size distributional data and the statistical parameters for the mean (*M*), weight mean (WM), heterogeneity index (HI) and the electron microscope surface area (EMSA).[136]

Mean particle size, number average:

$$d_{\text{mean}} = \sum_{n}^{i=1}\left[f_i d_i / f_n\right] \tag{3.6}$$

Weight mean particle size, midpoint of weight distribution:

$$d_{\text{weightmean}} = \sum_{n}^{i=1}\left(f_i d_i^4\right) / \sum_{n}^{i=1}\left(f_i d_i^3\right) \tag{3.7}$$

The heterogeneity index, HI, indicator of the breadth of particle size distribution:

$$\text{HI} = d_{\text{weightmean}} / d_{\text{mean}} \tag{3.8}$$

The electron microscope surface area or EMSA is calculated as follows:

$$\text{EMSA} = 6000 / \left(d_{\text{surfacemean}} \times \rho\right) \tag{3.9}$$

Surface mean particle size, the size of the particle whose surface area, when multiplied by the total number of particles, gives the surface area of the material.

$$d_{\text{surfacemean}} = \sum_{n}^{i=1}\left(f_i d_i^3\right) / \sum_{n}^{i=1}\left(f_i d_i^2\right) \tag{3.10}$$

The term ρ is the density of carbon black taken as 1.8 g cm^{-3}.

The mean particle size and EMSA data for a wide range of carbon blacks was determined using the TEM/AIA mean-chord model (Table 3.5). Note the correlation of EMSA with NSA for these mostly non-porous carbon blacks, indicating the TEM/AIA method can provide a reasonable measurement of the carbon black particle size and its distributional nature.

Transmission electron micrographs (Figure 3.24) can give a visual indication of the differences in mean particle size amongst carbon blacks with widely varying mean particle size and surface area, and the particle size distributions from TEM/AIA can graphically show differences in size and distributional range and width (Figure 3.25). Note that the particle size range for N234, a high surface area product used in tyre tread compounds for good abrasion resistance, resides well below that for the number distribution for an N772 (Figure 3.25), a product used in non-tread or carcass compounds in a tyre, such as inner liners, where abrasion resistance is not important, rather higher compliance (flexibility) and heat build-up are desirable. This example demonstrates the degree of engineering that carbon blacks are subjected to in order to optimise its morphology for the many different compounds in a tyre, as well as the many other applications in which carbon black is used.

Table 3.5 Representative particle size distributional data for a wide range of carbon blacks. Reproduced/adapted from ref. 2 with permission from Marcel Dekker, Inc., Copyright 1993.

Carbon black	d_{mean}/nm	$d_{\text{weight mean}}$/nm	Heterogeneity index, HI	EMSA/m^2 g^{-1}	NSA/m^2 g^{-1}
N121	19	30	1.6	131	132
N234	20	31	1.6	124	125
N220	21	31	1.5	121	116
N339	26	40	1.6	95	95
N326	27	39	1.4	87	85
N550	56	94	1.7	41	41
N660	67	110	1.7	35	38
N762	107	168	1.6	24	26
N990	285	464	1.6	9	9

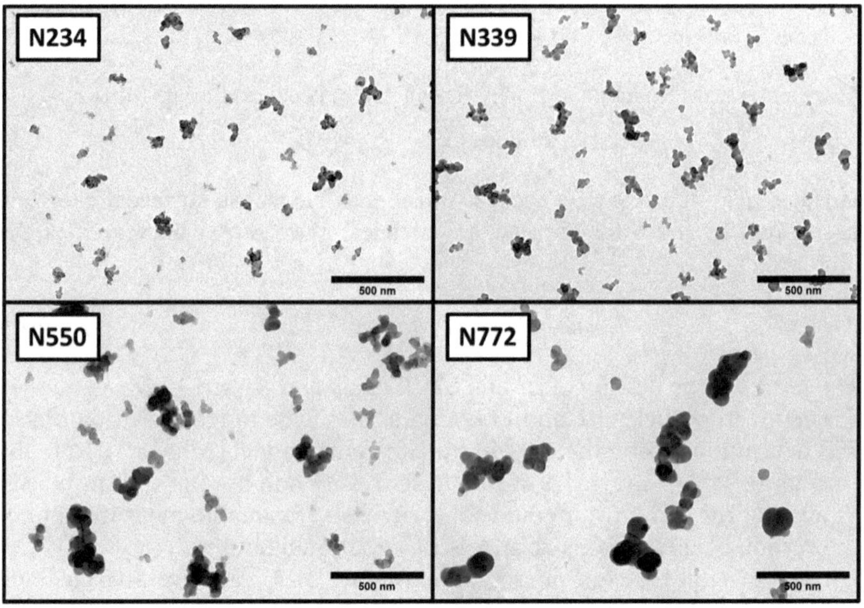

Figure 3.24 TEM micrographs of carbon blacks of widely varying particle size and surface area. Reproduced with permission from Birla Carbon USA, Inc.

3.5.2 Carbon Black Aggregate Size

As mentioned, the carbon black aggregate is the true, primary unit of carbon black, therefore characterisation of the carbon black aggregate size distribution is critical. First it provides an understanding of one of the fundamental properties of carbon black and, secondly, the knowledge gained provides insight into avenues to alter the aggregate size distribution for design and development of new products.

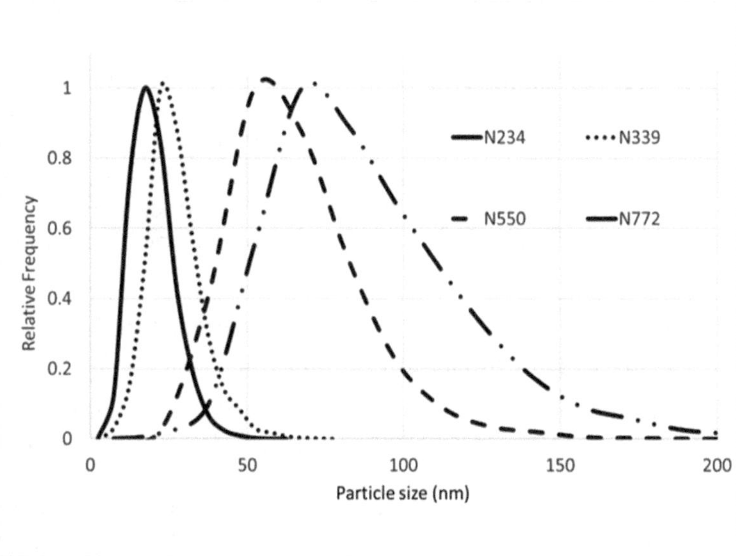

Figure 3.25 Particle size distributions for carbon blacks of widely varying particle size and surface area. Reproduced with permission from Birla Carbon USA, Inc.

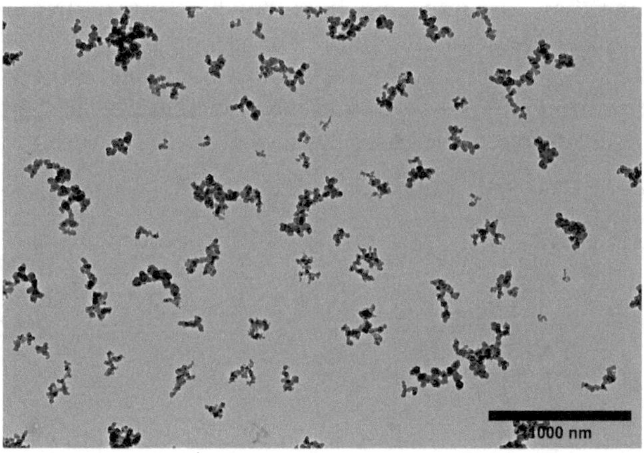

Figure 3.26 TEM micrograph of a collection of carbon black aggregates showing their diverse size and shape. Reproduced with permission from Birla Carbon USA, Inc.

Fundamentally it is difficult to accurately measure the size or diameter of a single carbon black aggregate simply due the fact that the individual aggregates are of varied structure, some simple and nearly spherical, ellipsoidal and even linear in nature, and some complex with a high degree of branching (Figure 3.26). Deciding upon a single measurement of its diameter is simply not possible; however, various techniques have been developed that provide ways in which to describe the aggregate 'size'.

The two most commonly employed techniques for measurement of the aggregate size distribution of carbon black are: disc centrifuge photosedimentometry, DCP, which utilises Stokes' Law[137] modified for centrifugal sedimentation; and the other is transmission electron microscopy with automated image analysis, TEM/AIA.[136] Recently a laser diffraction technique was also reported for evaluation of carbon black aggregate size distribution.[138]

3.5.2.1 Aggregate Size Distribution Measurement by Disc Centrifuge Photosedimentometry

In the DCP technique, (see ISO-certified method ISO 15825:2017),[139] a disc spins at a high rate of speed, 5000–20 000 rpm, and a spin fluid is added to the disc along with a small aliquot of carbon black dispersed using ultrasonication in a water–ethanol mixture with some amount of surfactant. As the carbon black elutes due to centrifugal forces, the larger aggregates are of course eluted first followed by the smaller aggregates, and the amount of aggregates in each size range is measured by signal intensity variation of an LED light due to scattering or absorption of the visible light by the carbon black, ultimately generating an aggregate size distribution (ASD) plot (Figure 3.27). The DCP technique uses Stokes' Law modified for centrifugal sedimentation[140] and assumes that the carbon black aggregates are spherical in nature. Typical ASDs for three different carbon blacks, N234, N330 and N660 (Figure 3.27), reveal that the ASDs for carbon black show a log-normal type of distribution and the mode and mean increase in size as the particle size of the three products increases (surface areas are lower). Structure also

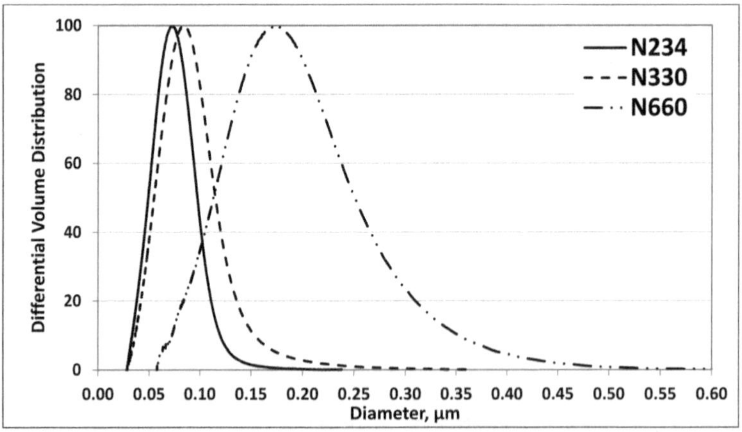

Figure 3.27 Carbon black aggregate size distributions for N234, N330 and N660 as measured by DCP. Reproduced with permission from Birla Carbon USA, Inc.

Table 3.6 DCP representative aggregate size distributional data for a wide range of carbon blacks (Birla Carbon USA, Inc.)

Carbon black	STSA[a]/m^2 g^{-1}	OAN[a]/cc per 100 g	Mean/nm	Mode/nm	ΔD50	ΔD50/ mode
N134	127	137	70	63	56	0.883
Broad ASD	129[b]	130[b]	79	53	79	1.50
N234	112	125	76	72	48	0.674
N220	106	114	83	77	58	0.758
N330	75	102	94	83	60	0.724
N326	72	76	85	79	57	0.722
N550	39	121	210	174	152	0.875
N660	34	90	199	175	137	0.788
N772	30	65	188	143	144	1.01
N990	8	43	513	355	333	0.937

[a]ASTM target values.
[b]Non-ASTM product.

plays a secondary role in determining the aggregate size *via* DCP and will be discussed later.

A wide range of ASTM-type carbon blacks with varied surface area and structure were measured *via* DCP (Table 3.6). Note this technique reports a mean, mode, ΔD50, and ΔD50/mode, among other parameters. In particular, the ΔD50 and ΔD50/mode represent the breadth of the aggregate size distribution. The ΔD50 value represents the width of the ASD peak at 50% half height or full width at half maximum, FWHM. In general, this data reflects the basic particle size/specific surface area and structure of the carbon black, where a larger mean and mode are measured as the specific surface area decreases, going from the N100 to N900 series products. In some cases, where surface area is similar, as for N330 to N326, the mean and mode are smaller for N326, mostly driven by the lower structure of the N326. Thus, when evaluating aggregate size data, both the particle size/specific surface area and the structure must be considered.

In terms of the ΔD50 and ΔD50/mode, the ΔD50 is of course again driven by size properties, that is particle size/specific surface area and structure, but when divided by and normalised by the mode, this parameter can serve as an indicator of the breadth of the aggregate size distribution. For the ASTM products below, the ΔD50/mode for most products ranges from about 0.65 to 0.80, which places these products in a narrow to normal ASD range typical of ASTM type products. Three of the products show a ΔD50/mode >0.85, which begins to place these products into a broader ASD category.

Carbon blacks with modified ASDs have been made in the lab or produced in pilot and commercial-scale reactors, mostly making broader ASDs to lower tyre compound hysteresis,[141,142] but in some cases, generating narrower ASDs for improved abrasion resistance.[143] In these instances, the DCP has been used to measure the ASDs for feedback in the development process, in-rubber validation of the design changes, and for publications of papers and filing of patents.

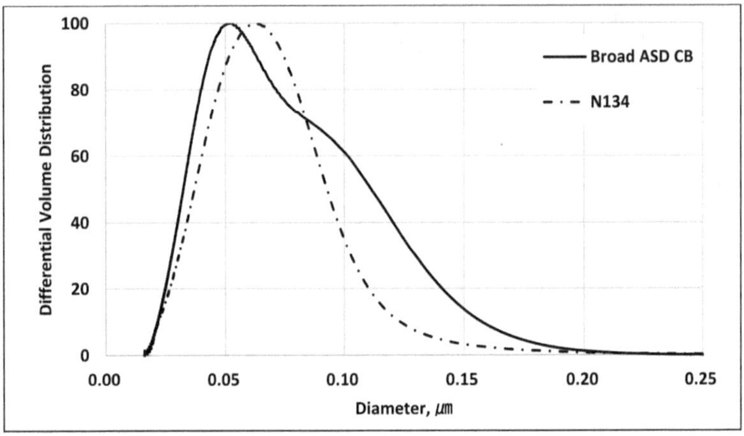

Figure 3.28 Broad ASD carbon black for reduced heat build-up and lower rolling resistance tyre tread compounds. Both carbon blacks have similar surface area and structure. Reproduced with permission from Birla Carbon USA, Inc.

The ASD plot for a broad ASD product normally shows a broadening, mostly toward the larger aggregate diameters (Figure 3.28), which can be accomplished through manipulation of the reactor process setup. Compared to an ASTM product (N134, Figure 3.28) of the same surface area and structure, the broad ASD carbon black has a broader ΔD50 and longer tail in the larger aggregate diameters. Again, such a variation may be undertaken to produce rubber compounds with differentiated behaviour, and in this case, the broader ASD carbon black will yield a rubber compound with lower heat build-up,[141,142] and provide for a tyre tread with lower rolling resistance and reduced CO_2 output. Note that these two carbon blacks have similar surface area and structure, so the ASDs indeed represent fundamental differences in the aggregate morphology and distribution.

The ability of broad ASD carbon blacks to lower heat build-up in a rubber compound is thought to occur due to diminished carbon black aggregate–aggregate contact or flocculation. In general, this aggregate–aggregate flocculation is known as networking, where the carbon black essentially forms a jammed network.[144] In practical terms, this behaviour was described by Payne,[145–147] in which low strain dynamic moduli are enhanced as a result of the jammed network acting as one large structure, while at higher strains, this jammed network is broken and the aggregates are well separated from one another. Such networking as described by Payne[145–147] can be quantified by calculating the $\Delta G'$ (dynamic elastic modulus) as the difference of G' at low strain (0.5–1%) minus the high strain (50% or more) G', with a larger difference indicating more significant networking and higher heat build-up and *vice versa*.

Broad ASD carbon blacks reduce the Payne effect ($\Delta G'$ low–high strain) since the network is not developed due to the larger average interaggregate spacing of broad ASD carbon blacks. In this case, the broad ASD reduces the low strain G', which is more influenced by the degree of networking, *versus* the high strain G', where the network is broken and now the impact of the carbon black morphology dominates, particularly the structure. Thus the basic hypothesis behind the broad ASD approach is that the hysteresis of carbon blacks in rubber compounds arises from breaking and re-formation of the carbon black network during application of dynamic stress and strain, as in a tyre tread during driving; hence by increasing the average interaggregate spacing *via* broadening of the ASD, this friction is reduced, and it lowers heat build-up in the rubber compound.

The hysteresis and heat build-up by carbon blacks is known to increase as surface area increases and indeed the interaggregate spacing can be shown to be inversely correlated to the carbon black surface area *via* the following equation developed by Wang, Wolff and Tan:[148]

$$\delta_{aa} = 6000 / \rho S \left(k\Phi^{-\frac{1}{3}} \beta^{-\frac{1}{3}} - 1 \right) \beta^{1.43} \tag{3.11}$$

where δ_{aa} = average aggregate distance between aggregates; ρ = density of carbon black; S = specific surface area of the reinforcing fine particle; k = constant (0.86, used for random packing); Φ = volume fraction of carbon black in the compound; β = ratio of effective volume fraction of carbon black aggregates to the volume fraction of carbon black in the compound.

This equation shows that the average interaggregate distance is primarily a function of the carbon black specific surface area, S, and the volume fraction, Φ, with structure (β) playing a minor role. In this instance, δ_{aa} is inversely proportional to surface area and volume fraction as higher levels of both indicate a higher number of aggregates per unit volume that leads to a jammed network and smaller average interaggregate spacing and hence higher hysteresis and heat build-up. This equation is of course aligned with practical observations of the impact of various carbon blacks on rubber compound heat build-up and hysteresis.[142]

3.5.2.2 *Aggregate Size and Structure Measurement by TEM/AIA*

In TEM/AIA measurement of aggregate size distributions, the aggregate diameter is simply taken as the diameter of a circle with an equivalent two-dimensional projected area of the individual carbon black aggregate. The procedure for this technique is described in detail in ASTM D3849[136] and in this instance, since the size of the true fundamental unit of carbon black is desired, the aggregate measurements are made for dispersions of carbon black in the dry state or as-manufactured state. TEM/AIA ASD plots as bar chart histograms typical for N234, and N660 (Figure 3.29) show these ASDs are similar as for the DCP and both techniques show carbon black to primarily follow a log-normal distribution.

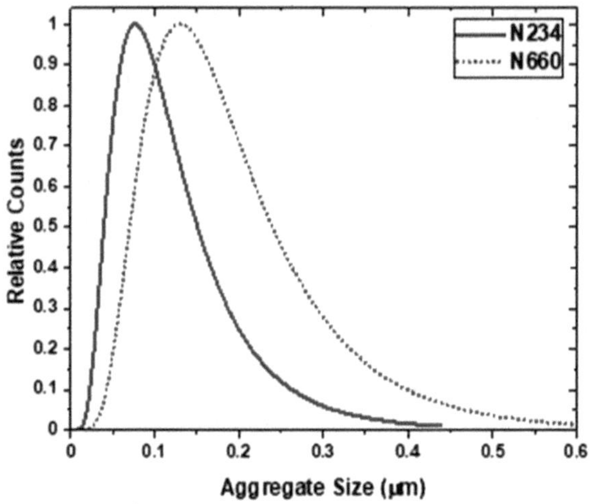

Figure 3.29 Aggregate size distributions for N234 and N660 as measured by TEM/AIA.

The disadvantages of the TEM/AIA technique are cost of equipment and maintenance, and specialised personnel to run these microscopes; however, the TEM/AIA methodology provides a significantly richer and deeper data set for describing and understanding carbon black morphology beyond a simple equivalent-circular diameter. These advanced methods for evaluation of carbon black morphology are reviewed in the sections that follow.

As mentioned, the TEM/AIA technique can be costly to implement, maintain and utilise daily, *versus* a DCP, but the TEM/AIA technique is key in providing additional insight into carbon black morphology. The additional morphological properties extractable from TEM/AIA data sets include: the particle size distribution, the aggregate structure distribution, the aggregate shape distribution, and the three-dimensional size and shape distribution *via* TEM tomographic reconstruction.

Several researchers have used the TEM/AIA technique to study carbon black to move beyond basic colloidal measurements and elucidate the true nature of carbon black morphology through image analysis and measurement of individual carbon black aggregates. Some of the first significant research attempts to describe the range of carbon black aggregate structures and shapes were undertaken in the late 1960s and early 1970s by Medalia[149–152] and by Hess.[153,154] These researchers went beyond simple size distributions and moved forward the concept of occlusion capacity and void volume of carbon black aggregates, as well as shape descriptors and shape classification.

The concept of carbon black aggregate occlusion capacity and aggregate void volume arises from the idea of carbon black structure, which results from an open architecture or void space between aggregate branches. It is within this void space or void volume, between aggregate branches, that vehicle systems, such as oils, elastomers, and plastics, can be occluded and essentially shielded from external stress. The net result is that the compound properties will be impacted as if the compound contains an apparent higher volume or higher effective volume fraction of carbon black. This higher effective volume fraction is most notable in higher structured products, for example when considering its impact on modulus development as discussed by Medalia.[155] This concept is very important in carbon black reinforcement as carbon black structure and occlusion capacity have a direct influence on numerous rubber compound properties, such as hardness and modulus,[155–157] tear strength and abrasion[158] (although to a lesser degree than specific surface area), and die swell.[159]

Medalia[155,159] determined an equation for calculation of the void volume for a given carbon black product based on its number of particles per aggregate and its TEM two-dimensional projected area, along with basic assumptions of the absorption of oil in the OAN test. The final equation is a shown below:

$$\frac{\text{Occluded Volume of Carbon Black}(\Phi' - \Phi)}{\text{Volume of Carbon Black}(\Phi)} = \frac{\text{OAN} - 21.5}{68.26} \tag{3.12}$$

where Φ' = volume fraction of carbon black plus occluded rubber; Φ = volume fraction of carbon black in the compound.

In this instance, the occlusion capacity represents the amount of oil, elastomer, or vehicle system that can be occluded by the collection of carbon black aggregates, taking into consideration that in addition to the intra-aggregate void volume, some additional void volume is due to inter-aggregate void space. Additionally, it has been considered that for correlation to rubber properties, the occlusion of elastomer is only partially effective.[159]

McDonald and Hess[160] in a study of carbon black morphology in rubber compounds, adopted the models of void volume developed by Medalia[155,159] to TEM/AIA measurements using the following calculation scheme to determine the void volume, V'_{agg}/V_{agg}, of an individual carbon black aggregate as follows:

$$\frac{V'_{agg}}{V_{agg}} = \frac{(V_{es} - V_{agg})}{V_{agg}} \tag{3.13}$$

where the volume of the aggregate (V_{agg}) and volume of the equivalent sphere (es), (V_{es}) are calculated as follows:

$$V_{agg} = (8/3)A^2/P \tag{3.14}$$

$$V_{es} = (4/3)\pi r^3 \tag{3.15}$$

where again A is the two-dimensional projected area of the aggregate and P is the two-dimensional projected perimeter of the aggregate as determined by TEM/AIA (Figure 3.30).

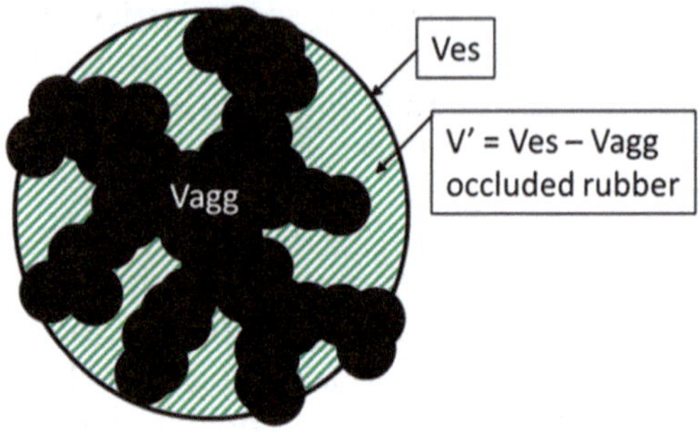

Figure 3.30 Void volume concept for an individual carbon black aggregate, showing the aggregate area, V_{agg}, and the equivalent circular diameter, V_{es}, the difference between the two representing the void volume, V'_{agg}, of the aggregate. Reproduced with permission from Birla Carbon USA, Inc.

As described, the void volume of an aggregate is an indication of its occlusion capacity where a void volume or V'_{agg}/V_{agg} value of 2 means that the individual aggregate can occlude up to twice its own volume. The V'_{agg} is calculated as the difference of the two-dimensional projected area of the aggregate as denoted by the shaded aggregate less the area of the equivalent circular diameter, taken as the average of the diameter of the equivalent aggregate area circular diameter and the average Feret diameter (Figure 3.30).

Once the V'_{agg}/V_{agg} for each individual aggregate is measured from a collection of aggregates for a given carbon black product as measured *via* TEM/AIA, then it is also possible to calculate the void volume for an aggregate distribution (as opposed to a single aggregate) to obtain a weighted average for a statistically meaningful void volume distribution. So again, if a weighted void volume for a given carbon black product is 2.3, it means this product can occlude up to 2.3 times its own volume.

Transmission electron micrographs of four N300 series carbon blacks (Figure 3.31), similar in surface area, that range in OAN (structure) from 72 to 150 cc per 100 g,[161] show that the aggregates move from a high degree of branching and open architecture with a large amount of void volume between aggregate branches in the high structure N358, to a progressively lesser amount of branching or structure in going from N351 to N330 to N326, where the aggregates become ellipsoidal, spherical, and linear in shape, with very little branching and available void volume. This is reflected in the OAN values for these products (Figure 3.31).

The void volume distribution for these four N300 series products (Figure 3.32) shows the mode and breadth of the void volume distributions decreases to smaller values as the OAN and structure decreases.[161] It is the ability of TEM/AIA to measure the individual aggregate size, structure and shape that provides the distributional construct of these parameters that gives the TEM/AIA technique an advantage over any other particle (aggregate) size measurement techniques known in the field of particle size measurement.

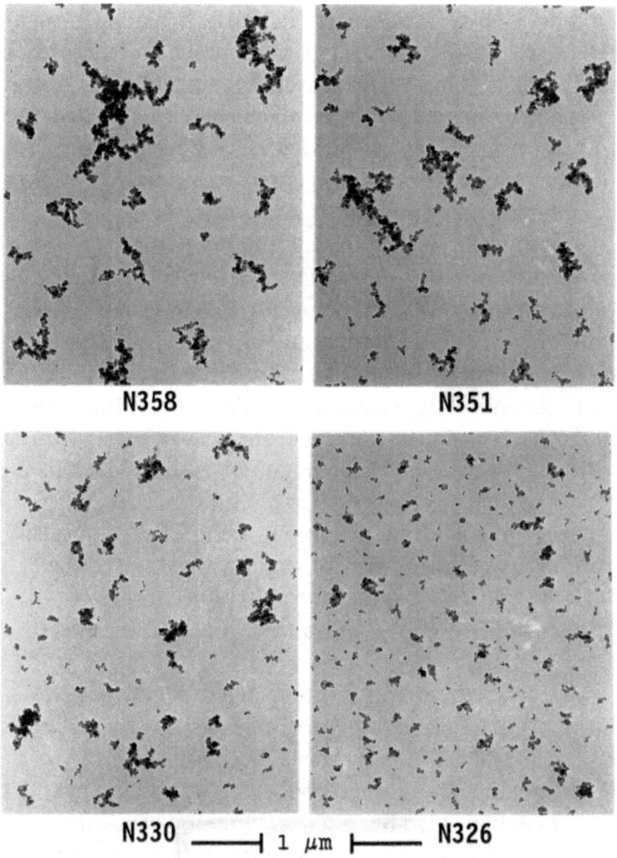

Figure 3.31 TEM micrographs of N300 series of similar surface but widely varying structure: N358 (OAN = 155 cc per 100 g); N351 (OAN = 120 cc per 100 g); N330 (OAN = 100 cc per 100 g); N326 (OAN = 72 cc per 100 g). Reproduced from ref. 153 with permission from American Chemical Society, Copyright 1994.

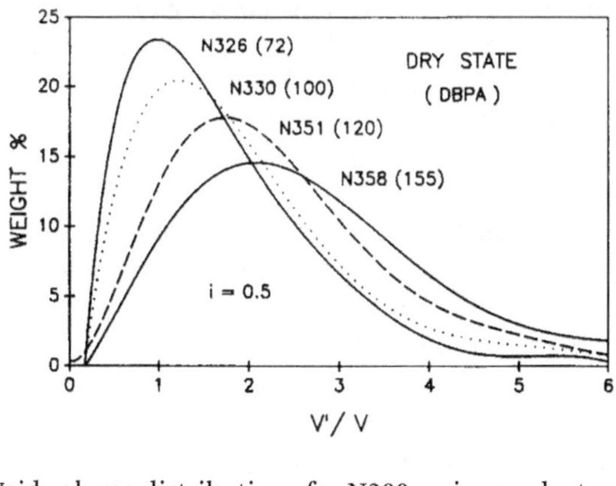

Figure 3.32 Void volume distributions for N300 series products over a wide range of structure. Reproduced from ref. 153 with permission from American Chemical Society, Copyright 1992.

Inherent in the idea of measuring the carbon black aggregate structure *via* TEM/AIA is that carbon black aggregates have a structure or shape distribution. It is clear that high-structured aggregates have a high degree of branching and occlusion capacity as discussed in the last section, while low-structured aggregates have a low degree of branching and low occlusion capacity. When viewing the range of structures in a carbon black aggregate distribution, one can imagine and observe a range of aggregate *shapes*, as already mentioned when describing aggregate structure. It was this observation, along with advances in computer power, automated image analysis and high-resolution camera technology for TEMs starting in the 1960s and 1970s, that has driven the advancement in the use of *shape* descriptors to move beyond simple size descriptors and provide deeper insight into carbon black aggregate morphology. This and the drive to relate carbon black morphology to the aggregate formation environment in the carbon black reactor and to improve the in-rubber performance of carbon black to meet the ever expanding performance envelopes of tyres and other rubber goods.

Over time several researchers developed various aggregate shape descriptors for use in shape classification of carbon black. Medalia[162] and Hess[163] were some of the first to consider various shape descriptors and Hess and co-workers[163] devised protocols *via* pattern recognition and discrimination analysis.

Some of the basic shape descriptors used by Medalia[162] and Hess[163] included number of particles per aggregate, anisometry, bulkiness and form factor among others (Table 3.7). Newer shape descriptors as described by Herd *et al.*[164] using advanced image analysis parameters, such as number of aggregate branches, have also been used for shape classification (Table 3.7).

Shape factors, such as anisometry and bulkiness, are related to fitted ellipses based on centre of gravity of the two-dimensional projection (Figure 3.33), where rounder, ellipsoidal-shaped aggregates fill the ellipse and have lower bulkiness factors and a lower degree of anisometry while higher structure aggregates with a more irregular shape have higher levels of bulkiness (more open architecture) and higher degree of anisometry.

More specifically, the spheroidal shaped aggregate on the left (Figure 3.33), based on the definition of bulkiness (Table 3.7), has a low bulkiness factor of 1.08 as it nearly fills the ellipsoid. Note by definition, a bulky aggregate, one mostly roundish with little branching, will have a low bulkiness factor. The same is true for the anisometry factor, where its definition (Table 3.7) means this more spherical aggregate will have a lower anisometry factor, and in this case, it is 1.37. Conversely, the aggregate on the far right (Figure 3.33) that is higher in structure and elongated, by definition (Table 3.7), has a higher bulkiness factor of 1.24, and since it is an elongated aggregate, its anisometry factor is also higher at 3.01. These results demonstrate how such shape parameters can be useful in describing the carbon black aggregate shape.

However one or two basic shape descriptors do not provide for meaningful shape classification, as the problem of shape classifying carbon black aggregates, and any other fine particle, is that while human beings can visualise

Table 3.7 Shape factors used for carbon black aggregate shape descriptors and shape classification[162–164]

Shape factor name	Symbol or factor	Function	Comments
Area[162,163]	A/nm	Size	Used in plots of shape factor *vs.* size to delineate size *vs.* shape
Number of particles per aggregate, number and weight[162]	$(N_p)_n$ and $(N_p)_w$	Assumed to be related to structure	The higher the N_p usually the more branched or complex the aggregate shape
Anisometry[162]	K_a/K_b	Separates, spherical, ellipsoidal and linear aggregate shapes	K_a and K_b are the larger and smaller radii of gyration of a radius equivalent ellipse as fit to the aggregate
Form factor, F[163]	L/W	Separates, spherical, ellipsoidal and linear aggregate shapes	Also an anisometry factor; in this instance L is the longest Feret and W is the shortest Feret diameter
Bulkiness, B[162] (area of ellipse/area of silhouette)	$(4\pi K_a K_b)/A$	Related to structure or branching	Higher bulkiness means spherical or ellipsoidal shape
Shortest Feret/particle size	W/d^2	Linearity	Good discriminator for linearity of aggregates
Longest Feret/particle size	L/d^2	Anisometry/branching	Higher values tend toward higher structure
Number of aggregate branches[164]	$N_{branches}$	Branches counted after skeletonisation	Skeletonisation is an image analysis technique that can quantitatively count the degree of branching
Irregularity factor[164] perimeter²/area	P^2/A	Related to structure and N_p	A higher value indicates higher structure

an aggregate or fine particle and 'classify' its shape in their mind based on shape factors, such as bulkiness, anisometry and degree of branching, and overall 'complexity' or lack thereof, it becomes much more difficult when image analysis parameters have to be applied in a computer algorithm to do the same. In fact, it requires a wide range of descriptors as listed in Table 3.7 and the application of sophisticated mathematical processes, such as pattern recognition or discrimination analysis, to categorise and classify aggregates into different and distinct shape categories. In the future it appears that artificial intelligence may provide a new and exciting pathway for fine particle shape classification.

As mentioned, McDonald *et al.*[163] and Herd *et al.*[164] developed strategies for shape classification of carbon black aggregates using pattern recognition and discrimination analysis. In the original work undertaken in shape classification, nine different shape categories were identified, but Herd *et al.*[164] simplified this scheme to four shape categories that included spheroidal, ellipsoidal, linear and branched aggregate shapes (Figure 3.34).

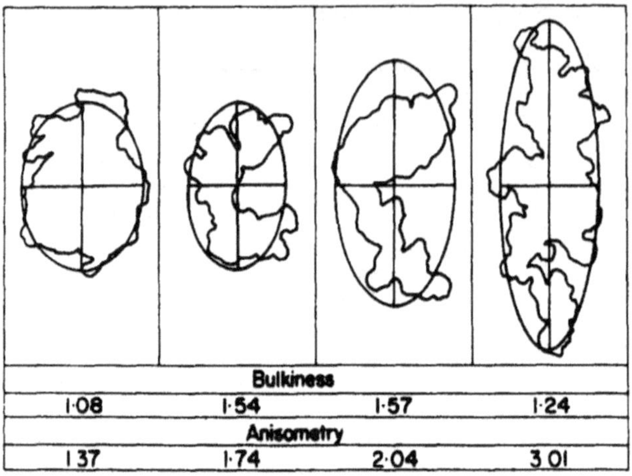

Figure 3.33 Bulkiness and anisometry shape descriptors. Reproduced from ref. 154 with permission from American Chemical Society, Copyright 1992.

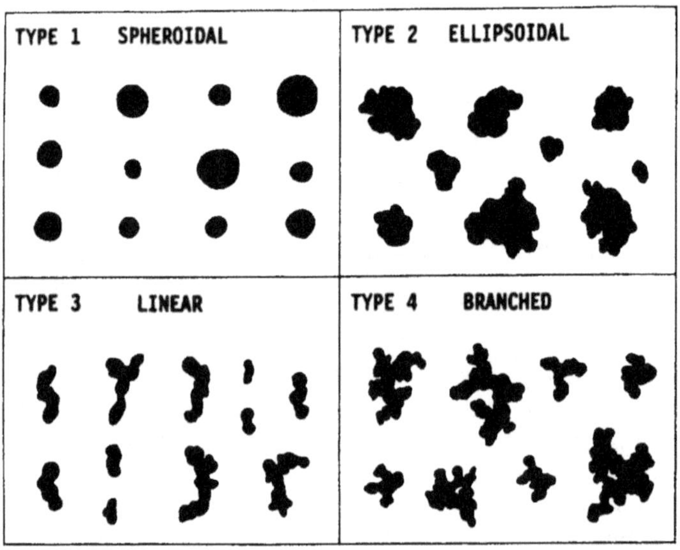

Figure 3.34 Four shape categories implemented for shape classification of carbon black aggregates. Reproduced from ref. 164 with permission from American Chemical Society, Copyright 1993.

Additionally, to provide an improved method of shape classification, an image analysis algorithm called skeletonisation was utilised and applied to shape classification of carbon black aggregates.[164] This technique can identify the branches of an aggregate and yield the count of the number of branches and even yield an indication of their magnitude (short, long). This technique

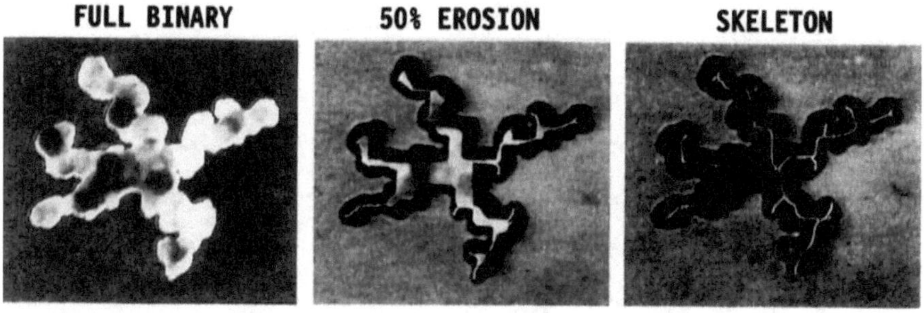

Figure 3.35 Skeletonisation progressive erosion of a single carbon black aggregate in going from a full binary image to its skeleton. Reproduced from ref. 164 with permission from American Chemical Society, Copyright 1993.

provides additional parameters not previously available through shape descriptors to enhance the shape classification of carbon black aggregates.

The skeletonisation process (Figure 3.35) utilises a binary image from TEM/AIA and erodes pixels from the boundary inward until a skeleton of the object is achieved. From the final skeleton the number of branch ends and hence number of branches can be determined.

This technique has proven sensitive in differentiating the degree of aggregate branching and can provide aggregate branch distributions for products of widely varying OAN, where it can be seen in a bivariate plot of number of aggregate branches *versus* OAN or structure (Figure 3.36), that as the structure of the carbon black product increases, so does the mode and breadth of the aggregate branch distribution. This result is consistent with plots of the aggregate void volume distributions described previously (Figure 3.32).

For shape classification, the number of branches in an aggregate (B), as well as B^2/L or number of branches divided by the longest Feret diameter, P^2/A and L/W (as described previously), among others, were used as shape descriptors in a discrimination analysis program to shape classify the distribution of carbon black aggregates within a given carbon black product for a wide range of tread and carcass products of varied OAN (Table 3.8).

As can be seen, as the OAN drops in going from BC2005 to N990, the weight% of branched (Type 4) aggregates decreases, but some reversal is seen for the carcass products, N550 and N660, and this is most likely due to differences in reaction time and collision dynamics (ballistic *versus* diffusion limited, as described by Schaefer and Hurd[165]). Also, as the OAN decreases, the ellipsoidal (Type 2) aggregates increase and reach their highest level for the very, very low structure N990, where the weight% of spherical (Type 1) and ellipsoidal (Type 2) aggregates, *i.e.* low structure aggregates with minimal branching, approaches 80% of the four possible shape types in N990.

Skeletonisation and shape classification has also been used to examine the impact on the carbon black shape types upon mixing in rubber and cellulose acetate butyrate (or CAB chips), for extensive aggregate breakdown and

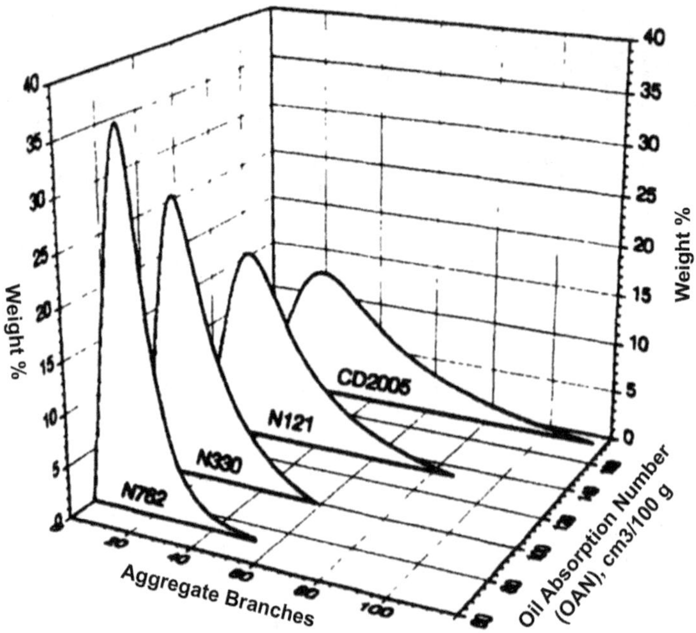

Figure 3.36 Aggregate branch distributions from skeletonisation for N762, N330, N121 and CD2005 showing progressively larger mode and broader branch distribution as OAN increases. Reproduced from ref. 164 with permission from American Chemical Society, Copyright 1993.

Table 3.8 Shape classification *via* skeletonisation of carbon black products[164]

Carbon black	OAN/cc per 100 g	NSA/m^2 g^{-1}	Weight% (Type 1) Spheroidal	Weight% (Type 2) Ellipsoidal	Weight% (Type 3) Linear	Weight% (Type 4) Branched
BC2005	174	120	0.1	4.8	17.8	77.3
N121	134	132	0.4	8.8	28.7	62.1
N550	120	39	0.6	13.8	45.3	40.3
N330	100	83	0.2	10.2	44.1	45.5
N660	91	36	0.4	15.4	52.5	31.7
N762	67	26	2.5	22.4	47.7	27.4
N990	35	9	44.9	34.8	14.4	5.9

improved dispersion on TEM grids. For example, it was observed that the changes in the aggregate shapes in the dry state or as-manufactured state *versus* those upon mixing in rubber and the CAB chip (Figure 3.37), that the branched Type 4 aggregates, that are susceptible to breakdown, decrease in amount, going from about 77% in the dry state to 53% and 43% in rubber and the CAB chip, respectively. Consequently, the branches broken from the branched Type 4 aggregates appear to increase the amount of linear Type 3 and ellipsoidal Type 2 aggregates, as might be envisioned.

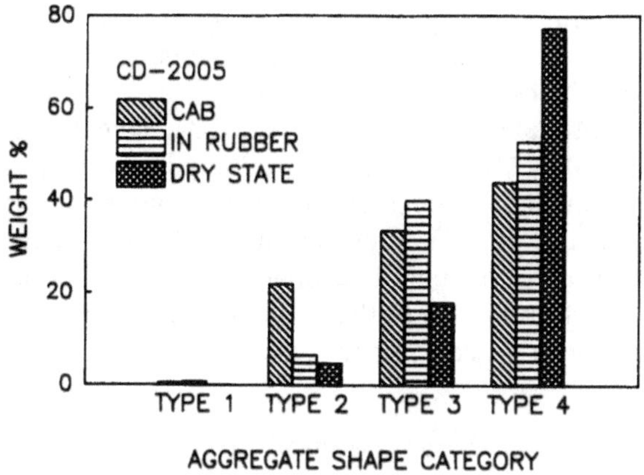

Figure 3.37 Aggregate branch distributions from skeletonisation for BC2005 in different forms showing changes in shape distribution as a function of breakdown. Reproduced from ref. 164 with permission from American Chemical Society, Copyright 1993.

These results show the usefulness of approaches, such as shape classification, in gaining a deeper understanding of carbon black morphology and how it might be impacted by aggregate formation mechanisms and its performance in rubber compounds, among many other applications.

The vast majority of the TEM/AIA research conducted in the past 50 years has been on TEM two-dimensional images. Such studies provided great insight into carbon black morphology, as well as all other fine particles studied in a similar manner; however, inclusion of the third dimension will provide a more complete description of the morphology of fine particles. The following section describes results from past and present studies on the three-dimensional structure of carbon black aggregates.

3.5.3 Three-dimensional Characterisation of Carbon Black Aggregates

All of the early research conducted on the size and shape of carbon black aggregates involved TEM two-dimensional projected areas or images of carbon black aggregates and missing, of course, is the third dimension or three-dimensional spatial arrangement of the aggregate. It might be considered that the third dimension is the aggregate height in the *z*-direction, with the *x*–*y* plane representing the two-dimensional projection. This fact, *i.e.* the lack of the third dimension in TEM/AIA morphological studies, was often discussed in the various papers published by researchers conducting the size and shape analyses on carbon black aggregates. Obviously, the motivation for three-dimensional characterisation of carbon black aggregates is to not only provide a visualisation of its true structure, but a true quantitative measurement of its size, surface area, volume, void volume, and shape.

Certainly, a common stereoscopic technique was used in the past in various labs to be able to visualise the three-dimensional structure of carbon black and carbon black in rubber matrices, using two TEM or SEM images taken at different tilt angles and visualised using a stereoscopic viewer. This technique was used by Mercer *et al.*[166] to not only view the three-dimensional character of carbon black aggregates, but also to construct models using theory from Boyde *et al.*[167,168] This method was used to calculate the heights of two points as a result of the parallax between two points as normally viewed by the eyes, giving human beings the ability to see in three-dimensions. The equation used from this concept to determine differences in height is as follows:

$$k = p/(2M\sin\Delta\Phi) \tag{3.16}$$

where k = the difference in height; p = the parallax; M = the magnification; $\Delta\Phi$ = half of the tilt angle.

Mercer was able to reconstruct aggregate models using this technique and determine aggregate volumes from three-dimensional models and found that volumes using two-dimensional projected areas and heights was significantly larger than the model aggregates generated using this parallax construct.

Later, other researchers, initiated studies into the three-dimensional nature of carbon black[169–171] and devised techniques for generating computer simulated three-dimensional models that could then be evaluated in detail using image analysis techniques and shape parameters, but now in three-dimensions. First and foremost, the visual appearance of these three-dimensional models is impactful and conveys significant information about the carbon black aggregate three-dimensionality, with one of the major findings that carbon black aggregates are very anisotropic in their volume spatial distribution, generally shorter in the *z* direction than the *x*–*y* dimension.

Gruber *et al.*[169,170] analysed 84 aggregates of N220 and 20 aggregates of N660 using a trigonometric method to determine the spatial distribution of particles within an aggregate relative to a reference particle on the rotation axis. In this way, three-dimensional models could be reconstructed to allow viewing of the three-dimensional aggregate at various tilt angles and orientations. The authors defined a flatness index, I_f, where I_f is defined as follows:

$$I_f = 2T/(L+W) \tag{3.17}$$

where T = the greatest thickness in the normal direction of the longest dimension; L = longest in-plane (x–y) dimension; W = the aggregate dimension perpendicular to L.

The authors found the aggregates to be relatively 'flat' with I_f values ranging from ~0.2 to ~0.5 for N220 aggregates and from ~0.4 to ~0.7 for N660 aggregates. A value of 1.0 would indicate a completely isotropic aggregate.

Later, Xu *et al.*[171] performed three-dimensional characterisation of carbon black aggregates using atomic force microscopy across a wide range of ASTM grades and an experimental product. In this instance the authors also used a flatness index, I_f, adjusted for the AFM tip size, which measures particles broader in the *x*–*y* dimension than their height or *z* dimension. Again, the authors found carbon black aggregates of various grades to be relatively flat. An experimental grade was briefly described as being made on an experimental reactor that was isotropic in nature. Both of these studies confirmed the anisotropic nature of carbon black aggregates with a *z*-direction of lesser magnitude than the *x*–*y* dimension, considering the *x*–*y* dimension as the plane on the surface of the TEM grid support of the carbon black during TEM or AFM imaging. These results also confirmed that the broadest *x*–*y* dimension was almost always parallel to the supporting surface, meaning that two-dimensional imaging in the TEM or AFM would provide the most informative view concerning the structure level or degree of aggregate branching and aggregate occlusion capacity.

Additional studies of the three-dimensional character of carbon black aggregates were carried out by Gruber and Herd,[172,] and Gruber,[173] where a range of aggregate morphological parameters were measured on N300 and N600 series products, measuring approximately 2000 aggregates to obtain the three-dimensional distributional variation of fundamental dimensional parameters. These studies again confirmed the anisotropic nature of carbon black aggregates and that the widest dimension is normally the *x*–*y* plane parallel to the TEM grid support surface.

The variation in the two-dimensional aggregate projection for a high structured carbon black aggregate as a function of the TEM goniometer tilt angle, shows that the two-dimensional projected area and perimeter are at a maximum at 0° tilt angle (Figure 3.38), and decrease similarly in rotation of the TEM goniometer between 0° and ±60°. From the actual aggregate tilt series, a three-dimensional model was reconstructed using the method of Gruber *et al.*[169,170] The three-dimensional model was then used to view the aggregate from 0° to ±90° in 10° increments and compared visually and quantitatively to the two-dimensional aggregate area and perimeter projections. The results showed that the changes in the aggregate area and perimeter moved in a very similar manner between the model and the actual aggregate as measured in the TEM two-dimensional projection. Note how the anisotropic nature of the carbon black can be seen at the 90° angle from the model, where the change in the two-dimensional projected area and perimeter reduce by approximately −40% and −60%, respectively.

It is also instructive to view the four basic shape types in three dimensions[172] using the same techniques described above in Gruber *et al.*[169–171] (Figures 3.39 and 3.40). The three-dimensional aggregate images, both real and from a model, as a function of the TEM goniometer rotation angle, for a spheroidal Type 2 aggregate and a linear Type 3 aggregate, reveal that their three-dimensional morphology confirms their shape classification.

The above results confirm that the largest dimension of carbon black aggregates is measured at the 0° projection and that models generated using the technique of Gruber *et al.*[169–171] indeed reproduce the actual three-dimensional construct of carbon black aggregates and that the aggregates

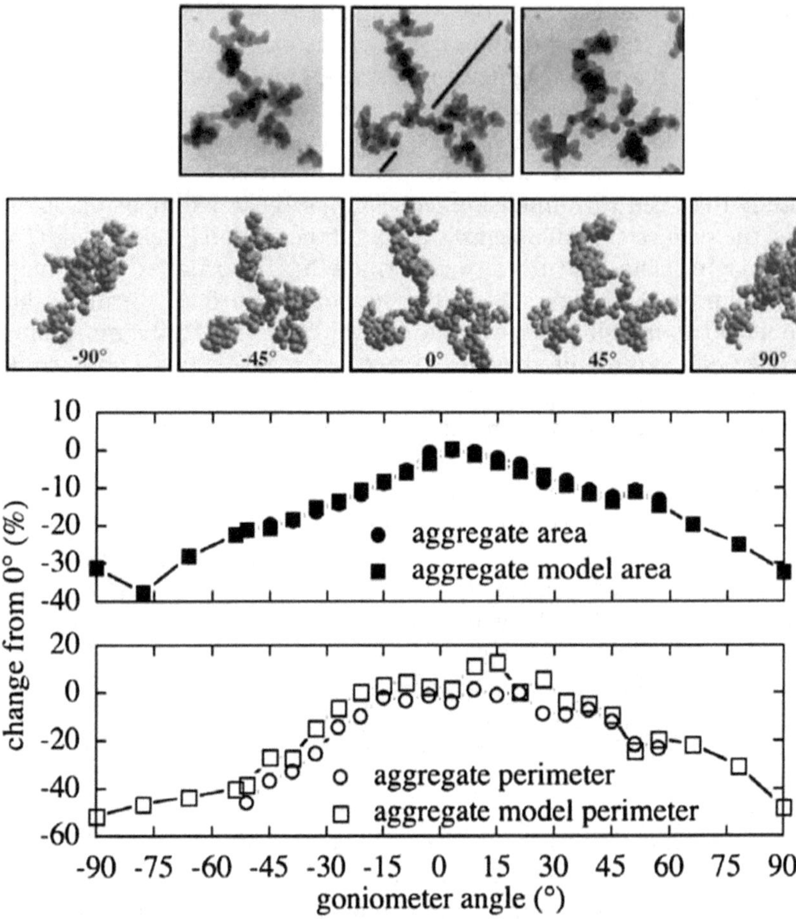

Figure 3.38 Type 4 branched aggregate area and perimeter of two-dimensional projection measured at various TEM goniometer tilt angles (0° to ±60° every 5°). Line in top middle image indicates axis of rotation in the TEM goniometer. Reproduced from ref. 172 with permission from American Chemical Society, Copyright 1997.

are anisotropic in nature, being shorter in the *z*-direction compared to the 0° TEM projection or its *x*–*y* dimension.

The above techniques are time consuming as each particle in the aggregate has to be tracked and identified in the images, but another technique, electron tomography, has been utilised to generate three-dimensional images of carbon black aggregates, where only the aggregates themselves need to be tracked as a function of the TEM goniometer tilt angle. The next section describes the past, present and future initiatives in the use of electron tomography for the three-dimensional characterisation of carbon black aggregates. This technique has generally been applied for characterisation

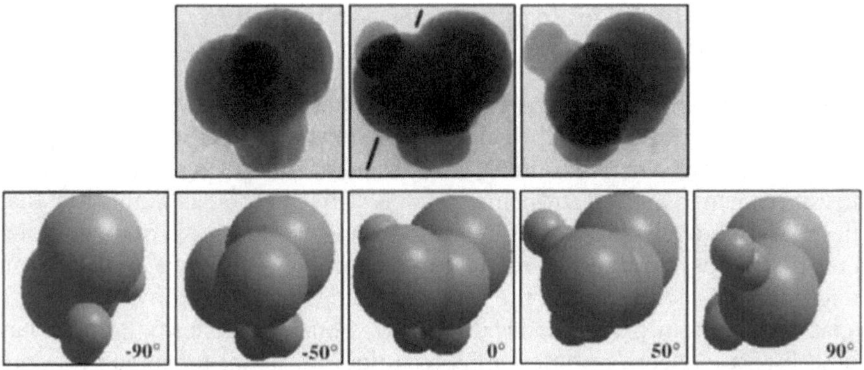

Figure 3.39 Type 2 ellipsoidal aggregate shape of two-dimensional projection displayed at various TEM goniometer tilt angles (0° to ±60° every 5°). Line in top middle image indicates axis of rotation in the TEM goniometer. Reproduced from ref. 172 with permission from American Chemical Society, Copyright 1997.

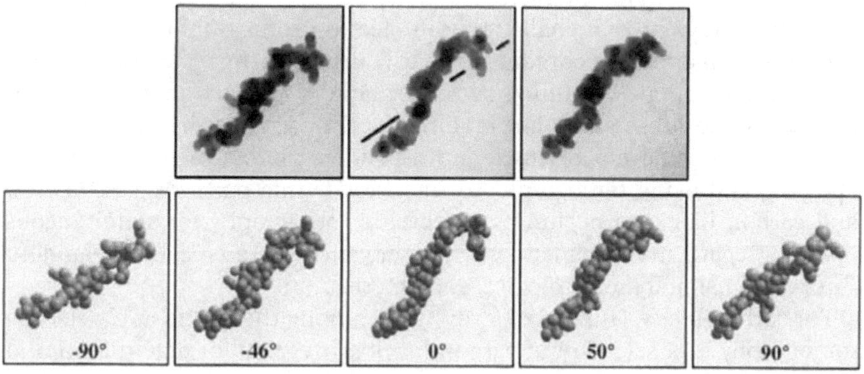

Figure 3.40 Type 3 linear aggregate shape of two-dimensional projection displayed at various TEM goniometer tilt angles (0° to ±60° every 5°). Line in top middle image indicates axis of rotation in the TEM goniometer. Reproduced from ref. 172 with permission from American Chemical Society, Copyright 1997.

of a single fine particle for each tilt series, but electron tomography needs to be applied to the measurement of multiple aggregates at a time to greatly increase the tomographic throughput for development of data sets descriptive of the aggregate distributions for individual carbon blacks. In this way, a fuller, richer and complete description for carbon black morphology can finally be obtained, making these data sets useful for determination of truly three-dimensional quantitative morphological parameters and for use in models of carbon black reinforcement in rubber compounds at the aggregate level never before possible.

3.5.4 Electron Tomographic Reconstruction of Carbon Black Aggregates

Electron tomographic reconstruction of materials and biological entities is a technique utilised for the three-dimensional reconstruction of objects, as derived from a series of images taken at various goniometer tilt angles in the transmission electron microscope. This method is similar to that described above where researchers used images taken at various goniometer tilt angles, but the reconstruction technique is different and relies on alignment of each image in the tilt series and transformation of the image spatial domain into a frequency domain *via* Fourier transform. Frank[174] has described the electron tomography technique in detail for biological entities for examining three-dimensional structures in the cell. Obviously, the goal of any three-dimensional reconstruction technique is to yield a realistic, three-dimensional structure for a more thorough and representative description of a given material or fine particle as opposed to its two-dimensional projection. Hence the electron tomography approach has merit to be developed using a statistically significant number of carbon black aggregates to yield a meaningful characterisation of its size and shape distribution. This has always proven challenging in electron tomography as it is also a very time-consuming technique, primarily driven by image capture time at numerous tilt angles, defining suitable markers for each image in the tilt series, and the subsequent time required for the computer-based reconstruction. Yet strides have been made in first understanding the applicability of applying and using tomography to characterise fine particles, such as soot and carbon black, to beginning to develop tomography for simultaneous, multiaggregate measurement, thereby advancing statistically meaningful three-dimensional morphological characterisation.

For carbon black, Hunt and Herd[175] were among the first to apply electron tomography to a selection of carbon black aggregates for determination of their three-dimensional surface area and volume, where it was found there was good agreement between the two-dimensional and three-dimensional surface area and volume measurements; however, the degree of thresholding of the various images did impact the measured surface area and volumes. This result was different than that found by Mercer,[166] as described above, for SEM evaluation and use of stereology, where differences greater than 50% in these values were found.

Also, from the study conducted by Hunt and Herd,[175] the grey-scale image of the reconstructed carbon black aggregate was sliced to show the general three-dimensional arrangement of the aggregate mass and particles, for a more complex branched, Type 4 aggregate (Figure 3.41).

Such a series of images shows the three-dimensional complexity of the carbon black aggregate and aids in the visualisation of the degree of interconnectedness or lack thereof of the various particles that form the aggregate, which is due to the inherent anisometry of the carbon black aggregates and the high number of particles and branches that compose the aggregate.

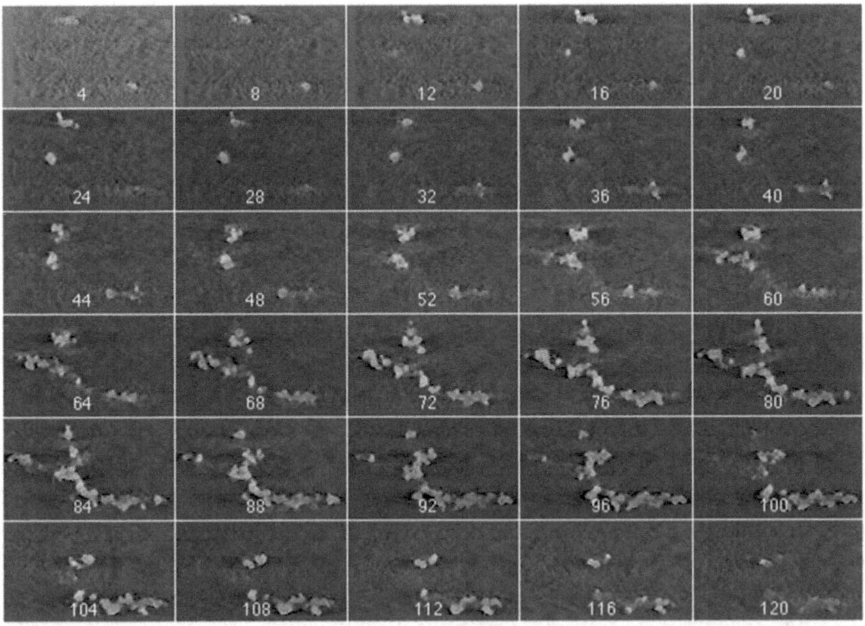

Figure 3.41 A complex branched Type 4 aggregate sliced to show the spatial distribution of mass and particles forming the aggregate structure. Reproduced with permission from Birla Carbon USA.

Orhan *et al.*[176] studied soot particles, from flame-generated soot and soot-in-oil, using electron tomography (transmission) and found deviations in volume for flame-generated soot as high as 38% compared to two-dimensional measurements, yet for soot-in-oil, the deviation was much lower at only 4% deviation. The differences were explained in the overall complexity of the two different sources of soot where the flame-generated soot had a more complex and three-dimensional or isotropic morphology, while the soot-in-oil was more two-dimensional and highly anisotropic in its mass distribution. All of these early studies showed the importance of advancing three-dimensional aggregate morphological characterisation for a better and more in-depth understanding of the aggregate morphology. These advances reveal insight into aggregate formation in many different temperature, pressure and chemical environments, as well as providing better aggregate structures for modelling and understanding such phenomena as rubber reinforcement and rheology.

As mentioned, moving toward statistically meaningful three-dimensional aggregate characterisation requires sampling of numerous soot or carbon black aggregates at a more rapid pace, without sacrificing the quality of the reconstruction. Hafiner-Staton *et al.*[177] conducted research into methods for increasing the throughput of tomographic analysis of soot and carbon black aggregates. They evaluated three different approaches for improving throughput that included the type of tomographic reconstruction program, the optimum number of images from a tomographic series and the interpolation range. They found through visual and quantitative comparison of these various factors that a range of ±60° with an angle increment of 5° or a total of 41 images, along with the use of a weighted-back-projection algorithm that takes less than one minute of computational time, that they were able to get very similar results (within about 3 to 4%) for the measurement of aggregate volume, surface area, mass fractal dimension, and radius of gyration *versus* the most thorough tomographic reconstruction method (±80° and angle increments of 1°, with use of a different algorithm that involves sequential interpolation and takes 5 to 6 hours to complete a single reconstruction). Overall, they were able to reduce the time to reconstruct a single aggregate with acceptable quality by ~70% (or a total time of ~1.5 hours). It should be noted that they identified the most difficult barriers to good quality tomographic reconstruction which were: use of the widest tilt range possible to eliminate elongation of features in the *z* direction, difficulty of discerning the mass at the interior of an aggregate (usually leads to an over-estimation of volume), use of fiducial markers, and identification of the best segmentation or thresholding boundary (due to the low density of carbon-based materials in the TEM).

Other researchers, namely, Okyay and co-workers,[178] also examined the three-dimensional geometry of soot aggregates using scanning electron microscopy (SEM), developing a tomographic technique to increase the speed of tomographic reconstruction.

Expanding this concept, Gruber and co-workers[179] significantly increased the throughput of electron tomography as applied to carbon black aggregates, developing a technique that can gather data on, and reconstruct, three-dimensional structures of up to 30 to 50 aggregates per image field, moving significantly beyond acquiring the tomographic reconstruction of a single aggregate per tilt series. This approach will make tomographic reconstruction and morphological analysis of up to 250 aggregates in 40 hours of attended time possible and yield statistically meaningful distributional data for three-dimensional aggregate reconstructions. Shown below in Figure 3.42 is an image field containing 63 carbon black aggregates with the aggregates suitable for tomographic reconstruction (no overlap at high tilt angles) circled in the image field.

This technique was also developed such that fiducial markers were not required, by utilising pattern recognition modules in ImageJ where, across tilt angle increments of 1–10°, any one aggregate can be identified in each successive tilt angle. Secondly, this technique has been devised to use a particle-based reconstruction algorithm instead of a voxel-based reconstruction algorithm, as described in the various papers referenced in this section. The particle-based method works by first determining the number of particles in an aggregate. This is done by determining the aggregate particle size using the mean-chord model as described previously,[135] and then dividing the aggregate volume from two-dimensional projections as $8/3\ (A^2/P)$ (Janzen and Goodarz-Nia)[180] by the aggregate particle size. These particles are then

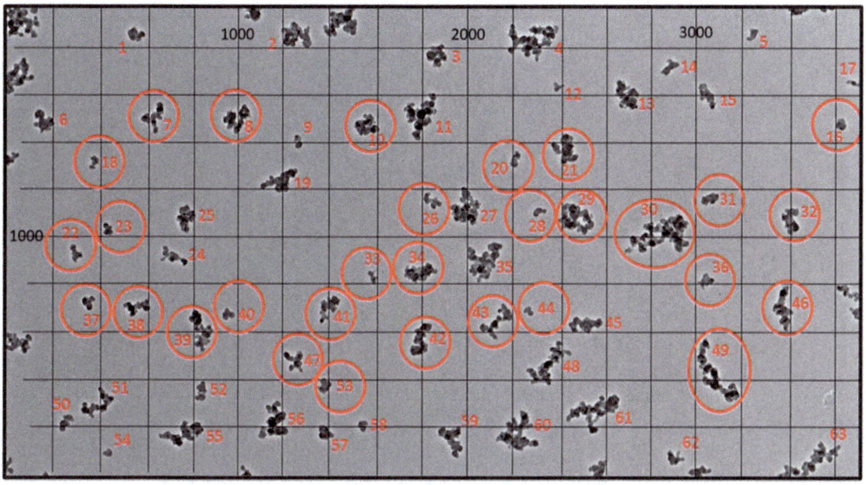

Figure 3.42 Field of N330 carbon black aggregates used for generation of tomographic image series from ±70° using JEOL TEM at 120 kV on a 2672 × 4008 pixel camera.

used to form the general shape of the aggregate in the various two-dimensional projections, adjusting the particle diameter and x, y and z locations in three-dimensional space in an iterative process until the aggregate shape is reproduced for all tilt angles. The resulting three-dimensional reconstruction is then overlaid as a binary image on each two-dimensional projection at the various tilt angles of the binary image of the actual aggregate and compared. The differences in shape are counted as pixels and termed the cost function. This cost function is minimised to provide the most accurate reconstruction.

Three different images of the same carbon black aggregate that were generated using different algorithms are compared (Figure 3.43): first is the TEM two-dimensional phase-contrast image, second is the two-dimensional image slice of the tomographic reconstruction of the aforementioned aggregate, performed using a conventional voxel-based reconstruction algorithm, and the third is the particle-based tomographic reconstruction. Both the voxel-based and particle-based reconstructions qualitatively appear to give a good match to the two-dimensional projection, but the voxel-based method does suffer from segmentation or threshold accuracy, leaving some parts of the aggregate weak in their density and hence overall reconstruction. The particle-based method is relatively new and further refinement is currently being undertaken at the time of writing and one area of improvement is the resolution of the interior volume of the aggregate, as described previously as an area of improvement for tomographic reconstruction.[177]

Of interest of course is how well the particle-based tomographic reconstruction method compares to the morphological properties, such as volume and surface area, of the two-dimensional projection. Comparing the results

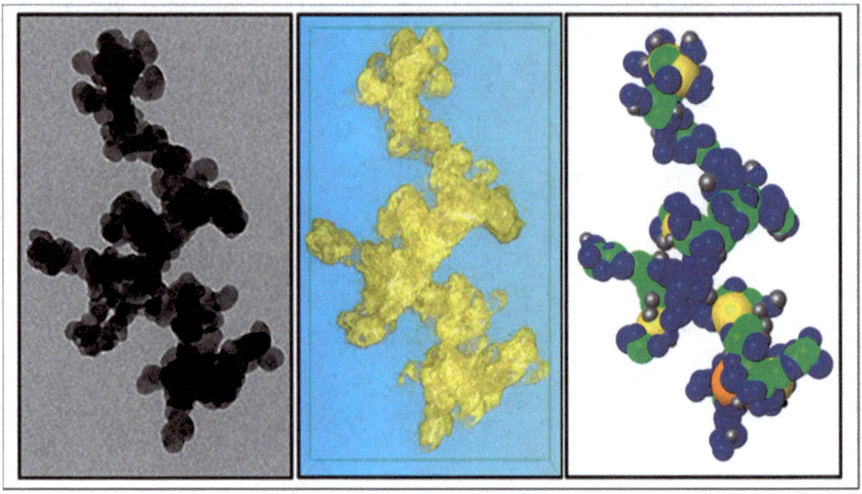

Figure 3.43 Comparison of TEM phase contrast image of a carbon black aggregate (left), voxel-based reconstruction method (centre), and particle-based reconstruction method (right).

of the particle-based three-dimensional volume and the two-dimensional projection volume (as $8/3(A^2/P)$) showed that for the smaller volumes, for aggregates generally lower in structure, the agreement is well aligned, but for larger volumes, which may be higher structure and more anisotropic, especially in the z axis, the differences in volume become larger,[179] as has been reported previously.[176]

The realm of morphological characterisation of fine particles and of carbon black in particular, has yielded a very rich and useful basis of knowledge that drives progress from understanding aggregate formation mechanisms to rubber reinforcement along with information for development of new morphologies; however, the surface activity of carbon blacks that drives the interaction of carbon black with its surroundings, whether it is an elastomer, plastic, resin or curatives, is equally important and discussed in the next section.

3.6 Surface Activity of Carbon Black

The surface activity of carbon black (and the numerous other forms of carbon) is a very broad topic and has been studied in great detail over the history of carbon black.[1–4] As described previously, the carbon black surface appears to be composed of overlapping layers of lamella arranged in a tile-like fashion with exposed edge sites. Fortunately, the surface, as produced, has a relatively high number of edge sites and these exposed lamella edges and edge sites seemingly play a large role in driving reinforcement of rubber (Donnet *et al.*[112–114] and Robertson[77]), and the interaction with any organic vehicle system in which carbon black is dispersed. Overall, the carbon black surface is primarily hydrophobic and organophilic in nature making it very compatible with elastomeric systems as a reinforcing fine particle.

As already discussed, the carbon black lamellas are mobile and carbon black can be graphitised upon heating under an inert atmosphere. In this way, the number of edge sites can be lowered through graphitisation, essentially creating a surface with properties more similar to graphite, and even more hydrophobic, resulting in poor interaction with elastomers and other organic vehicle systems. This 'negative result' appears to lend credence to the idea that carbon black edge sites at the surface are important in carbon black reinforcement in rubber and with other organic-based vehicle systems.

In another approach to alter the carbon black surface activity, functional groups *via* oxidation (treatment with ozone, nitric acid, or naturally in a flame as with the lampblack and impingement and gas black process) can be added to the surface, making it significantly more hydrophilic and acidic in nature. Thus, it might be considered that carbon black can have three basic surface characteristics and chemistry, which include as-produced, oxidised, and graphitised (Figure 3.44).

Note that graphitisation is irreversible, but carbon black can be heated under a nitrogen atmosphere from about 700–950 °C to release the oxygen-based functional groups, although normally the surface activity will not be similar to its pre-oxidised, highly reinforcing state.

Overall, much effort has been put forth in altering the carbon black surface *via* heat treatment and surface oxidation as already mentioned, but other surface treatments can include gasification, reactions of specific molecules with specific functional groups, and even grafting of monomers – all done in an attempt to improve performance and dispersibility of carbon black in a wide range of applications. However, even today, surface-modified carbon blacks are only sold into specialty applications, such as inks and coatings. Little or none are being sold into rubber applications, except where slower

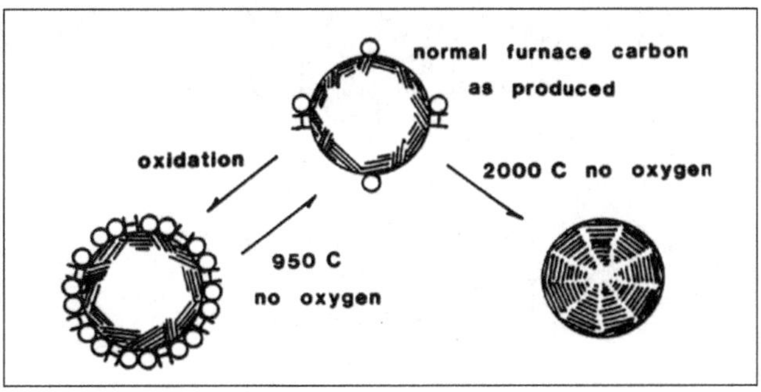

Figure 3.44 Carbon black surface activity; (1) as-produced; (2) post-oxidation and post-volatilisation (950 °C, no oxygen); (3) post-graphitisation (2000 °C, no oxygen). Reproduced with permission from Birla Carbon USA, Inc.

cure kinetics are favoured through the use of an oxidised carbon black, as may happen in larger, thicker rubber components in an attempt to achieve a more homogeneous crosslink density.

3.6.1 Carbon Black Surface Functional Groups and Their Characterisation

As mentioned, the carbon black surface contains a range of functional groups at the edge sites of the lamella at the surface, primarily composed of hydrogen and oxygen, as a result of the carbon black formation environment. Studies[181–186] of the carbon black surface have shown that a variety of oxygen-based functional groups are present, including carboxyl, phenolic, lactone, quinone, and ether and anhydrides (Figure 3.45).

Boehm wrote one of the first substantial essays on the characterisation of surface groups of carbons and other fine particles[181–183] utilising a series of bases of increasing strength, classically known as the Boehm titration. These bases, in order of increasing basicity, include sodium bicarbonate ($NaHCO_3$), sodium carbonate (Na_2CO_3), and sodium hydroxide (NaOH). In this manner, it has been possible to make assumptions about the various functional groups that react in a strong-acid : weak-base to weak-acid : strong-base range and use the differences in the moles of each functional group detected by the corresponding base, to ascertain the amount of each functional group.[185,186] This technique generated some of the first data that allowed researchers to begin to quantify the different functional groups present on the carbon black surface using classical wet chemistry methods. The Boehm titration method has also been applied to other carbonaceous materials including diamond,[182] activated carbon, multiwall carbon nanotubes, and graphenes.[184–186]

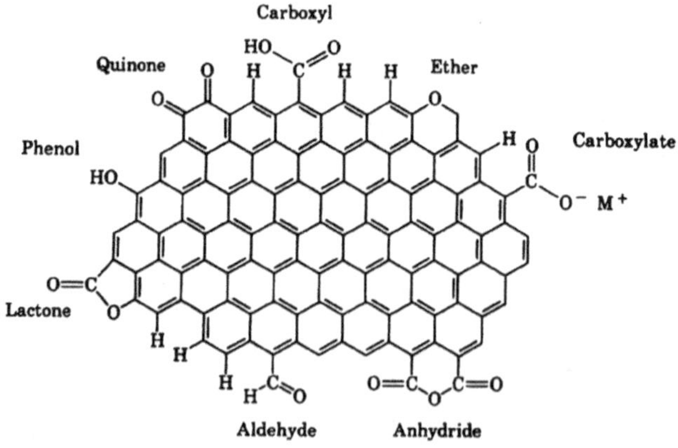

Figure 3.45 Representative surface lamella with typical hydrogen and oxygen-based functional groups. Reproduced with permission from Birla Carbon USA, Inc.

Carbon black functional groups demonstrate increased acidity in going from ether to the quinone, lactone, phenolic and carboxylic acid functional groups. The carboxylic, phenolic, and lactone-type functional groups are the primary functional groups on the surface of carbon black and other carbonaceous materials.[186] Carbon black surfaces that have been oxidised *via* treatment with ozone or nitric acid or other inorganic-based oxidisers (such as $KMnO_4$) can become highly acidic ($pH < 6$) with a preponderance of carboxylic acid and phenolic type functional groups on the surface.

Other techniques used to evaluate the surface functional groups on carbon black and other carbonaceous materials include temperature programmed desorption (TPD) with mass spectrometry or TPD-MS[184,187] also thermogravimetric analysis or TGA-MS, falling within the TPD category, pyrolysis gas chromatography or GC-MS,[188] and X-ray photoelectron spectroscopy (XPS).[184,187]

The thermal desorption method works by heating the carbon black (or other carbonaceous materials) under helium from 25 °C to 1000 °C, which provides the required thermal or translational energy required to break bonds of the functional groups on the surface. Typically, it is observed that the resulting volatile components are comprised of hydrogen, carbon monoxide and carbon dioxide. It has also been reported that the monoatomic oxygen-based functional groups, such as phenolic, carbonyl, ether and quinone, generate the CO volatile component; while the polyatomic oxygen-based functional groups, such as carboxylic, lactone and the carboxylic anhydride, generate the CO_2 volatile components.[187] Quantification of the amount of each volatile component *via* thermal desorption is possible and can be related back to the amount of specific functional groups based on collective knowledge reported in the literature, but this method is semi-quantitative at best for each specific functional group.

XPS, which is also known as electron spectroscopy for chemical analysis or ESCA, is a method that detects specific surface elements, but also provides chemical state information of the element. The probe depth is of the order of 10–100 Å (1–10 nm), which for carbon black, would mean a surface depth of approximately 3–35 lamella layers. Hence, XPS probes a much deeper volume than the TPD or Boehm titration, which are much more surface specific, but XPS will detect elements in the lamella itself, which could be oxygen, nitrogen and sulfur (Table 3.9).

Practically XPS works by bombarding the surface of a compressed disc of carbon black (or other carbonaceous or non-carbonaceous material) with monochromatic X-rays, which results in the ejection of electrons with a binding energy specific to the hybridisation of the various elements present in the material, which allows identification of its chemical state. For carbons with oxygen-based functional groups this would enable identification of the various functional groups (Table 3.9).[185]

In summary, the identification and characterisation of the surface functional groups on carbon black has evolved in its level of sophistication and knowledge over the years as more advanced techniques have been applied.

Table 3.9 XPS binding energies for carbon and oxygen and their respective chemical states[185]

Binding energy	Element, functional group
283.8	C, vacancy carbon
284.9	C, conjugated carbon
285.9	C, non-conjugated carbon
286.6	C–O, phenols, lactones, ethers
287.7	C=O, quinones, pyrones
288.8	O–C=O, carboxyl groups, lactones
291.4	π–π*

Overall, this information has proven useful in understanding the impact of utilisation of various surface treatments, mainly oxidation agents, but obviously these techniques can be applied to analysis of other surface moieties, that allows insight into the effectiveness and optimisation of surface treatments to improve performance of carbon blacks in a wide range of applications. Another aspect of the carbon black surface, besides edge-site functional groups, is its nanoscale surface roughness and the dislocations that result from that, which determine the surface energy distribution of the carbon black surface. This aspect of the carbon black surface activity is discussed in the next section.

3.6.2 Carbon Black Surface Energy and Its Characterisation

The surface activity of carbon black, besides its chemical composition, is also strongly influenced by its surface energy, as a result of its nanoscale surface characteristics due to lamella layering and resulting exposed edges and dislocations in the lamella present on the surface. Amorphous carbon (sp^3) and porosity, or pores, also contribute to the surface energy and surface-active sites as previously described in this chapter, considering the gas adsorption work of Klüppel *et al.*[120,189] Other techniques besides gas adsorption have been employed to evaluate the surface energy of carbon black and include inverse gas phase chromatography (IGC)[190–193] and calorimetry.[194]

One of the most comprehensive studies of carbon black using IGC was carried out by Wang, Wolff and Donnet[190] and Wang and Wolff.[191] In IGC, a typical gas chromatography set-up is used in which the carbon black is now the stationary phase in the GC column and a gaseous probe molecule is utilised to determine the adsorption characteristics of the carbon black surface. Of course probes of various chemistries and polarities can be used to yield a rich array of information on the surface energy of carbon black and the molecular characteristics of the probe molecules, often very low molecular weight monomer analogues of polymers, can be varied to gain insight into the characteristics of the various elastomers that drive the carbon black elastomer interaction.

Information, derived from IGC, primarily centres around the measurement of the change in the free energy of adsorption, $\Delta G°$, and the determination of the total surface energy of the filler, γ_s, which is composed of a dispersive, γ_s^d (London forces or non-specific forces), and polar component, γ_s^{sp} (polar, hydrogen bonding or acid–base),[4] as follows:

$$\gamma_s = \gamma_s^d + \gamma_s^{sp} \tag{3.18}$$

For example, both the dispersive and specific components can be calculated from the change in the free energy (obtained from the net retention volume) from a plot of $\Delta G°$ for a series of *n*-alkanes of varied length, to yield the $\Delta G^2{}_{CH2}$ or free energy of adsorption of the methylene group. The polar component is determined from the specific interaction, I^{sp}, which is determined from the $\Delta\Delta G°$ of a polar probe *versus* the *n*-alkane $\Delta G°$ plot.[190–192]

The results of various IGC-based investigations[189] using a wide range of carbon blacks (Table 3.10) showed that both the dispersive and polar surface components decreased with decreasing surface area or increasing mean particle size.

As can be seen in the table, there is a correlation of the dispersive component *versus* surface area, with a decreasing dispersive surface energy as the surface area decreases, which is an intuitive result if it is considered that the crystallite dimensions of the larger particle size carbon blacks are larger, reducing the amount of edge sites and unsaturated charge.[190] Note that for the graphitised N375, which has significantly larger crystallite dimensions and many fewer edge sites, the dispersive surface energy drops by approximately 30%.

To further highlight the impact of the L_c or crystallite dimension on the dispersive surface energy, the γ_s^d was plotted *versus* the L_c (Figure 3.46) and shows that the dispersive component, decreases with increasing L_c.

The dispersive component relates to the general London forces and represents a general surface affinity, and in this case, for hydrocarbons. The

Table 3.10 Surface energy, dispersive and polar, for a wide range of carbon blacks.[182]

Carbon black	NSA/m^2 g^{-1}	γ_s^d/mJ m^{-2}	I^{sp}, benzene/mJ m^{-2}	I^{sp}, MeCN/mJ m^{-2}
N115	137	412	105	169
N121	121	362	102	207
N234	126	403	105	197
N220	111	396	113	190
N375	93	268	98	172
N375, graphitised	93[a]	189	60	100
N347	87	274	97	177
N330	78	276	99	186
N550	39	190	88	163
N660	35	133	67	137
N772	36	129	66	150

[a]Not listed in the reference, but assumed to be similar to the original N375.

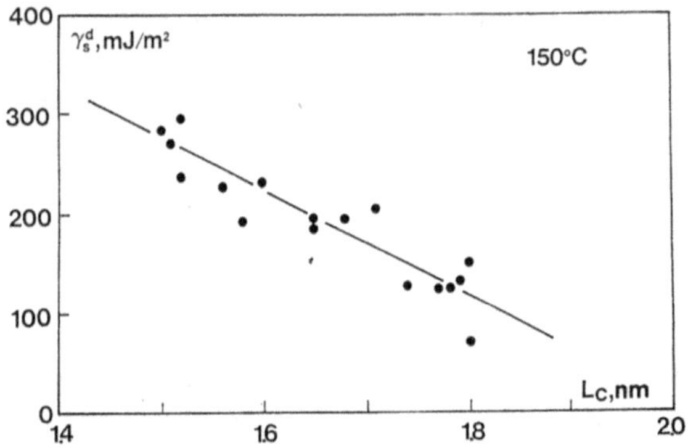

Figure 3.46 Dispersive surface energy *versus* average crystallite stacking height, L_c (angstroms). Reproduced from ref. 190 with permission from American Chemical Society, Copyright 1991.

specific component relates to polar interactions with the surface and in the case of carbon black can be driven by the aromatic π character of the surface and functional groups. For example, the benzene molecule was used as one probe to measure the polar impact of the π-electron cloud of the carbon black surface (Table 3.10), while acetonitrile was used as a means to reflect hydrogen bonding and surface polarity due to the polarity of the nitrile group.[190] Again the specific interaction parameter shows some dependency on surface area, with smaller particle size showing a higher I^{sp}, but this factor is also strongly influenced by the surface chemistry, and at a given carbon black surface area, probes, such as acetonitrile, will have a higher I^{sp}, if the surface is oxidised.

Wang, Wolff and Donnet[190] also studied a series of small molecules that represent polymer analogues and included simple linear and branched alkanes and alkenes (C4 through C8), cyclohexane, cyclohexene, cyclohexadiene, benzene, and acetonitrile. The results showed that the highest energy of adsorption was found for linear and branched alkenes and acetonitrile, and decreased steadily in going from the higher unsaturated cyclic C6 molecules, to the saturated alkanes and the C6 cyclic molecule, hexane. The authors concluded that these results were aligned with previous work, in which the affinity of carbon black for various elastomers was established, where it was concluded that carbon black has the highest affinity for NBR and SBR, then decreases on going to NR, EPDM and finally butyl (IIR) rubber.[189]

In terms of the carbon black affinity for various elastomers, surface modification has been studied as a way to increase the carbon-black–elastomer interaction while simultaneously decreasing the carbon-black–carbon-black interaction, primarily as a means to lower the compound hysteresis.[195–197] Other surface modifications have been implemented for use of carbon blacks in liquids and plastics systems to achieve improved interaction and to stabilise the dispersion (prevent flocculation).[198] The literature for carbon black surface modification is voluminous and it is not possible to review it in its entirety, but it should be noted that surface modified carbon blacks are not sold in any significant volume to tyre or rubber goods customers today, but they do find extensive use in inks and coatings. In this instance, all of these carbon blacks are oxidised during their formation or during a post-treatment, imparting a significant number of oxygen-based functional groups to the surface as described previously.

3.7 Porosity in Carbon Black

Porosity is often considered a fundamental property of carbon black and indeed the performance of certain carbonaceous materials, such as activated carbon, is primarily determined by their porosity or pore profile, among other properties. However, the overwhelming majority of carbon blacks do not contain any significant amount of porosity, and if they do, it is a relatively small amount imparted to the product as a side gasification reaction during the production process, and normally observed only for very high surface area or very small particle size carbon black products (<15 nm), where reaction temperatures are the highest used in any of the carbon black production processes.

There are however certain carbon blacks that are purposely made with a significant amount of porosity *via* gasification with water vapour, CO, CO_2, or small amounts of O_2, typically at temperatures of 900 °C and higher, and generating surface areas >400 $m^2\ g^{-1}$ and up to and greater than 1000 $m^2\ g^{-1}$, as measured by the NSA.[199] One may gauge the amount of porosity by the difference between the NSA and STSA, but for highly porous products the STSA model may not work as well as for non-porous carbon blacks.

Carbon blacks with such high levels of porosity are known as extra-conductive or super-conductive and are made by several companies including Engineered Carbons, Orion, Cabot Corporation and Unipetrol. They are produced in the same way as non-porous carbon blacks in their first step of production (incomplete combustion in a furnace reactor), but a second post-treatment step may be utilised for gasification and generation of the high degree of porosity and an apparent significant enhancement in electrical conductivity. There is one extra-conductive product line, Ketjenblack, made by Nouryon, that is a by-product of the shell gasification process where coal tar or other heavily aromatic hydrocarbon residues are injected into the gasification tower with oxygen and steam to form syngas. A highly porous carbon black is produced as a side product, and it is subsequently collected and sold into the extra-conductive application market. In most cases, however, the production of solid carbon black by the gasification process is minimised and it is seen as an unwanted product.

All of these products find applications in numerous conductive applications in all major carbon black market sectors (rubber, plastics, inks, coatings), but are more recently in high demand for lithium-ion batteries, essentially creating a market shortage.

From a technical standpoint. the interior of the carbon black particle is preferentially gasified, which makes the particles hollow with pores and open

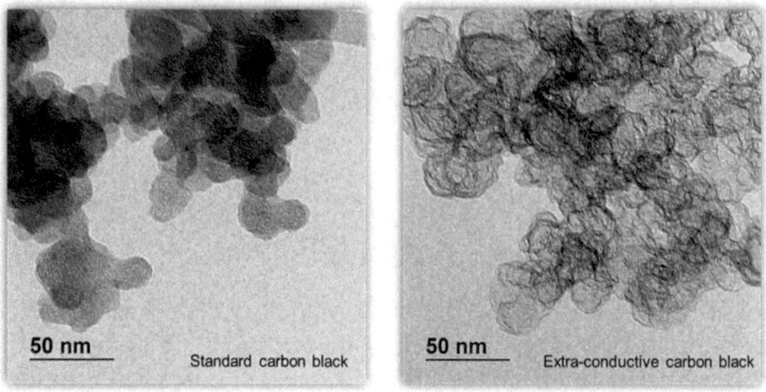

Figure 3.47 Standard carbon black *versus* extra-conductive, highly porous carbon black. Reproduced with permission from Birla Carbon USA, Inc.

channels in the exterior shell of lamella (Figure 3.47). It appears that the interior lamella of any carbon black are preferentially gasified due to shorter and more numerous lamella edges that are very reactive to gasification once the outer shell is breached.

As already noted, the degree of porosity is normally taken as the difference between the NSA and the STSA, where the NSA measures the exterior shell surface area available to the polymer or elastomer, as well as the pores and porosity and even perhaps the interior shell surface area, depending upon the severity of the porosity. The resulting extra-conductive carbon blacks can have NSAs as high as 1000 $m^2\ g^{-1}$, or even higher.

The impact of very high levels of porosity, such as seen in superconductive carbon blacks, results from a much lower specific gravity than for a normal non-porous carbon black, perhaps as much 30% lower density. This is easily observed in a simple experiment done by placing equal weights of a normal and a superconductive carbon black in a small jar and observing the volume difference.

The much lower specific gravity, yet much higher volume, of an extra- or super-conductive-type carbon black has many implications when tested for colloidal properties and when used in various applications or compounds. First almost all colloidal tests specify given sample weights when testing such properties as surface area (iodine number, NSA, STSA) and structure (OAN) and thus a much higher volume is actually used, *versus* testing of normal carbon blacks with a much higher specific gravity. The net result is that a test, such as structure or OAN, yields a value that is much higher than the true structure of the superconductive carbon black. So, caution has to be used when interpreting colloidal values for these highly porous products. Secondly the same can happen in a plastic or rubber compound, where the term extra- or super-conductive most likely originated, since when a super-conductive carbon black is added to the compound, a much lower conductive percolation threshold is achieved at a much lower weight, making the carbon black appear to be 'superconductive'. In actuality, the volume fraction of the super-conductive carbon black can be the same as a normal carbon black hence achieving a through-going electrical network at a much lower weight loading. That is the super-conductive carbon black is fundamentally not more conductive itself, and primarily gives good compound conductivity at lower loadings, due to its much lower specific gravity *versus* normal carbon blacks.

3.8 Summary and Future Outlook

Carbon black, as has been shown, continues to be a key ingredient in numerous applications that provide significant societal benefit. The mere fact that it is the key ingredient in tyres, that gives them outstanding durability, allows modern societies to function, learn and grow through sustainable mobility. The societal benefits of this performance feature of carbon black are numerous.

The latest trends in the tyre industry are development of sustainable products. This fact is driving the carbon black industry toward new production processes for carbon black that significantly reduce its CO_2 footprint. There are also moves toward the use of recycled or sustainable feedstocks, such as renewable tall oils from the pulping of pine wood or recycled-tyre pyrolysis oils for use as feedstocks to produce carbon blacks. The yields from use of such feedstocks are diminished *versus* current feedstocks, so research continues to not only bring sustainable characteristics to carbon blacks and tyres, but to also improve their performance to the levels of current materials. Additionally recycled materials, such as sustainable carbonaceous materials (SCMs) from pyrolysis of end-of-life tyres are being developed and commercialised. These SCMs are primarily composed of the original carbon black and precipitated silica content of the original tyre, yet they are not considered carbon blacks nor silicas, and the performance of the SCMs is below that of the original reinforcing fine particles due to the pyrolysis process. Ultimately the producers of tyres want all raw materials used in their production to be 100% sustainable by 2050.

Other trends in the carbon black industry, as mentioned, involve development of processes with significantly lower CO_2 output and water usage, as well as continued use of the excess heat for production of electricity or steam, making the carbon black process an energy positive process. Additional challenges are being faced in pollution control mechanisms to reduce and eliminate SO_x and NO_x emissions as mandated by regional governing bodies. Lastly new products of carbon black are still very important since 80% of the environmental impact of a tyre is during its use on a vehicle that emits CO_2, and use of low rolling resistance carbon black technologies helps to reduce this impact.[200] Newer developments in tyre abrasion regulations are also now under consideration as wear particles are conjectured to impact the environment.[200] Tyres with improved wear profiles along with lower rolling resistance, while maintaining the tyre's traction, are very much desired, but this type of performance balance has proven difficult to achieve.

Now, sustainable raw material content is added to the performance balance in addition to the main three tyre performance properties described above.

The future of mobility holds the key for carbon black, and whether it is carbon blacks for tyres for lower rolling resistance and improved wear performance, or for carbon blacks in lithium-ion batteries for electric vehicles, carbon black will continue to play a large role in supporting sustainable mobility and the benefits this concept brings to our modern societies. The next 50 years will require vigorous research and development for new processes and materials to meet the ever-changing needs of our modern, sustainability-focused world.

References

1. J. B. Donnet and A. Voet, *Carbon Black Physics, Chemistry and Elastomer Reinforcement*, Marcel Dekker, New York, New York, 1976.
2. W. M. Hess and C. R. Herd, Microstructure, Morphology and General Physical Properties (of Carbon Black), in *Carbon Black Science and Technology*, ed. J.-B. Donnet, R. C. Bansal, and M.-J. Wang, Marcel Dekker, New York, New York, 1993.
3. R. Taylor, Carbon Blacks: Production, Properties and Applications, in *Introduction to Carbon Technologies*, ed. H. Marsh, E. A. Heinz, and F. Rodriguez-Reinoso, Universidad de Alicante, Secretariado de Publicaciones, 1997.
4. M. J. Wang and M. Morris, *Rubber Reinforcement with Particulate Fillers*, Carl Hanser Verlag GmbH & Co. KG an Chemical Industry Press Co., Ltd., P.R. China, 1st edn, 2021.
5. D. Tracy, *Jalopnik Reviews*, Jalopnik, https://jalopnik.com/heres-why-tires-are-black-1828925177, accessed 15 November 2018.
6. *Annual Book of ASTM Standards*, ASTM International, West Conshohocken, PA, 2021, **vol. 09.01**, Section 9.
7. ASTM D1765-19, in *Annual Book of ASTM Standards*, ASTM International, West Conshohocken, PA, 2021, vol. 09.01, Section 9, p. 383.
8. Notch Consulting Inc., *Carbon Black World Data Book 2022*, Amherst, MA, September 2022.
9. The Cary Company Online, https://www.thecarycompany.com/media/pdf/specs/orion-what-is-carbon-black.pdf, accessed November 2022.
10. Columbian Chemicals Company, *US Pat.*, 5256388, 1993.
11. Degussa – Hüls Aktiengesellschaft, *US Pat.*, 6251983B1, 2001.
12. Cabot Corporation, *US Pat.*, 6485693B1, 2002.
13. Cabot Corporation, *US Pat.*, 6548036, 2003.
14. J. G. Mini, *US Letters Pat.*, 3821, 1844.
15. W. G. W. Laeger, *Letters Pat.*, 11331, 1854.
16. E. Binney, *US Pat.*, 453140, 1891.
17. A. Schuch, *Rubber World*, 2003, 18.
18. Columbian Carbon Company, *US Pat.*, 1510485, 1924.
19. Columbian Carbon Company, *US Pat.*, 1838316, 1931.

20. Degussa, *FR Pat.*, FR881442A, 1943.
21. Degussa, *DE Pat.*, DE845683C, 1952.
22. Shawinigan Chemicals Limited, *US Pat.*, 2453440, 1948.
23. Tennessee Eastman Corporation, *US Pat.*, 2475282, 1949.
24. R. E. Brownee and R. F. Uelinger, *US Pat.*, 1478730, 1923.
25. RJ Lee Group, Inc, *US Pat.*, 7341646B2, 2008.
26. Black Bear Carbon, *US Pat.*, 10119031B2, 2018.
27. Pyrolox AG, *US Pat.*, 10184081B2, 2019.
28. Frederick Taylor, *US Pat.*, 10329489B2, 2019.
29. Bolder Industries, *US Pat.*, 2020/0123390A1, 2020.
30. Kvaerner Engineering AS, *Canadian Pat.*, CA2117329C, 1998.
31. Monolith Materials, *US Pat.*, 10808097B2, 2020.
32. Seerstone LLC, *US Pat.*, 8679444 B2, 2014.
33. Ekona Power, *WO Pat.*, WO2020/118417A1 Filed, 18 June 2020.
34. Carbon X, *Danish/(DK) Pat.*, 3475367T3, 2022.
35. A. J. W. Watson and D. Parkinson, *Ind. Eng. Chem.*, 1955, **47**(5), 1053.
36. ASTM D1510 – 19a, Standard Test Method for Carbon Black – Iodine Adsorption Number, in *Annual Book of ASTM Standards 2021*, ASTM International, West Conshohocken, PA, 2021, vol. 09.01, Section 9, p. 312.
37. S. Brunauer, P. H. Emmett and E. Teller, *J. Am. Chem. Soc.*, 1938, **60**, 309.
38. ASTM 6556 – 19a, Standard Test Method for Carbon Black – Total and External Surface Area by Nitrogen Adsorption, in *Annual Book of ASTM Standards 2021*, ASTM International, West Conshohocken, PA, 2021, vol. 09.01, Section 9, p. 1120.
39. R. W. Magee, *Rubber Chem. Technol.*, 1995, **68**, 590.
40. A. I. Medalia, *J. Colloid Interface Sci.*, 1970, **31**, 115.
41. ASTM 2414 – 19, Standard Test Method for Carbon Black – Oil Absorption Number, in *Annual Book of ASTM Standards 2021*, ASTM International, West Conshohocken, PA, 2021, vol. 09.01, Section 9, p. 456.
42. ASTM 3493 - 20, Standard Test Method for Carbon Black – Oil Absorption Number of Compressed Sample (COAN), in *Annual Book of ASTM Standards 2021*, ASTM International, West Conshohocken, PA, 2021, vol. 09.01, Section 9, p. 603.
43. ASTM 7854 - 18, Standard Test Method for Carbon Black –Void Volume at Mean Pressure, in *Annual Book of ASTM Standards 2021*, ASTM International, West Conshohocken, PA, 2021, vol. 09.01, Section 9, p. 1269.
44. R. E. Dollinger, R. H. Kallenberger and M. L. Studebaker, *Rubber Chem. Technol.*, 1967, **40**, 1311.
45. G. A. Joyce, W. M. Henry, A. W. Thornton and J. C. Hodgin, *Rubber World*, 2011, **244**, 21.
46. G. A. Joyce and W. M. Henry, *Rubber Chem. Technol.*, 2006, **79**, 735.
47. F. Grunert, A. Wehmeier and A. Blume, *Polymers*, 2020, **12**, 567.
48. A. Medalia and L. Richards, *J. Colloid Interface Sci.*, 1972, **40**, 233.
49. ASTM 3265 – 19b, Standard Test Method for Carbon Black – Tint Strength, in *Annual Book of ASTM Standards 2021*, ASTM International, West Conshohocken, PA, 2021, vol. 09.01, Section 9, p. 567.

50. ASTM 1618 - 18, Standard Test Method for Carbon Black – Carbon Black Extractables – Transmittance of Toluene Extract, in *Annual Book of ASTM Standards 2021*, ASTM International, West Conshohocken, PA, 2021, vol. 09.01, Section 9, p. 357.
51. ASTM 1512 – 15b, Standard Test Method for Carbon Black – pH Value, in *Annual Book of ASTM Standards 2021*, ASTM International, West Conshohocken, PA, 2021, vol. 09.01, Section 9, p. 326.
52. ASTM 1511 – 12, Standard Test Method for Carbon Black – Pellet Size Distribution, in *Annual Book of ASTM Standards 2021*, ASTM International, West Conshohocken, PA, 2021, vol. 09.01, Section 9, p. 322.
53. ASTM 5230 - 19, Standard Test Method for Carbon Black – Automated Individual Pellet Hardness, in *Annual Book of ASTM Standards 2021*, ASTM International, West Conshohocken, PA, 2021, vol. 09.01, Section 9, p. 913.
54. ASTM 1509 – 18 (reapproved 2020), Standard Test Method for Carbon Black – Heating Loss, in *Annual Book of ASTM Standards 2021*, ASTM International, West Conshohocken, PA, 2021, vol. 09.01, Section 9, p. 309.
55. ASTM 1506 – 15 (reapproved 2020), Standard Test Method for Carbon Black – Ash Content, in *Annual Book of ASTM Standards 2021*, ASTM International, West Conshohocken, PA, 2021, vol. 09.01, Section 9, p. 302.
56. ASTM 1514 – 15 (reapproved 2020), Standard Test Method for Carbon Black – Sieve Residue, in *Annual Book of ASTM Standards 2021*, ASTM International, West Conshohocken, PA, 2021, vol. 09.01, Section 9, p. 333.
57. Phillips Petroleum Company, *US Pat.*, 4213957, 1980.
58. H. Bockton, *Soot Formation in Combustion*, Springer, Berlin, Heidelberg, Springer Series in Chemical Physics #59, 1994.
59. M. Frenklach and A. M. Mebel, *Phys. Chem. Chem. Phys.*, 2020, **22**, 5314.
60. A. S. Semenikhin, A. S. Savchenkova, I. V. Chechet, S. G. Matveev, M. Frenklach and A. M. Mebel, *Phys. Chem. Chem. Phys.*, 2020, **22**, 17196.
61. M. Frenklach, A. S. Semenikhin and A. M. Mebel, *J. Phys. Chem. A*, 2021, **125**, 6789.
62. A. I. Medalia, D. Rivin and D. R. Sanders, *Sci. Total Environ.*, 1983, **31**(1), 1.
63. A. Y. Watson and P. A. Valberg, *Am. Ind. Hyg. Assoc.*, 2001, **62**, 218.
64. G. Porter, *Proc. Combust. Inst.*, 1953, **4**, 248.
65. S. J. Harris and A. M. Weiner, *Annu. Rev. Phys. Chem.*, 1985, **36**, 31.
66. M. Singh, A. Gharpure, R. L. Vander Wal, J. Kollar and C. R. Herd, *Appl. Sci.*, 2022, **12**, 2569.
67. B. E. Warren, *J. Chem. Phys.*, 1934, **2**, 551.
68. J. Biscoe and B. E. Warren, *J. Appl. Phys.*, 1942, **13**, 364.
69. C. E. Hall, *J. Appl. Phys.*, 1947, **18**, 588.
70. C. E. Hall, *J. Appl. Phys.*, 1948, **19**, 198.
71. W. M. Hess and L. L. Ban, *Proceedings of the Sixth International Congress for Electron Microscopy*, Norelco Reporter, Kyoto, Japan, 1966, vol. 13, p. 102.

72. R. D. Heidenreich, W. M. Hess and L. L. Ban, *J. Appl. Crystallogr.*, 1968, **1**, 1.
73. W. M. Hess, K. A. Burgess, F. Lyon and V. E. Chirico, *Kautsch. Gummi Kunstst.*, 1968, **21**, 689.
74. L. L. Ban and W. M. Hess, *Proceedings of the 26th Annual EMSA Meeting*, 1968, p. 256.
75. D. F. Harling and F. A. Heckman, *Mater. Plast. Elastomeric*, 1969, **35**, 80.
76. S. Ergun, *Carbon*, 1968, **6**, 141.
77. C. Robertson and N. J. Harman, *Polymers*, 2021, **13**, 538.
78. P. Delhaes, *Graphite and Precursors in World of Carbon*, Overseas Publishers Association, published by License under the Gordon and Breach publisher's imprint, 2001, vol. 1.
79. W. Wang, R. Lamba, C. Herd, D. Tandon and C. Edwards, *Rubber World*, 2005, 22.
80. W. H. Bragg and W. L. Bragg, *Proc. R. Soc. London, Ser. A*, 1913, **88**, 428.
81. P. Scherrer, *Göttinger Nachrichten Gesell*, 1918, vol. 2, p. 98.
82. M. A. Short and P. L. Walker Jr., *Carbon*, 1963, **1**, 3.
83. T. Ungár, *Scr. Mater.*, 2004, **51**, 777.
84. Z. Q. Li, C. J. Lu, Z. P. Xia, Y. Zhou and Z. Luo, *Carbon*, 2007, **45**, 1686.
85. R. E. Franklin, *Proc. R. Soc. A*, 1951, **209**, 196.
86. L. E. Alexander and S. R. Darin, *J. Chem. Phys.*, 1955, **23**, 594.
87. A. E. Austin, *Proceedings of 3rd Conference on Carbon*, University of Buffalo, New York, 1958, p. 389.
88. W. V. Kotlensky and P. L. Walker Jr., *Proceedings on the Fourth Conference on Carbon*, Buffalo, New York, 1960, p. 423.
89. M. A. Short and P. L. Walker Jr., *Carbon*, 1963, **1**, 3.
90. J. Mering and J. Marie, *Les Carbones*, 1965, **1**, 129.
91. T. Ungar, J. Gubicza, G. Ribarik, C. Pantea and T. W. Zerda, *Carbon*, 2002, **40**, 929.
92. C. V. Raman, *Indian J. Phys.*, 1928, **2**, 387.
93. N. Nakamizo, R. Kammereck and P. L. Walker Jr., *Carbon*, 1974, **12**, 259.
94. P. Lespade, A. Marchand, M. Couzi and F. Cruege, *Carbon*, 1984, **22**, 375.
95. K. Angoni, *Carbon*, 1993, **31**, 537.
96. L. Bokobza, J. L. Bruneel and M. Couzi, *C*, 2015, **1**, 77.
97. T. Gruber, T. W. Zerda and M. Gerspacher, *Carbon*, 1994, **32**, 1377.
98. T. Jawhari, A. Roid and J. Casado, *Carbon*, 1995, **33**, 1561.
99. T. W. Zerda and T. Gruber, *Rubber Chem. Technol.*, 2000, **73**, 284.
100. M. Pawlyta, J. N. Rouzaud and S. Duber, *Carbon*, 2015, **84**, 479.
101. F. Tuinstra and J. L. Koenig, *J. Chem. Phys.*, 1970, **53**, 1126.
102. R. L. Vander Wal, A. J. Tomasek, K. Street, D. R. Hull and W. K. Thompson, *Appl. Spectrosc.*, 2004, **58**, 230.
103. R. L. Vander Wal and A. J. Tomasek, *Combust. Flame*, 2003, **134**, 1.
104. M. Singh, C. K. Gaddam, J. P. Abrahamson and R. L. Vander Wal, *Aerosol Sci. Technol.*, 2019, **53**, 207.
105. K. O. Johansson, M. P. Head-Gordon, P. E. Schrader, K. R. Wilson and H. A. Michelsen, *Science*, 2018, **361**, 997.

106. K. Ono, Y. Matsukawa, K. Dewa, A. Watanabe, K. Takahashi, Y. Saito, Y. Matsushita, H. Aoki, K. Era and T. Aoki, *Combust. Flame*, 2015, **162**, 2670.
107. M. Thomson and T. A. Mitra, *Science*, 2018, **361**, 978.
108. J. W. Martin, M. Salamanca and M. Kraft, *Prog. Energy Combust. Sci.*, 2022, **88**, 100956.
109. M. Frenklach and A. M. Mebel, *Phys. Chem. Chem. Phys.*, 2020, **22**, 5314.
110. G. Binning and H. Rohrer, *Helv. Phys. Acta*, 1982, **55**, 726.
111. M. J. Wang, S. Wolff and B. Freund, *Rubber Chem. Technol.*, 1994, **67**, 27.
112. J. B. Donnet and E. Custodero, *Carbon*, 1992, **30**, 813.
113. J. B. Donnet and R. Y. Qin, *Carbon*, 1992, **30**, 787.
114. J. B. Donnet, W. Wang, A. Vidal and M. J. Wang, *Carbon*, 1994, **32**, 199.
115. W. Niedermeier, J. Stierstorfer, S. Kreitmeier, O. Metz and D. Goritz, *Rubber Chem. Technol.*, 1994, **67**, 148.
116. W. Niedermeier, H. Raab, J. Stierstorfer, S. Kreitmeier and D. Goritz, *Kautsch. Gummi, Kunstst.*, 1994, **47**, 799.
117. J. B. Donnet, E. Custodero and T. K. Wang, *Kautsch. Gummi, Kunstst.*, 1996, **49**, 274.
118. H. Raab, J. Frohlich and D. Goritz, *Kautsch. Gummi, Kunstst.*, 2000, **53**, 137.
119. Degussa – Huls Aktiengesellschaft, *US Pat.*, 6251983B1, 2001.
120. M. Klüppel, *Adv. Polym. Sci.*, 2003, **164**, 1.
121. M. Knoll and E. Ruska, *Z. Phys.*, 1932, **78**, 318.
122. P. J. F. Harris, *C*, 2018, **4**(1), 4.
123. U. Hofman and W. Höper, *Naturwissenschaften*, 1944, **32**, 225.
124. J. H. L. Watson, *J. Appl. Phys.*, 1946, **17**, 121.
125. Columbian Carbon Company, *Colloidal Carbon (Internal Publication)*, 1940, vol. II.
126. Columbian Carbon Company, *Colloidal Carbon (Internal Publication)*, 1942, vol. II.
127. W. B. Wiegand and W. L. Ladd, *Rubber Age*, 1942, **50**, 431.
128. W. B. Wiegand and W. L. Ladd, *Rubber Chem. Technol.*, 1942, **15**, 644.
129. Endter and H. Gebauer, *Optik*, 1956, **13**, 97.
130. B. Schubert, F. P. Ford and F. Lyon, *Encyclopaedia of Industrial Chemical Analysis*, John Wiley and Sons, 1969, 8, p. 179.
131. W. M. Hess, L. L. Ban and G. C. McDonald, *Rubber Chem. Technol.*, 1969, **42**, 1209.
132. K. A. Burgess, C. E. Scott and W. M. Hess, *Rubber Chem. Technol.*, 1971, **44**, 230.
133. W. M. Hess, G. C. McDonald and E. Urban, *Rubber Chem. Technol.*, 1973, **46**, 204.
134. L. L. Ban, W. M. Hess and L. A. Papazian, *Rubber Chem. Technol.*, 1974, **47**, 858.
135. W. M. Hess and G. C. McDonald, *Rubber Chem. Technol.*, 1983, **56**, 892.
136. ASTM 3849- , Standard Test Method for Carbon Black – Morphological Characterization of Carbon Black Using Electron Microscopy, in

Annual Book of ASTM Standards 2021, ASTM International, West Conshohocken, PA, 2021, vol. 09.01, Section 9, p. 657.
137. G. G. Stokes, *Trans. Cambridge Philos. Soc.*, 1851, **9**(part ii), 8.
138. T. R. Mohanty, S. Ramakrishnan, T. R. Gopalakrishnan, S. K. P. Amarnath, D. Lorenzetti and P. K. Mohammed, *Rubber Tire Digest*, April/May 2020.
139. ISO 15825:2017 Rubber Compound Ingredients – Carbon Black – Determination of Aggregate Size Distribution by Disc Centrifuge Photosedimentometry.
140. I. H. Herron, S. H. Davis and F. P. Bretherton, *J. Fluid Mech.*, 1975, **68**, 209.
141. C. J. Stacy, P. H. Johnson and G. Kraus, *Rubber Chem. Technol.*, 1975, **48**, 538.
142. M. D. Morris and T. Al, *Rubber World*, 2010, 15.
143. F. D. Werner, W. Niedermeier and L. Bradley, *Rubber World*, 2016, 34.
144. C. G. Robertson, S. R. Vaikuntam and G. Heinrich, *Polymers*, 2020, **12**, 190.
145. A. R. Payne, *Mater. Res. Stand.*, 1961, **1**(12), 942.
146. A. R. Payne, *J. Appl. Polym. Sci.*, 1962, **6**, 57.
147. A. R. Payne, *J. Appl. Polym. Sci.*, 1962, **6**, 368.
148. M. J. Wang, S. Wolff and E. H. Tan, *Rubber Chem. Technol.*, 1993, **66**, 178.
149. A. I. Medalia, *J. Colloid Interface Sci.*, 1967, **24**, 393.
150. A. I. Medalia and F. A. Heckman, *Carbon*, 1969, **7**, 567.
151. A. I. Medalia, *Powder Technol.*, 1970–71, **4**, 117.
152. A. I. Medalia, *J. Colloid Interface Sci.*, 1970, **32**, 115.
153. W. M. Hess, L. L. Ban and G. C. McDonald, *Rubber Chem. Technol.*, 1969, **42**, 1209.
154. W. M. Hess, G. C. McDonald and E. Urban, *Rubber Chem. Technol.*, 1973, **46**, 204.
155. A. I. Medalia, *Rubber Chem. Technol.*, 1972, **45**, 1171.
156. G. Kraus, *Rubber Chem. Technol.*, 1971, **44**, 199.
157. R. W. Sandbrook, *J. Inst. Rubber Ind.*, 1970, **4**, 210.
158. G. Kraus, *J. Appl. Polym. Sci.*, 1971, **15**, 1679.
159. A. I. Medalia, *J. Colloid Interface Sci.*, 1970, **32**, 115.
160. G. C. McDonald and W. M. Hess, *Rubber Chem. Technol.*, 1977, **50**, 842.
161. C. R. Herd, G. C. McDonald and W. M. Hess, *Rubber Chem. Technol.*, 1992, **65**, 107.
162. A. I. Medalia and F. A. Heckman, *Carbon*, 1969, **7**, 567.
163. W. M. Hess, G. C. McDonald and E. Urban, *Rubber Chem. Technol.*, 1973, **46**, 204.
164. C. R. Herd, G. C. McDonald, R. E. Smith and W. M. Hess, *Rubber Chem. Technol.*, 1993, **66**, 491.
165. D. W. Schaefer and A. J. Hurd, *Aerosol Sci. Technol.*, 1990, **12**, 876.
166. H. M. Mercer, A. H. Boyer, P. L. Brusky and M. L. Deviney, *Rubber Chem. Technol.*, 1976, **49**, 1068.

167. A. Boyde, in *Scanning Electron Microscopy*, ed. O. C. Well, McGraw Hill, Inc, New York, 1974, ch. 11, p. 277.
168. A. Boyd, *Proceedings of the Third Annual Scanning Electron Microscope Symposium*, Chicago, April 1970, p. 105.
169. T. C. Gruber, T. C. Zerda and M. Gerspacher, *Carbon*, 1993, **31**, 1218.
170. T. C. Gruber, T. W. Zerda and M. Gerspacher, *Rubber Chem. Technol.*, 1994, **67**, 280.
171. W. Xu, T. W. Zerda, H. Raab and D. Goritz, *Carbon*, 1997, **35**, 471.
172. T. C. Gruber and C. R. Herd, *Rubber Chem. Technol.*, 1997, **70**, 727.
173. T. C. Gruber, *Kautsch. Gummi, Kunstst.*, 1998, **3**, 194.
174. J. Frank, *Electron Tomography: Methods for Three-Dimensional Visualization of Structures in the Cell*, Springer-Verlag, 2nd edn, 2006.
175. E. M. Hunt and C. R. Herd, *Microsc. Microanal.*, 2001, **7**(supplement 2), 82.
176. G. O. Orhan, E. Haffner-Staton, A. La Rocca and M. Fay, *Tribol. Int.*, 2016, **104**, 272.
177. E. Hafiner-Staton, A. La Rocca and M. W. Fay, *J. Microsc.*, 2018, **270**, 272.
178. G. Okyay, E. Héripré, T. Reiss, P. Haghi-Ashtiani, T. Auger and F. Enguehard, *J. Aerosol Sci.*, 2016, **93**, 63.
179. T. C. Gruber, K. Cauffiel and R. Waldheim, *Presented at the 200th Technical Meeting of the Rubber Division*, ACS, Pittsburgh, PA, October 5–7, 2021.
180. J. Janzen and I. Goodarz-Nia, *J. Colloid Interface Sci.*, 1979, **69**, 476.
181. H. P. Boehm, E. Diehl, W. Heck and R. Sappok, *Angew. Chem., Int. Ed. Engl.*, 1964, **3**, 669.
182. H. P. Boehm, *Adv. Catal.*, 1966, **16**, 179.
183. H. P. Boehm, *Carbon*, 1994, **32**, 759.
184. J. Schönherr, J. R. Buchheim, P. Scholz and P. Adelhelm, *C*, 2018, **4**(2), 21.
185. J. Schönherr, J. R. Buchheim, P. Scholz and P. Adelhelm, *C*, 2018, **4**(2), 22.
186. H. Ren, E. Cunha, Q. Sun, Z. Li, I. A. Kinloch, R. J. Young and Z. Fan, *Nanoscale Adv.*, 2019, **1**, 1432.
187. J. L. Figueiredo, M. F. R. Pereira, M. M. A. Freitas and J. J. M. Orfão, *Carbon*, 1999, **37**, 1379.
188. J.-H. Lin, *Carbon*, 2002, **40**, 183.
189. A. Schuster, M. Klüppel, R. H. Schuster and J. Heidberg, *Carbon*, 2002, **40**, 207.
190. M. J. Wang, S. Wolff and J.-B. Donnet, *Rubber Chem. Technol.*, 1991, **64**, 714.
191. M. J. Wang and S. Wolff, *Rubber Chem. Technol.*, 1992, **65**, 890.
192. M. J. Wang and S. Wolff, *Rubber Chem. Technol.*, 1991, **64**, 559.
193. J. B. Donnet, E. Custodero, T. K. Wang and G. Hennebert, *Carbon*, 2002, **40**, 163.
194. J. A. Ayala, W. M. Hess, G. A. Joyce and F. D. Kistler, *Rubber Chem. Technol.*, 1993, **66**, 772.

195. M. J. Wang, K. Mahmud, L. J. Murphy and W. J. Patterson, *Kautsch. Gummi, Kunstst.*, 1998, **5**, 348.
196. T. Mathew, R. N. Datta, W. K. Dierkes, J. W. M. Noordermeer and W. J. van Ooij, *Plasma Chem. Plasma Process.*, 2008, **28**, 273.
197. Columbian Chemicals Company and Arlanxeo, *EP Pat.*, EP2470604B1, 2019.
198. Cabot Corporation, *US Pat.*, 5851280, 1998.
199. G. Kühner and M. Voll, Manufacture of Carbon Black, in *Carbon Black Science and Technology*, ed. J.-B. Donnet, R. C. Bansal, and M.-J. Wang, Marcel Dekker, New York, New York, 1993.
200. A. Scott, *Chem. Eng. News*, 2023, **101**(17), 28.

CHAPTER 4

Carbon Dots

KHURRAM ABBAS[a], LEI JIANG[a], YUAN LI[b], ZIJIAN LI[b] AND HONG BI[b*]

[a]School of Chemistry and Chemical Engineering, Anhui University, Hefei, Anhui 230601, P. R. China; [b]School of Materials Science and Engineering, Anhui University, Hefei, Anhui 230601, P. R. China
*E-mail: bihong@ahu.edu.cn

4.1 Introduction

In recent decades, carbon-based nanomaterials (CNMs) have attracted increasing attention because of their rich physico-chemical properties and extensive applications.[1–4] Traditional CNMs, such as graphene, fullerene, and carbon nanotubes, are commonly non-fluorescent due to the population of non-radiative recombination caused by an infinitely large sp^2 domain. This situation has changed dramatically since the discovery of fluorescent carbon nanoparticles by Xu *et al.* in 2004.[5] These researchers found that massive carbon nanostructures could be oxidatively cut to produce luminous materials. In 2006,[6] the photoluminescent nanoparticles were given the name carbon dots (CDs) because of their ultra-small size and quasi-spherical shape, which endow CNMs with intriguing photoluminescent features. Since then, there has been a surge in interest in the synthesis, photophysics and applications of CDs.[7,8]

Carbon dots are zero-dimensional carbon nanomaterials with a size of less than 10 nm. The core/shell model is commonly used to describe the structure of CDs, with the core consisting primarily of sp^2- and sp^3-hybridised carbon atoms and the shell composed of different functional moieties (*e.g.*,

Inorganic Materials Series No. 15
Elemental Carbon
Edited by Peter J. F. Harris

Published by the Royal Society of Chemistry, www.rsc.org

amino, hydroxyl, and carboxyl groups). The synthetic routes of CDs can be classified into two categories: 'top-down' and 'bottom-up' methods.[9,10] In earlier studies, graphitic materials were broken down extensively by using the 'top-down' strategy to obtain CDs. In these cases, CDs were considered as tiny graphene fragments, with their emission originating from discrete sp^2 graphene domains. Unlike 'top-down' synthetic pathways, which are limited to graphitic structures (*e.g.*, carbon fibre, graphene oxide, and fullerene), 'bottom-up' synthetic approaches allow for the inclusion of a wide range of precursors, such as small molecules, polymers, and biomass materials. These precursors contain amino, carboxyl, and hydroxyl groups, which help to develop the sp^2 domain during solvothermal or microwave heating *via* intermolecular dehydration/condensation processes. By varying the synthetic methods and starting materials, it might be possible to create CDs with different structures and compositions that could be used later to modulate their properties.

As luminescent carbon nanomaterials, CDs have a number of merits, combining the biocompatibility and stability of carbon materials with bright photoluminescence, tuneable emission wavelength, and high colour purity. Moreover, the presence of various functional groups on the surface of CDs introduces good solubility and provides reactive sites for further modification reactions.[7,11] On account of these combined attributes, CDs have been extensively employed in energy conversion, optoelectronic devices, and biomedical areas.[12–14] Along with advances in materials science development, the underlying photoluminescence (PL) mechanism of CDs has also been explored.[15] Several studies have suggested that the emission of CDs could originate from the carbon core, the surface state, or both. However, slight modifications in precursors or minor changes in reaction conditions could result in CDs with distinct structural compositions and PL mechanisms. Clarifying the exact emission origin of CDs brings both challenges and opportunities.

The fast speed of research and publication rate have caused a few issues that affect the classification, purity criteria, and research of PL mechanisms of CDs, as Đorđević *et al.* addressed in one of their recent review articles.[13] Nevertheless, terms, such as 'carbon nanodots' (CNDs) or 'CDs', are widely used. Indeed, 'CDs' is a general term to denote a variety of small, nanometre-sized fluorescent carbon materials with photoluminescent behaviour,[14] including more carbonised graphene quantum dots (GQDs), carbon quantum dots (CQDs) and CNDs or less carbonised polymer dots (CPDs).[13,15] Regarding the structure of CDs and the connection between that structure and their properties, there is some ambiguity and controversy due to the variety of precursors and synthesis techniques used, as well as inconsistency in nomenclature. However, nomenclature issues have not hindered the widespread use of CDs in optoelectronics, energy conversion and storage, nanomedicine (for diagnostics, imaging, and therapies, such as photothermal therapies), and many other applications.[16]

This chapter will start by summarising the synthesis of CDs categorised by the type of precursor used (Section 4.2). It is worth noting that bottom-up syntheses are more versatile and, therefore, the main focus of this review. The structural characteristics of CDs, as well as standard characterisation techniques, are briefly discussed in Section 4.3 to assist readers in understanding what structural information these techniques may provide. Section 4.4 is devoted to the optical properties of CDs including a discussion of both structural analysis and mechanism, to better explain the structure–attribute connection. Following the overview of these intriguing properties, the applications of CDs with a focus on the latest research in photo/electro-catalysis, energy storage/conversion, optoelectronic devices, bio-imaging, phototherapy and nanozymes are highlighted (Section 4.5). Finally, some current challenges and possible future directions are provided which may benefit the rational design and practical uses of CDs (Section 4.6).

4.2 Synthesis of Carbon Dots

Carbon dots can be prepared by ‘top-down’ and ‘bottom-up’ methods. The ‘top-down’ approach for producing CDs involves electrochemical or physical methods for cleaving or breaking down sheet-like or bulky carbon materials. This can be done in a number of ways. During the formation process, the bond between carbon atoms must be broken.[17] Researchers typically use waste carbon sources, such as candle soot, natural gas soot, and tyre soot, as precursors in the green synthesis of CDs.[18] Pyrolysis or carbonisation of small organic molecules is involved in ‘bottom-up’ synthesis. The ‘bottom-up’ strategy is probably the most widely used mechanism for the green synthesis of CDs. Because most biomass materials are made up of small organic molecules, the ‘bottom-up’ technique can carbonise them into CDs at high temperatures.[19] Despite a seemingly endless number of new synthetic methods, some fundamental problems remain unresolved.[19] Research in this field is not only limited to the exploration of synthesis strategies but also aims to find more scientific, controllable, and greener synthesis methods.

4.2.1 Small Organic Molecules

In the first stage of synthesis from small organic molecules, the precursors can form chain linking intermediates through a condensation reaction, such as amidation,[18,20] aldol condensation,[21,22] Schiff base condensation[23] or free radical reaction.[24] Second, the intermediates are further polymerised or aggregated with certain polymers (also known as carbon clusters) through covalent, noncovalent, or other interactions to form polymer-like CDs.[25] Third, the polymer-like carbon clusters are further carbonised to generate the carbon core, especially at high temperatures.[26–28] Finally, the remaining precursors can be modified to passivate the surface of CDs to improve the luminescence efficiency. Currently, small organic molecules, such as aromatic rings, citric acid (CA), glucose, cellulose, and phenolic compounds are very popular for the preparation of CDs.

Citric acid is a natural organic acid found in citrus fruits. Plants and animals both produce it as a metabolite. It is established that CA produces fluorescent carbon nanostructures by thermal pyrolysis or carbonisation.[29,30] Because of the direct procedure and high quantum yield (QY), this reagent has become a popular starting material for the synthesis of CDs *via* bottom-up carbonisation pathways.[31–34] Many groups have used pyrolysis, hydrothermal techniques, or microwave-assisted heating methods to make CDs from CA and amine-containing molecules, and in these cases, the obtained CDs are usually doped by nitrogen or have various nitrogen-containing groups on their surface.

Schneider *et al.*[35] investigated three kinds of CDs formed by the hydrothermal reaction of CA with three amine precursors, identified as ethylenediamine (e-CDs), hexamethylenetetramine (h-CDs) and triethanolamine (t-CDs) (Figure 4.1). They found differences in the reactivities of different molecular precursors, resulting in simultaneous chemical reactions under hydrothermal conditions that facilitate the parallel formation of CDs and

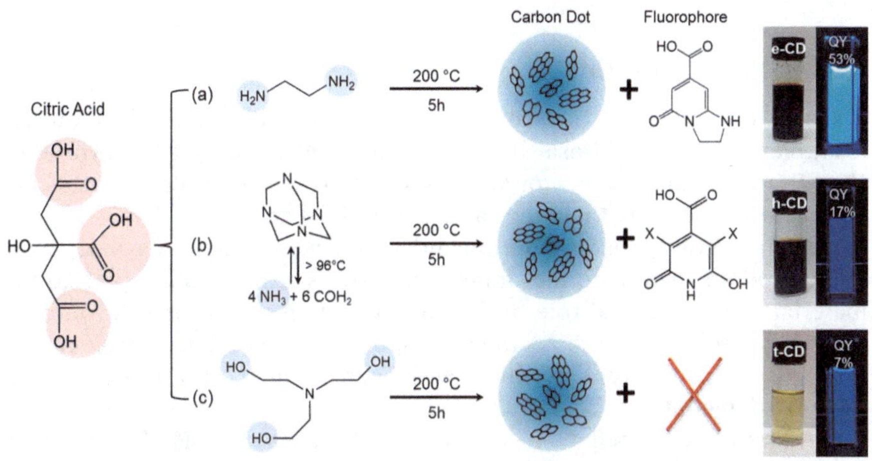

Figure 4.1 Synthesis conditions of citrus acid-based CQDs using three different nitrogen-containing precursors (a) reaction of CA and ethylenediamine, resulting in e-CQDs and the fluorophore IPCA. (b) The reaction of CA with hexamethylenetetramine, resulting in h-CQDs and citrazinic acid and/or 3,5-derivatives due to hexamethylenetetramine decomposition into ammonia and formaldehyde at temperatures above 96 °C. (c) Reaction of citric acid and triethanolamine, resulting in t-CQDs and no derivatives of citrazinic acid, since the tertiary amine prohibits their formation. Images on the right show the purified reaction products under ambient light and the corresponding diluted solutions under UV light excitation reveal blue emission with photoluminescence quantum yields as shown. Reproduced from ref. 35 with permission from American Chemical Society, Copyright 2017.

molecular species likely to be bound to the CDs. Structural discrepancies between the three species have been addressed.

Song *et al.* successfully prepared CDs with strong yellow fluorescence by a simple microwave-assisted method using *o*-phenylenediamine (*o*-PD) as the carbon precursor and named them *o*-CDs.[36] The fluorescent properties of the CDs prepared from *m*-phenylenediamine (*m*-PD) and *p*-phenylenediamine (*p*-PD) differed significantly from those of *o*-CDs. A fluorescence imaging experiment with SW480 cells showed that the *o*-CQDs, thus produced, can be efficiently taken up by cancer cells, showing a good imaging effect.[37] Xu *et al.* have demonstrated that sp^3 compartmentalisation can be achieved by boosting sp^3 bonding fractions during hydrothermal synthesis of UVB-emissive CDs, using *o*-PD and glacial acetic acid (AcOH) precursors. This suggests that the formation mechanism of CDs is not yet fully understood.[38] Nitrogen-doped carbon dots (N-CDs) had significant solid-state emission quenching and solution-state photoluminescence emission in the visible range. However, when AcOH was introduced during the hydrothermal process, the emission of CDs was significantly shifted to the UVB area. This can be attributed to AcOH sp^3 compartmentalisation, which increases the proportion of sp^3 hybridisation and reduces the size of sp^2 conjugating units in UVB-emissive CDs.

4.2.2 Natural Products

The use of renewable, low-cost, and environmentally friendly biomass resources can help address the immediate need for large-scale synthetic CDs. Examples of methods for synthesising CDs include hydrothermal carbonisation, microwave heating, and pyrolysis. The prepared CDs have a consistent particle size distribution and a high yield, and it is advantageous to incorporate heteroatomic doping into the synthesis process. Natural materials have a critical role in the energy system due to their favourable environmental properties.[39] A range of approaches have been developed to prepare CDs from natural products, with hydrothermal carbonisation being the most frequent.

Bandi *et al.* developed a one-step method for the synthesis of fluorescent CDs, using onion waste as the precursor, and a simple autoclave.[40] This method not only offers a solution to the disposal of onion waste but also produces fluorescent CDs that exhibit outstanding properties such as high aqueous dispersibility, good chemical and photostability, strong fluorescence emission with good QY (28%), and excellent biocompatibility. The production yield of this method is also significant (6.06%), making it economical and scalable. Furthermore, these synthesised CDs were successfully applied as fluorescent probes for the sensitive and selective detection of Fe^{3+} ions and have potential as multicolour imaging agents in bioimaging studies with HeLa cells. Xu *et al.* reported a simple green hydrothermal method for processing potatoes to prepare CDs with a particle size distribution between 4 and 8 nm as well as a high QY of 15% in aqueous solution.[41]

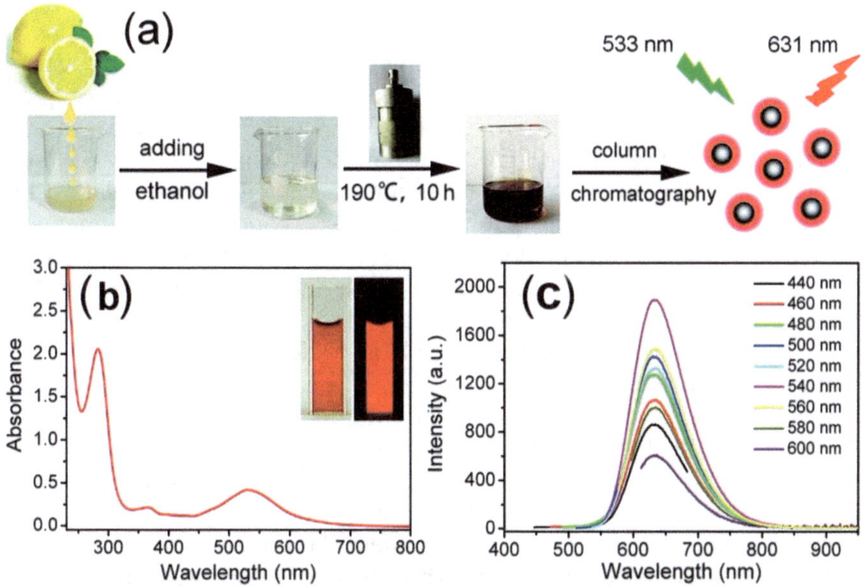

Figure 4.2 (a) Illustration of the preparation process of the R-CDs from lemon juice by solvothermal treatment. (b) UV-visible absorption spectrum and (c) fluorescence spectra of the as-prepared CD mixtures excited by different wavelengths of light. Reproduced with from ref. 42 with permission from the Royal Society of Chemistry.

Ding *et al.* obtained excitation-independent luminescent CDs emitting at 631 nm through heating an ethanol solution of pulp-free lemon juice followed by purification with silica column chromatography (Figure 4.2a–c).[42] The resulting red-emitting carbon dots (R-CDs) are mono-dispersed with an average diameter of 4.6 nm, exhibiting a high QY of 28% in water, which was demonstrated to originate from their surface states and nitrogen-derived structures.

Peanut shells and peanut skins can also be used as carbon sources for pyrolysis to produce CDs.[43,44] In a simple pyrolysis reaction, peanut shells were used as a carbon source to produce spherical CDs with an average diameter of 1.62 nm and a diameter distribution ranging from 0.4 to 2.4 nm. Konjac (root vegetable) flour was used as a carbon source to produce quasi-spherical N-doped CQDs with an average diameter of 3.37 nm. The fluorescence QY of N-doped CQDs was high (up to 22%), but the product yield was as low as 3%.[45]

Plant leaves were also used by Zhu *et al.* as a carbon source.[46] In this work the CDs' fluorescence intensity was greatly improved through plasma and microwave-assisted heating treatments. The resulting CDs were utilised to create fluorescent inks that were useful for printing luminescent patterns. These patterns may be used in a variety of applications, including anti-counterfeit and optoelectronic applications.

4.2.3 Synthetic Polymers

Zeng *et al.* prepared highly fluorescent polyethylenimine (PEI) CDs using a hydrothermal method. The team managed to achieve a fluorescence quantum yield of 38%. Interestingly, the fluorescence intensity decreased with increasing 6-thioguanine (6-TG) concentration, while interference substances had minimal impact. The experiment yielded a linear relationship between fluorescence intensity function and 6-TG concentration under optimal conditions, with a detection limit of 1.33 μM. The method was successfully applied to determine 6-TG in human serum and urine samples. The PEI-CDs fluorescence probe has excellent potential for practical application due to its inexpensive precursors, simple operation, low toxicity, and excellent biocompatibility.[47]

Chaudhary *et al.* conducted a study in which they used cups and polyethylene bags and subjected them to simple heating to synthesize fluorescent CDs. The obtained CDs displayed absorption peaks at 260 nm and had a size range of between 5 and 30 nm. The process of upcycling caused structural changes in plastic waste, which in turn affected the optical properties of the CDs. The emission properties of the three types of plastic waste used as precursors were excellent, and give a peak at 422 nm. Furthermore, the CDs generated from plastic polybags, cups, and bottles (P-CDs, C-CDs, and B-CDs) respectively, had quantum yield (QY) values of about 62%, 65%, and 64%. The researchers also tested the toxicity of the CDs *via* multi-assay biocompatible activities, such as antibacterial and antifungal activities. The study explored the potential use of CDs derived from plastic waste in analytical applications, specifically in selective copper metal ion sensing in aqueous media. The study results highlighted the importance of preserving the environment and addressing the issue of plastic waste.[48]

4.3 Purification, Structure, and Characterisation of Carbon Dots

As a new type of zero-dimensional carbon-based nanomaterial, CDs are usually spherical nanoparticles with a size of less than 10 nm, mainly composed of a graphite-like carbon core with abundant functional groups on the surface. As already noted, CDs can be classified as carbonised polymer dots (CPDs), carbon nanodots (CNDs), carbon quantum dots (CQDs), or graphene quantum dots (GQDs).[49] Carbonised polymer dots contain small carbon cores and numerous polymer chains on the surface, which are typically generated by hydrothermal/solvothermal processes that result in inadequate carbonisation of polymer clusters. Scheme 4.1 shows two core–shell structures for CDs derived from different precursors. Carbon quantum dots, unlike CPDs and CNDs, have a conjugated carbon core, and their luminescence is influenced by the quantum confinement effect, resulting in a redshift of their emissive wavelength as the sp^2-conjugated domain increases. Graphene quantum dots usually present a two-dimensional layered morphology, so are

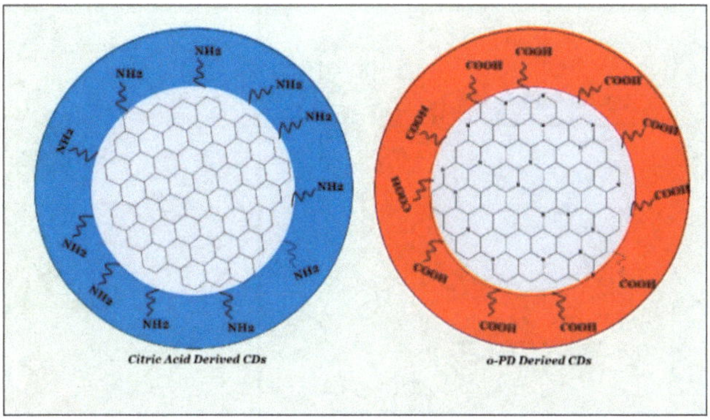

Scheme 4.1 Two basic core–shell structures of CDs derived from citric acid and *o*-phenylenediamine, respectively.

effectively graphene fragments with an average size of roughly 10 nm. Their properties are impacted by the quantum confinement effect while keeping a graphene layered structure, giving them new optical capabilities that graphene lacks.[49,50]

Since by-products (often formed during CDs synthesis) might alter their fluorescence, a purification process is necessary before characterisation. For the removal of large soot particles or low molecular weight compounds, standard centrifugation, dialysis, or filtration procedures are required. It is possible to isolate CDs with uniform properties as well as charge, mass, size, and hydrophobic or hydrophilic nature by combining techniques for fine separation (sucrose density centrifugation, gel electrophoresis, and chromatography) that are based on different physicochemical properties, such as density, charge, hydrophobicity *etc.* By carefully combining separation techniques based on the size, mass, and density of CDs with subsequent separation based on surface qualities, it will be possible to determine the morphological and structural characteristics of CDs, which are responsible for various emissive phenomena.[51,52]

Yuan *et al.* prepared high colour-purity NBE-t-CQDs from symmetric PG triangulogen with a QY of up to 54–72% and a full width at half maxima (FWHM) of 29/30 nm from blue to red (Figure 4.3).[53] Extensive theoretical calculations and structural and optical characterisations showed that the rigid triangular structure, molecular purity, crystallinity, and weak electron–phonon interaction are key factors in the high colour purity. The NBE-t-CQD-based multicoloured LEDs outperformed the inorganic quantum dots-based LEDs in terms of colour purity, luminance, and stability.

Xu *et al.* demonstrated that sp^3 compartmentalisation can be achieved by boosting sp^3 bonding fractions during the hydrothermal synthesis of CDs using *o*-PD and glacial AcOH (Figure 4.4).[38] Some aberration-corrected high-angle annular dark-field scanning transmission electron microscopy (HAADF-STEM) images are shown in the figure. In general, transmission electron microscopy is not easy to apply to carbon nanodots since lattice

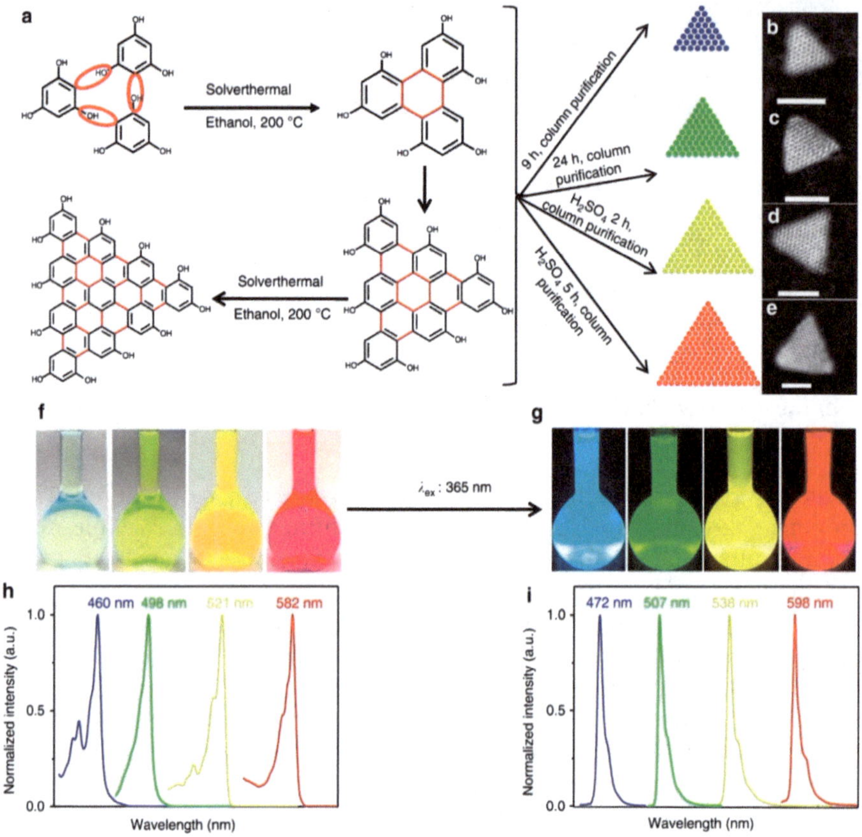

Figure 4.3 Design and synthesis of narrow bandwidth emission triangular CQDs. (a) Synthesis route of the NBE-t-CQDs by solvothermal treatment of PG triangulogen. Typical aberration-corrected HAADF-STEM images of B- (b), G- (c), Y- (d), and R-NBE-t-CQDs (e), respectively; scale bar, 2 nm. Photographs of the NBE-t-CQDs ethanol solution under daylight (f) and fluorescence images under UV light (excited at 365 nm) (g). The normalised UV-vis absorption (h) and PL (i) spectra of B-, G-, Y-, and R-NBE-t-CQDs, respectively. Reproduced from ref. 53 with permission from Springer Nature, Copyright 2018.

fringes are notoriously difficult to image in these tiny structures. The CD stock solution was purified by using column chromatography separation to yield UVB-emissive CDs with a high purity and uniform particle size.[38] These strategies reduce negative effects, such as spectrum broadening and aggregation-induced quenching, enabling high emission efficiency of UVB-CDs in solid state with a narrow bandwidth (PL QY = 20.20%, λ_{ex} = 265 nm) with a very narrow bandwidth (FWHM = 24 nm).

In other work, Wang *et al.* discovered that after purification, six distinct *o*-PD-based CDs showed nearly identical PL emission and absorption spectra.[54] Structural and spectral analysis revealed that *o*-PD-based CDs had

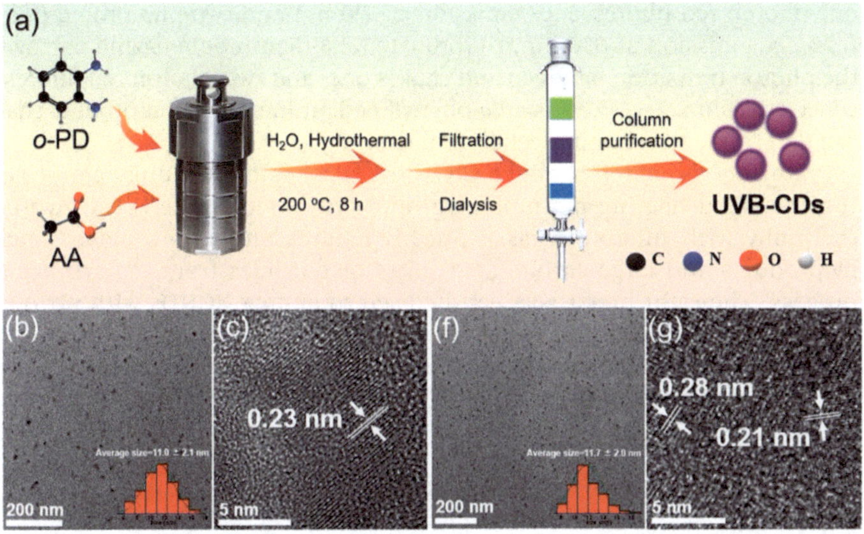

Figure 4.4 (a) Illustration of the synthesis and purification process of the UVB-emissive CDs. TEM, HRTEM images and size distribution (inset) of (b and c) UVB-CDs and (f and g) N-CDs. Insets show the corresponding dots diameter distribution histogram from 100 CDs. Reproduced from ref. 38 with permission from John Wiley & Sons, Copyright 2022.

similar carbon core structures, and single-molecule PL spectroscopy showed that the same PL centre was responsible for the multimodal emission. The 'spectral splitting' phenomena at low temperatures further demonstrated the uniqueness of these *o*-PD-based CDs. The carbon core structure and polymerisation mechanism were investigated by theoretical simulations. Carbon dots were characterised by TEM, Raman spectroscopy, X-ray photoelectron spectroscopy (XPS), and near-edge X-ray absorption fine structure spectroscopy (NEXAFS). UV-visible and fluorescence spectrophotometry were used to investigate the optical properties of CDs.[55] Zhu *et al.* used a one-pot hydrothermal approach to make multicolour emissive CDs and used time-resolved emission spectra to discriminate and identify the origin of multicoloured emission; this was the first time this had been done.[56] Park *et al.* used linear absorption, time-resolved photoluminescence, and femtosecond transient absorption to characterise the optical properties of CDs before and after laser induction.[57] Prato and his co-workers studied structural phenomena during the synthesis of CDs using small-angle X-ray scattering (SAXS) and wide-angle X-ray scattering (WAXS).[58]

He *et al.* synthesised a novel kind of carbon dot by using 2,7-naphthalenediol and dimethylformamide as precursors.[59] Molecular fluorophores and carbogenic nanodots were obtained through column chromatography separation and dialysis, with the molecular fluorophores being 8-ethoxy-3*H*-cyclopenta[*a*]naphthalen-3-one and the carbon dot being a carbon core with attached fluorophores. Zhang *et al.* developed a method to fabricate CDs with one-photon red fluorescence emission at 620 nm and two-photon red fluorescence emission at 630 nm.[60] Protonated 2,3-diaminophenazine narrows the photon transition bandgap and causes one- and two-photon red fluorescence emission. 2,3-DAPN is an *o*-phenylenediamine oxidation product that can determine CDs luminescence qualities.

Prato *et al.* prepared N-doped CNDs (NCNDs) using a multistep process that included heating arginine (Arg) and ethylenediamine (EDA) hydrothermally with microwave assistance.[61] It is essential to remove small by-products and large carbon aggregates or particles from the synthesis process. The right precursors can be used to endow NCNDs with various features, such as chromophore precursors, chirality, and quinones. The energy levels of the NCNDs were adjusted to make nanomaterials with the necessary properties for particular applications.

4.4 Optical Properties of Carbon Dots

4.4.1 Fluorescence

Carbon dots have remarkable optical characteristics. Fluorescence and phosphorescence are two important bioimaging and biosensing phenomena that can be used *in vitro* and *in vivo*, as discussed in Section 4.5. The capacity of carbon dots to emit light in the near-infrared (NIR) region is especially valuable since NIR light can penetrate into tissues and biological systems that are opaque to visible light.[62] A fluorescence induced by excitation is present in most CDs, but nitrogen and sulfur co-doped CDs exhibit an excitation-independent fluorescence.[63] Bao *et al.* used a well-controlled wet oxidative approach to make excitation-independent CDs with tuneable fluorescent colours. Carbon dots PL properties were primarily determined by their molecular weight and degree of surface oxidation.[64] Hu *et al.* have described luminous CDs with a fluorescence wavelength that can be controlled across the visible spectrum by altering the passivation/functionalisation agent, synthesis conditions, and precursor molar mass ratio.[65] Carbon quantum dot emissions can also be controlled using a variety of adaptable solvents.[66] All of these findings suggest that the excitation dependent/independent PL behaviour stems primarily from CDs' surface states.[67] At present, the PL mechanism of CDs is not very clear, but the following possible mechanisms are widely accepted by researchers in this area: the quantum size effect, electron–hole radiation rearrangement in the surface state, and the molecular state luminescence emission process.[68–71] Therefore, the preparation of monochromatic fluorescent CDs and the study of the fluorescence mechanism of CDs are important research directions for improving their applicability. Although CDs do not exhibit the same quantum size effects as classical semiconductor quantum dots, the effects still contribute to their PL properties.[30,72]

Li *et al.* developed a new and simple method for electrochemically synthesising high-quality CQDs. The CQDs displayed stable and strong PL, with a quantum yield of about 12%. These CQDs could be used as fluorescent markers and may also hold potential for high-efficiency catalyst design in bioscience and energy technology. The size-dependent optical properties and upconverted PL properties make them a promising avenue for further research in these fields.[73] Xiong *et al.* performed a silica gel column separation of CDs synthesised by hydrothermal treatment of *o*-PD and urea, and they obtained four components of varying sizes: B-CDs, G-CDs, Y-CDs, and R-CDs, which exhibited different fluorescent colours ranging from blue, green, yellow, and red, as shown in Figure 4.5.[74]

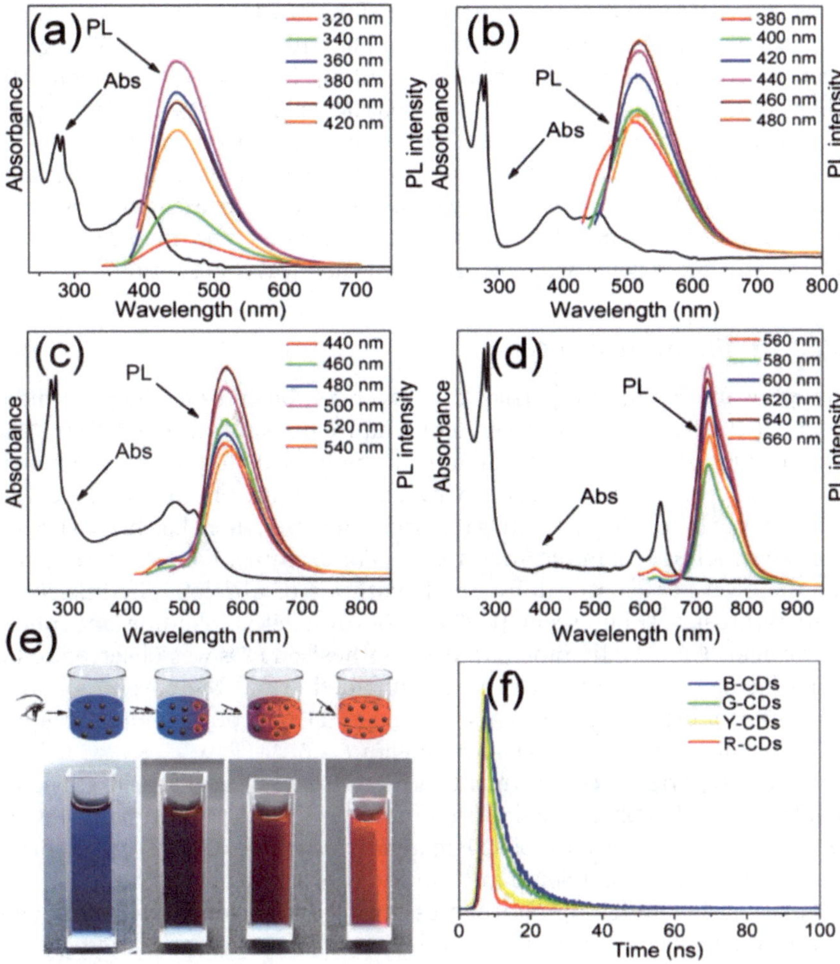

Figure 4.5 Absorption spectra and PL emission spectra of (a) B-CDs, (b) G-CDs, (c) Y-CDs, and (d) R-CDs under excitation of different wavelengths of light (see the inset legends). (e) An R-CDs solution exhibits different colours varying from blue to red when observed from different angles (upper), and the corresponding photos of the same R-CD solution under sunlight (bottom). (f) Time-resolved PL spectra of the four selected samples. Reproduced from ref. 74 with permission from John Wiley & Sons, Copyright 2018.

Surface functional groups of CDs influence surface energy levels and energy gaps significantly. Sun *et al.* were the first to employ PEG1500N to passivate CDs to boost fluorescence yield. The importance of CD surface passivation by N-containing organic moieties for improving CD fluorescence characteristics has been established in subsequent experiments.[6] Sciortino *et al.* synthesised two classes of CDs, one with low nitrogen content and one with high nitrogen content, each with different structural characteristics and optical emissions. The molar ratio of N-doping appears to be a key factor in defining the structural characteristics and optical emissions of CDs.[75] Tang *et al.* synthesised N-doped GQDs with broad emission (300–1000 nm) in the NIR region, but their absorption intensity was low, resulting in weak NIR emission.[76] In other work, Xiao *et al.* synthesised N-doped CDs exhibiting NIR emission with two photon luminescence.[77] Depending on the type of CDs, the emission can be either excitation-dependent or independent. The energy gap and size and structural homogeneity of CDs are well-defined if they are excitation independent. Excitation-dependent emission in CDs is frequently attributed to surface defects, size heterogeneity, and complicated chemical composition.[78]

4.4.2 Phosphorescence

The phosphorescent properties of CDs have been extensively studied. Before 2018, the majority of room-temperature phosphorescence (RTP)-CDs research was focused on embedded or immobilised CDs in specific matrices (*e.g.* polyurethane, zeolite, polyvinyl alcohol, or urea/biuret), and the RTP properties of bare CDs were rarely reported. However, there has been a breakthrough in RTP rCDs in recent years.[79–81] For example, Jiang *et al.* reported the microwave-assisted hydrothermal synthesis of ultralong-lifetime room temperature phosphorescent (URTP) CDs from ethanolamine and phosphoric acid. The RTP lifetime of the as-synthesised CDs was observed to be 1.46 s (actually more than 10 s to the unaided eye).[82] Moreover, a fluorescence/phosphorescence mechanism of the URTP CDs powder and the aqueous solution was deduced, as shown in Figure 4.6.[83]

Tao *et al.* reported URTP polymer CDs made from polyacrylic acid, ethylenediamine, and phosphoric acid, with an unexpected URTP lifetime of 1.39 s. Polymer CDs are a new form of RTP material that is free of metal and difficult to produce in an aqueous solution.[84] Shi and co-workers proposed a more general method for inhibiting phosphorescence by employing water molecules to form intermolecular hydrogen bonds between cyanuric acid and polymer CDs. This method significantly increased phosphorescence emission, but when the water content reached 98%, the phosphorescence decreased and eventually disappeared in a dissolved form. A solvothermal synthesis was used to create F,N-CDs with long-lived triplet excited states, which generated pH-stabilised blue fluorescence and pH-responsive green self-protective RTP.[85,86] The self-protective RTP of F,N-CDs arises from n–π^* electron transitions for C–N/C=N bonds with a narrow energy gap at room temperature. Tian and his co-workers developed a radical binary copolymerisation approach to prepare heavy-metal-free CDs from acrylamide and different phosphors. The

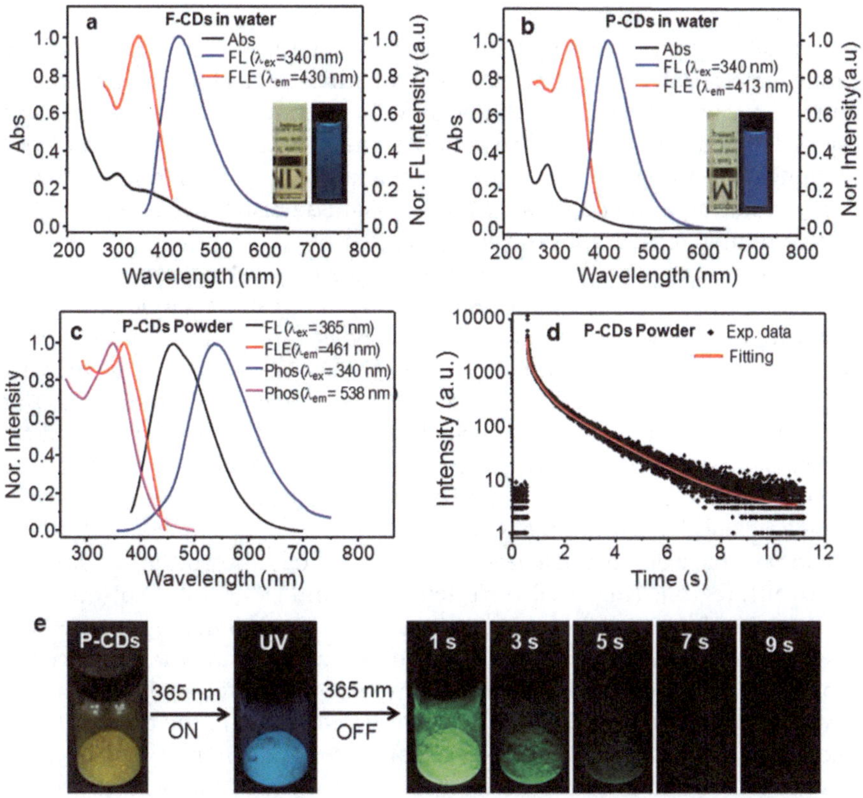

Figure 4.6 (a and b) The normalised UV–vis absorbance (black line), fluorescence emission (blue line), and excitation (red line) spectra of the F-CDs (a) and P-CDs (b) in water (insets showing the corresponding F-CD and P-CD water dispersions under daylight and UV lamp (365 nm)). (c) Normalised fluorescence (black and red lines) and phosphorescence (blue and pink lines) emission and excitation spectra of the P-CD powder. (d) Phosphorescence decay spectra of the P-CD powder (at 340 nm excitation and 540 nm emission). (e) Images of the P-CD powder under daylight, UV lamp (365 nm) ON and OFF with time. Reproduced from ref. 83 with permission from John Wiley & Sons, Copyright 2018.

cross-linked network of hydrogen bonding among polymeric chains immobilised phosphors to subdue non-radioactive transitions and endowed a micro-environment to shield quenchers.[87] Phosphorescent CDs are a fascinating research topic due to their high biocompatibility and low cytotoxicity, which could lead to novel applications in bioimaging or in solution.

4.4.3 Near-infrared Absorption/Emission

Carbon dots are ideal for biomedical applications, such as biosensing and tumour detection. Blue-green light has a shorter wavelength than red light, and therefore less energy. This makes it less able to penetrate deep

(1–2 mm) into the tissue, limiting its potential uses in biomedicine.[88] However, red/NIR-carbon dots with an absorption of 600–950 nm have a deeper tissue penetration depth, less interference with biological self-luminescence, and less tissue damage. They can be used for deep-tissue fluorescence imaging of living animals, photoacoustic (PA) imaging, and other kinds of bio-imaging. Functional groups allow paramagnetic metal ions to interact with them, resulting in metal-doped CDs for magnetic resonance imaging (MRI) and computed tomography (CT) imaging.[89–91]

The inner core of the CDs is a π–π conjugated sp^2 hybrid structure, which allows the CDs to capture photons efficiently in the short wavelength range, resulting in strong absorption in the ultraviolet range (230–320 nm).[92] Carbon dots can be doped with heteroatoms, such as N, O or S, to increase their absorption in the red/NIR range for high-efficiency phototherapy. This increases the amount of light absorbed by the CDs in the 'biological window', which can then be used for therapeutic purposes.[88] Doping with metal ions, such as Cu^{2+}, Mn^{2+} and Ni^{2+}, as well as with N, O or S, can also improve the long-wavelength absorption of CDs.[93–96]

Bao *et al.* have developed a new type of CD that has absorption bands in the red to NIR region. They used citric acid, urea, and DMSO *via* a solvothermal method, in which DMSO acts as both a solvent and an S-doping source. This resulted in high photothermal conversion efficiency and NIR fluorescence. Moreover, the CDs were found to be non-toxic and were renally excreted *in vivo*. After intravenous injection, the CDs accumulated in the tumour tissue and exhibited strong NIR fluorescence and PA signals *in vivo*. The tumours were eradicated after intravenous injection of the CDs followed by irradiation with a 655 nm laser. The CDs having absorption in the NIR region can be safely used as biomedical agents for tumour fluorescence imaging, PA imaging, and photothermal therapy (PTT).[93,94] The assembly of super-CDs is another effective technique for obtaining NIR light absorption, as demonstrated by Li *et al.*[97] Carbon dots have molecular surface layers that control their luminescence and are made up of conjugated carbon cores. The CDs' size effect, surface functional groups, and heteroatom doping are thought to have a significant impact on the luminescence properties of CDs. Red/NIR light-emitting carbon dots can be generated by altering the above features of CDs.

4.4.4 Non-linear Optical Properties

Single-photon fluorescence has a high detection speed and low instrumental requirements, but tissue penetration is weak and the higher the excitation light intensity, the more likely is tissue injury. Up-converting fluorescence overcomes some of the drawbacks of single-photon fluorescence by absorbing two long-wavelength photons and emitting a shorter wavelength photon when it returns to the ground state.[98]

Lin *et al.* synthesised CDs from glutathione using dimethylformamide as a solvent, and showed that these displayed NIR emission for two-photon excited fluorescence (TPEF) bioimaging.[98] Lan *et al.* reported the synthesis of S and Se co-doped CDs displaying maximum absorption at 526 nm with dual emission at 731 and 820 nm.[99] In addition, using a one-pot solvothermal method, Zhan *et al.* reported N-CDs with near-UV, Vis, and NIR emission, as well as excitation-independent NIR emission and excitation-dependent visible emission in the PL spectra.[77] Li *et al.* recently established a simple, high-throughput process for fabricating phosphorus-doped CDs (P-CDs), which demonstrated exceptional two-photon-induced NIR and three-photon-induced red emission in dimethyl sulfoxide (DMSO).[100] The first report of NIR-emitting CDs (centred at 710 nm, PL quantum efficiency of 26.28%) simultaneously demonstrated TPEF under 1200 and 1400 nm stimulation.[101] In the same report, the up-conversion PL of P-CDs was reported. The CDs' emission wavelengths were 450/550 nm when excited at 800/900 nm. Some researchers, however, question if the up-conversion performance claimed in some references is a fluorescence assay illusion.[101]

The grating's spectrum efficiency was a concern when TPEF CDs were activated by 800 nm light; it is combined with the frequency-doubled reflection peak of 400 nm as an excitation, but the CDs still show two-photon fluorescence at *ca.* 400 nm. In any event, more research is needed to ascertain whether all stated P-CDs' up-conversion performance is an illusion or not. A future development strategy is to investigate the theory of TPEF, examine the structure of two-photon fluorescent CDs, then improve the efficiency of the CDs TPEF and build an *in vivo* TPEF bioimaging approach.[102]

4.5 Applications of Carbon Dots

4.5.1 Photo/Electro-catalysis

4.5.1.1 Photocatalysis

Photocatalysis is a sustainable and environmentally friendly process in which a catalyst is activated by light to initiate a chemical reaction. It is widely used for environmental cleaning and energy conversion.[103] A photocatalyst is a substance that absorbs light and generates electron–hole pairs. Many catalysts, such as TiO_2, ZnO, WO_3, SnO_2, CdS, $CdInS_4$, and CeO_2, have been extensively documented in the literature and are frequently employed as photocatalysts.[104–106] Carbon quantum dots are a novel class of nanomaterial that has received a lot of attention because of their low cost and environmentally friendly nature. Their excellent light absorption, high PL QY, up-conversion, and great electron transport capability have made carbon quantum dots a good candidate for photocatalysis applications.[107]

Carbon quantum dots are excellent photosensitisers, and can be used, in combination with other semiconductors or photocatalysts, to enhance their photocatalysis. They absorb visible light with a longer wavelength and then relax, releasing photons with a shorter wavelength in the UV region. After photoexcitation, CQDs with an appropriate bandgap make it easier for photogenerated electrons to transfer from the surface of the photocatalyst.[108] Carbon quantum dots doped with other materials or heteroatoms can also be used for photocatalysis. Doping CQDs with heteroatoms or metals can adjust their band gaps, design and tailor surface groups, and improve electron transmission, as well as the catalytic performance.[109]

Kang *et al.* made CQDs through the electrochemical ablation of graphite that produced visible light-induced photons in solution and worked as acid catalysts. Carbon quantum dots were tested for their ability to generate light-induced protons *via* a variety of processes (aldol condensation, esterification, and Beckman rearrangement) by using acid as a catalyst. Carbon quantum dots catalysed the reactions under visible light for 10 h and achieved conversion efficiencies ranging from 34.7–46.2%. Other conditions remained unchanged. A conversion efficiency of about 5% was observed. Other researchers found that CQDs with acid catalysts are more catalytically active when exposed to visible light.[110] Miao *et al.* developed CQDs mesoporous TiO_2 composites (CQDs-mesoTiO_2) to investigate the degradation of methylene. The up-conversion and electron-withdrawing properties of CQDs resulted in a significant boost in photocatalytic activity, but only 5% of CQDs-TiO_2 were able to remove 98% of the methylene.[111]

Shi *et al.* have developed a new $CuBi_2O_4$ photocatalyst modification with nitrogen-doped carbon dots (N-CDs) that can perform water splitting under non-visible light (>400 nm). The N-CDs/$CuBi_2O_4$ composite produced more photocatalytic hydrogen than pure $CuBi_2O_4$ and CDs/$CuBi_2O_4$, demonstrating excellent photocatalytic activity. Additionally, N-CDs/$CuBi_2O_4$ composites have been demonstrated to show photocatalytic activity.[112] Hao *et al.* used a hydrothermal approach to make GQDs and an *in situ* photo-assisted strategy to anchor them to TiO_2 under UV-vis light irradiation. Their work showed that GQDs act as a solid-state electronic transfer reagent, trapping photoinduced electrons from TiO_2 and improving electron and hole separation efficiency. Additionally, π-conjugated GQDs may function as photosensitisers, providing electrons to TiO_2 that are photoexcited, thus increasing photocurrent density and H_2 evolution activity.[113]

Zhang *et al.* constructed a vis-NIR sensitive photocatalyst using $BiVO_4$ nanosheets and Ag-doped CQDs. Under visible and near-infrared light, the removal efficiency of tetracycline (TC) hydrochloride based on the $BiVO_4$/Ag@N-CQDs composite was studied. Under visible light, TC was degraded by 80.37% in 60 min and 65.67% in 180 min, while under NIR light, TC was degraded by 80.37% in 60 min. Molecular oxygen and surface plasma resonance were also activated in $BiVO_4$/Ag@N-CQDs (SPR).[114]

Carbon dots have unique photophysical and chemical properties that allow them to act as photocatalysts and photosensitisers. They are nonmetallic and chemically inert, making them attractive for photodegradation of contaminants in aquatic environments. They are useful in hybrids because they provide light absorption, photoluminescence, up-conversion, charge carrier separation, and sometimes photosensitisation.

4.5.1.2 Electrocatalysis

The development of sustainable, fossil-free pathways to produce fuels and chemicals of global significance could contribute to reducing carbon dioxide emissions while meeting our daily needs for feedstocks.[115] The electrocatalytic process can be applied in several clean energy conversion applications, such as fuel cells, water electrolysis, metal-air batteries, and CO_2 conversion. A major focus of research has been on the electrocatalytic activity of these reactions, including oxygen reduction reactions (ORR) and hydrogen oxidation reactions (HOR) that occur on the cathode and anode of hydrogen–oxygen fuel cells, respectively, and hydrogen evolution reactions (HER) and oxygen evolution reactions (OER) that also occur on the cathode and anode of hydrogen–oxygen fuel cells. Regeneration fuel cells rely on these reactions to determine their overall performance.

Zhu *et al.* explored the use of a CoP/CD composite as an electrocatalyst for OER in alkaline electrolytes. They found that adjusting the proportion of CDs in the composite had a significant impact on its OER activity. When the percentage of CDs was set at 28.79 wt%, the CoP/CD composite demonstrated superior OER activity in 1 M KOH with an overpotential of 400 mV compared to CoP alone. This improvement is due to the numerous functional groups, small size, excellent conductivity, and quick electron transfer of CDs. Its low cost, stability, and high efficiency make it a promising candidate as an electrocatalyst for OER applications.[116]

Bi *et al.* report on Ni-CDs with a low metal content, which can serve as efficient catalysts for urea catalysis. The low metal doping levels (1.1 at%) of Ni-CDs enhance the catalytic efficiency through the high activity of Ni sites. The catalytic onset potential of Ni-CDs at 100 mV cm^{-2} is as low as 1.38 V due to their abundant catalytic active sites and massive electrochemical active surface area. Additionally, the excellent stability of carbon nanomaterials and the high activity of Ni catalytic sites make Ni-CDs show good durability, with 96% retention after 23 h.[117]

According to Naqvi *et al.*, non-noble metal electrocatalysts can be used for both HER and OER. For water splitting, nitrogen doped carbon dots (NDCDs) were fabricated by facile methods in which CuCo alloy nanoparticles were substituted for nitrogen, resulting in a highly efficient and non-noble metal-based electrocatalyst.[118] Kou *et al.* fabricated highly fluorescent N-CDs with excellent selectivity and sensitivity, and developed a catalyst with high catalytic activity from N-CDs/NiO with a specific overpotential of 380 mV at 10 mA cm^{-2} and a Tafel slope of 57.96 mV dec^{-1}. This study demonstrated the possibility of using N-CDs in OER.[119]

Carbon-based materials, particularly CQDs, are promising for electrocatalytic reactions due to their small size, high conductivity, high surface area, and ease of surface functionalisation. Carbon quantum dots can improve the conductivity of electrode materials, optimise the geometric morphology and crystal phases of active materials, enhance the adsorption and desorption of intermediates, and provide rich active and defect sites for electrocatalytic reactions.

4.5.2 Bio-imaging

4.5.2.1 Fluorescence Imaging

The remarkable fluorescence properties of traditional semiconductor quantum dots and precious metal clusters has led them to be extensively investigated in the field of *in vitro* and *in vivo* bioimaging.[120,121] However, because of the heavy metal atoms found in semiconductor quantum dots, their practical application is limited due to biological safety concerns. Carbon quantum dots, on the other hand, are non-metals with similar fluorescence properties and excellent biocompatibility, making them interesting candidates for building new bioimaging probes.[6,122]

The unique properties of carbon quantum dots give them promise in a variety of biological applications, particularly biological imaging.[123,124] Diverse synthetic materials, reaction pathways, and purification processes, result in CQDs with different optical characteristics and topologies. Different types of cells or tissues, have distinct shapes and structures, and biomarkers in the cell membrane or cytoplasm correspond to particular reactions to foreign nanoparticles.[89,125] Carbon dots exhibit a variety of unique properties in cells (such as targeting membrane lipids, proteins, nucleus and other biomarkers).[126] The application of fluorescence imaging has been the principal research direction of CQDs since their discovery in 2004.[5] Sun and colleagues examined the bioimaging capabilities of CQDs by using human breast cancer MCF-7 cells to demonstrate the potential of CQDs passivated with poly(propionylethylenimine-*co*-ethylenimine) for two-photon fluorescence microscopy. Results showed high fluorescence in both the cell membrane and cytoplasm areas after incubation with CQDs for 2 h at 37 °C.[127] In addition, some studies have found that functionalised CQDs have specific targeting ability for a specific organelle in cells. For example, Gao *et al.*[128] prepared a series of CQDs with inherent mitochondrial targeting by solvothermal methods using glycerol and silane molecules as solvents (Figure 4.7). The green fluorescence of the CQDs coincides with the red fluorescence of the commercial Mito-tracker red dye, proving their excellent mitochondria-targeting ability.

Wu *et al.* reported a one-pot hydrothermal approach to make many novel multifunctional CQDs that resist lysosome/endosomal embedding and preferentially target the nucleus.[129,130] Yang *et al.* recently reported high-performance, deep-red-emitting CPDs fabricated from taxus leaves. Under excitation at 413 nm, the fluorescence QY of far-red emitting CPDs reaches 59%, while under excitation of the far-red fluorescent window, the recorded QY is 31%. The CPDs exhibit strong absorption and emission in the far-red region, low toxicity, and good biocompatibility, making them an efficient probe for *in vivo* fluorescence imaging.[131]

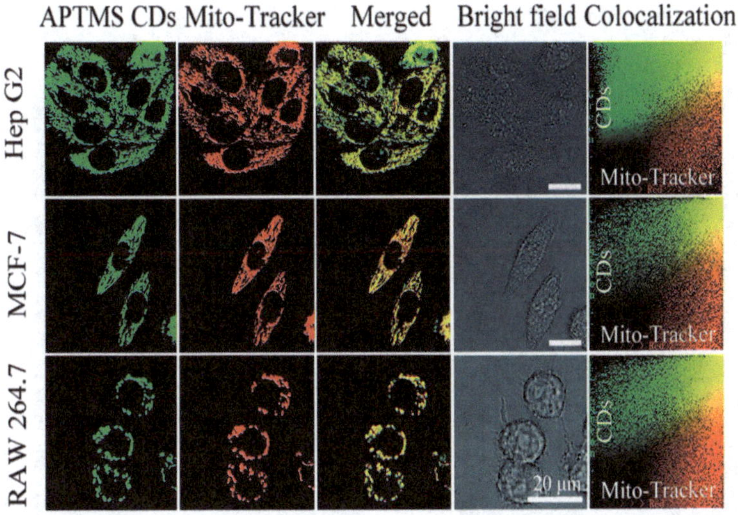

Figure 4.7 Mitochondrial localisation of (3-aminopropyl)trimethoxysilane (APTMS) CDs in different cells (Hep G2, MCF-7 and RAW 264.7). Cells were first incubated with APTMS CDs (30 μg mL^{-1}, Ex: 488 nm) for 30 min and then co-stained with Mito-Tracker (0.5 μM, Ex: 638 nm) for 20 min. The colocalisation coefficients were analysed by Image-Pro Plus software (Rockville, MD, USA). Scale bars: 20 μm. Reproduced from ref. 128 with permission from the Royal Society of Chemistry.

Bi *et al.* recently prepared a novel type of D–π-A conjugated fluorine and nitrogen-doped CQDs (N-CQDs-F) by using urea and NH_4F as dopants in a one-pot solvothermal synthesis procedure. The N-CQDs-F showed strong absorption across the entire UV/Vis-NIR spectrum, and excellent two-, three-, and four-photon excitation up-conversion fluorescence. The greatest advantage of the N-CQDs-F is their very bright far-red emission, which can be used for efficient *in vivo* NIR imaging (Figure 4.8).[132]

4.5.2.2 Phosphorescence Imaging

Optical probe imaging is widely used for monitoring complex biological processes, with great strides having been made in recent decades in improving their sensitivity, specificity, and achievable spatial and temporal resolution.[133] This is because the reporter signals and auto-fluorescence have similar decay times, but the reporter signal is stronger and its decay is more rapid. As a result, the reporter signal is more easily detected than the auto-fluorescence signal, and therefore it is possible to discriminate between the two.[134] Development of long-lived probes capable of distinguishing long-term auto-fluorescence from short-term auto-fluorescence is a highly desired goal in biological imaging applications. Phosphorescent materials are appealing in many applications and are beneficial in understanding light emission processes but preparing them remains a challenge.

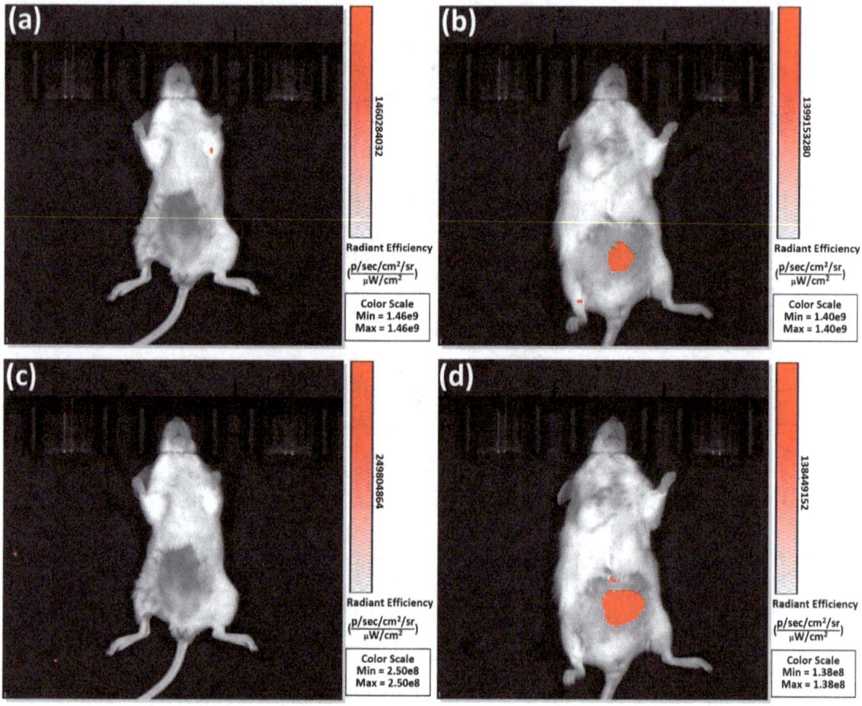

Figure 4.8 *In vivo* fluorescence images of nude mice treated with subcutaneous injection of (a) CQDs and (b) N-CQDs-F (λ_{ex} = 510 nm, λ_{em} = 630 nm). *In vivo* NIR fluorescence images of nude mice treated with subcutaneous injection of (c) CQDs and (d) N-CQDs-F (λ_{ex} = 735 nm, λ_{em} = 785 nm). The colour bars represent the fluorescence intensity. Reproduced from ref. 132 with permission from John Wiley & Sons, Copyright 2020.

Room-temperature phosphorescence (RTP) can be used to obtain high-quality images of biological tissue, but the materials must possess a sufficiently long phosphorescence lifetime in their aqueous environment to effectively eliminate auto-fluorescence. In recent years, RTP materials have been intensively studied and roughly classified into organometallic complexes and crystalline organic compounds.[135–138] The development of metal-free RTP materials with a long lifetime, low toxicity, good water dispersion properties, and inexpensive and simple production routes is highly desirable. Some attempts have been made to develop aqueous CQD-based RTP materials. These complexes usually contain noble metals and are expensive and cytotoxic, while crystalline organic compounds are fabricated by specific organic moieties with limited chemical diversity. This has limited their successful application for phosphorescence imaging. For example, Shi *et al.* successfully fabricated CQD-based long decay composites dispersed in water with a lifetime of 0.703 s, composed mainly of delayed fluorescence, which has a lifetime of 687 ms.[85]

Li *et al.* reported the first application of CQD-based RTP materials for practical imaging of biological samples.[139] They produced an ultra-long RTP from a CQD-based material (CQDs@SiO_2) by incorporating CQDs into a three-dimensional silica network. The system is non-toxic, readily penetrates cell walls, and provides an efficient means to discriminate against tissue autofluorescence. There is potential for future exploration of aqueous CQDs@SiO_2 as ultra-long-lived RTP reporters.

4.5.2.3 Magnetic Resonance Imaging

Recently, CD-based nanoparticles acting as T_1-weighted MRI contrast agents have attracted increasing interest because of their many advantages, such as good renal clearance capability, extremely low cytotoxicity, and strong fluorescence.[122,140,141] Furthermore, many functional groups on the surface of CDs can coordinate with metal ions and then increase the direct contact sites with surrounding water molecules to raise the longitudinal relaxivity (r_1) values.[142–144] Other than one report on metal-free boron-doped graphene quantum dots,[145] most of the previous reports focused on metal-doped CDs showing enhanced r_1 relaxivity and efficient MRI effects. In particular, Gd-doped CDs[146,147] and Gd-encapsulated graphene carbon nanotheranostics[148] endowed clearly enhanced T_1-weighted images of tumours and a relatively long circulation time in tumour-bearing mice, which can be potentially applied for imaging-guided cancer therapy. Moreover, either Gd-encapsulated or Mn-doped CDs afforded not only good r_1 relaxivity but also strong PL, making them appealing MRI/fluorescence dual-mode imaging probes.[141,149]

Carbon dots have been considered promising candidates in nanotheranostics for simultaneously imaging-guided cancer treatment due to their physiochemical and theranostic properties.[150,151] Gd^{3+}-based CDs have previously been produced specifically for multi-modal MRI/fluorescence probes for imaging-guided tumour treatment using various methodologies, such as Gd^{3+} doping and so on.[152–154] For example, Liu *et al.* used self-assembly to produce Gd^{3+} ion-induced CD-aggregates with a photosensitiser for improved fluorescence/MRI dual imaging and photodynamic treatment.[154] The obtained Gd@GCNs featured a strong longitudinal relaxation rate r_1 (16 mM^{-1} s^{-1}, 7 T) as well as a vivid red fluorescence.

Bi *et al.* demonstrated the successful preparation of single-atom Gd anchored onto graphene quantum dots (SAGd-GQDs) with a dendrite-like morphology (Figure 4.9). Interestingly, at a low Gd^{3+} concentration of 2 mol kg^{-1}, the as-fabricated SAGd-GQDs displayed a dramatically improved longitudinal relaxivity (r_1 = 86.08 mM^{-1} s^{-1}), which is 25 times higher than the commercial Gd-DTPA (r_1 = 3.44 mM^{-1} s^{-1}). *In vitro* and *in vivo* experiments reveal that the SAGd-GQDs developed are highly potent contrast agents for obtaining high-definition MRI, potentially paving the way for more precise clinical theranostics in the future.[155]

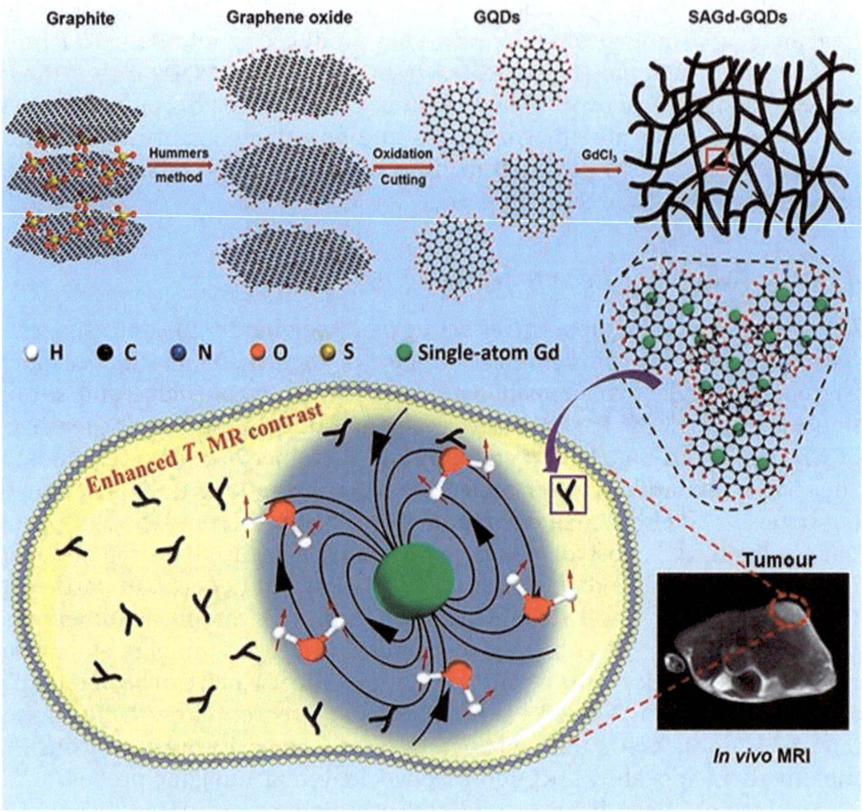

Figure 4.9 *In vivo* MRI of the tumour-bearing mice after intravenous injection of the SAGd-GQD and Gd-DTPA samples at different time intervals. Reproduced from ref. 155 with permission from American Chemica Society, Copyright 2021.

4.5.2.4 *Photoacoustic Imaging*

Photoacoustic imaging (PAI) is a promising biomedical imaging modality that combines the remarkable spatial and temporal resolution of ultrasonic (US) imaging with the excellent contrast of standard optical imaging.[156] Absorbing light energy causes an immediate thermoelastic expansion of a tissue structure, resulting in a broadband US wave that can be detected by a US transducer and converted to electric signals, which are then processed to provide a PA image. Photoacoustic imaging is a non-invasive imaging technique that relies on the imaged tissue's light absorption coefficient. Carbon dots are promising for PAI due to their simplicity of manufacture and functionalisation, despite having a lower molar-extinction coefficient than gold nanocrystals. However, they lack a distinct peak in the wide-band absorption spectrum, preventing them from being used in absorption wavelength-dependent PAI. Fortunately, by incorporating plasmonic metal nanoparticles or small molecules into carbon-based nanomaterials, the PA intensity can be enhanced much further.[157,158]

Ge *et al.* used another conjugated polymer, polythiophene phenylpropionic acid, as a precursor to make CQDs.[159] The synthesised CQDs showed a broad absorption band in the visible to near-infrared region from 500–800 nm with a red emission centred at 640 nm. The results show that red-emissive CDs can simultaneously act as fluorescent, photoacoustic, and thermal theranostics for cancer diagnosis and treatment in living mice at different excitation wavelengths: 350/600 nm for red fluorescence imaging and 671 nm for PAI and PTT.

Qu *et al.* explored supra-CNDs to explore their potential as contrast agents for PAI and photothermal agents for photothermal treatment.[160] They found that intratumour injection of supra-CNDs enabled *in vivo* tumour photothermal therapy under 655 nm laser irradiation. Furthermore, *in vivo* PAI demonstrated that supra-CNDs could accumulate in tumour tissue following intravenous injection. As a result of intravenous administration and subsequent supra-CND tumour accumulation, *in vivo* PTT was carried out, and the lifespan of the mice was extended as a result of the PTT-induced prevention of tumour growth. These highly desirable features indicate that supra-CNDs may be used as biological agents for both tumour therapy and PA imaging.

4.5.3 Phototherapy

4.5.3.1 Photodynamic Therapy

The anti-tumour effects of photodynamic therapy (PDT) and photosensitisers have long been established in treating several types of cancer. Photodynamic methods have evolved since dyes under light irradiation were shown to have a potential therapeutic effect at the beginning of the 20th century. As a result of research, off-target effects are being reduced, and pharmacokinetic properties are being improved. It is notable that the FDA authorised PDT as the first drug/device combo, reflecting the tremendous degree of interest and significant study on the subject.[161] Carbon quantum dots have attracted a lot of interest in connection with PDT, which is an alternative to surgical resection for almost all forms of cancer. This treatment reduces pain and postoperative problems while speeding up healing, but can lead to hair loss, mouth ulcers, and skin redness. Non-invasive PDT has become a recommended cancer treatment to improve patient care. In tumour therapy, PDT is an alternative to tumour ablation.[162–164]

4.5.3.2 Carbon Quantum Dots as Nanocarriers for Photosensitisers

Because of their numerous surface groups, CQDs have the potential to be used as a loading platform for a variety of compounds. Various functional elements, such as drug molecules, proteins, and aptamers, can be incorporated into CQDs through covalent or noncovalent interactions for a range of biomedical applications.[165]

Kim *et al.* prepared CQDs with cyclodextrin as the carbon source in the presence of H_2SO_4 and HNO_3 to study the π–π* stacking interaction.[166] The surfaces of the CQDs were then modified using NH_2-PEG-NH_2 and folic acid (FA), which improved their biocompatibility and tumour targeting abilities even further. Furthermore, through a π–π stacking interaction, zinc phthalocyanine (ZnPc) with a large π-conjugated system was loaded onto CQDs with abundant sp^2-hybridised carbon in their cores. The loading efficiency was estimated to be 6 wt% (Figure 4.10).

The electrostatic interaction between tetraplatinated porphyrin complex (PtPor) with a type of negatively charged CQD, was used by Zhu *et al.* to build a novel therapeutic drug. The CQDs transferred light energy to PtPor through the fluorescence resonance energy transfer (FRET) effect in the CQDs@PtPor system. Under xenon lamp irradiation, CQDs@PtPor produced more $^1O_2$ than pure PtPor, indicating that CQDs@PtPor had a better PDT impact than PtPor.[165]

Carbon quantum dots as nanocarriers can effectively improve the water solubility, biocompatibility, and therapeutic effects of photosensitisers. However, not all the energy from the light absorbed by the CQDs may be efficiently transferred to the photosensitisers through the FRET system. Tian *et al.* proposed several strategies to increase the utilisation efficiency of light and improve the efficiency of FRET by increasing the fluorescence QY of CQDs and shortening the distance between CQDs and PSs. They used acrylic acid, 1,2-ethanediamine and $Mg(OH)_2$ as starting materials to create

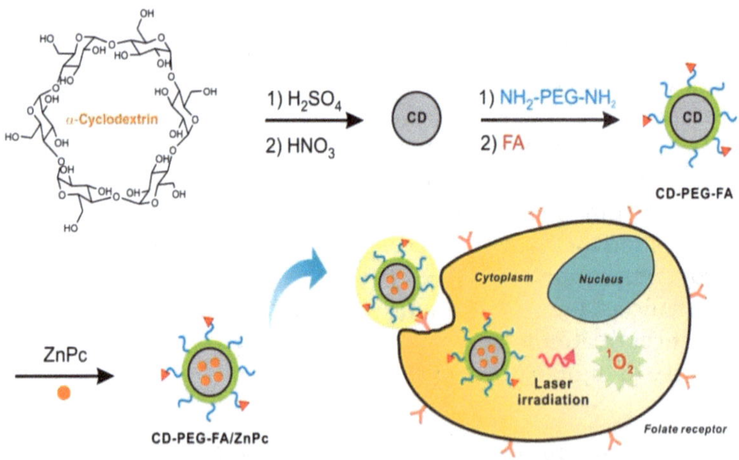

Figure 4.10 Schematic illustration of the preparation of carbon nanodots (CD) from α-cyclodextrin and targeted photodynamic therapy with folic acid functionalised carbon nanodots loaded with zinc phthalocyanine (CD-PEG-FA/ZnPc). Reproduced from ref. 166 with permission from John Wiley & Sons, Copyright 2014.

Mg,N-co-doped CQDs with an ultrahigh PL QY of 84.6%. Chlorin e6 (Ce6) was covalently coupled to the NH_2 group on the surface of CQDs, allowing most of the light energy to be transferred to Ce6. The short distance between CQDs and Ce6 significantly improved FRET efficiency.[167]

4.5.3.3 Carbon Quantum Dots as Photosensitisers

In addition to the excitation wavelength, $^1O_2$ quantum yield is an important parameter for evaluating photosensitiser performance.[168] Common methods for detecting reactive oxygen species (ROS) include UV-visible absorbance, fluorescence, and electron spin resonance (ESR) spectroscopy. Reactive oxygen species are difficult to detect directly because of their high activity and short lifetimes. In ESR, 2,2,6,6-tetramethylpiperidine (TEMP) and 5,5-dimethyl-1-pyrroline N-oxide (DMPO) are used as trapping agents for $^1O_2$ and $^{\bullet}OH$, respectively. The ROS produced by CQDs can react sensitively and selectively with these traps, causing changes in the ESR spectrum.[169] The synthesis of CQDs by Nie *et al.* using citric acid as a starting material and ethanol as a solvent was found to be a cost-effective and eco-friendly process that produced little waste. The CQDs were analysed using both physical and spectroscopic techniques, which confirmed their nanoscale size and photophysical characteristics. The CQDs were also shown to be effective photosensitisers for *in vitro* antibacterial photodynamic inactivation (aPDI). Under visible light illumination, both Gram-negative *Escherichia coli* 8099 and Gram-positive *Staphylococcus aureus* ATCC-6538 were inactivated (99.9999 +%) for 60 min. Further mechanistic studies revealed that the bacteria were inactivated by singlet oxygen, and SEM images of the bacteria incubated with the CQDs showed damage to cell membranes consistent with the action of reactive oxygen species. These results demonstrate that CQDs can be made easily and are efficient photosensitisers for *in vitro* aPDI, working through a Type II mechanism.[170]

In addition to its biological relevance, the cell nucleus is very sensitive to a variety of therapeutic treatments. As a result, nuclear tracking and nuclear-targeted photodynamic treatment are gaining popularity. Wu *et al.* synthesised multifunctional CQDs that effectively target the nucleolus using the carbon sources *m*-phenylenediamine and L-cysteine (Figure 4.11).[130] Carbon quantum dots can track both stationary and living cells in real-time, and when CQDs are combined with protoporphyrin IX (PPIX), the resulting PPIX-CQDs retain their nuclear targeting capacity and create $^1O_2$ directly in the nucleus, destroying genetic information. According to *in vivo* investigations, PPIX-CQDs rapidly concentrate in tumour locations due to an improved infiltration and retention effect, substantially inhibiting tumour growth after laser irradiation. PPIX-CQDs have a higher photokinetic efficiency than free PPIX.

The tumour microenvironment is usually in hypoxia. The therapeutic efficacy of PDT can be enhanced by increasing the oxygen content of the tissue. Wang and colleagues recently used hydrothermal methods to make unique Mn-CQDs containing manganese(II) phthalocyanine. Near-infrared fluorescence and T_1-weighted MRI (Figure 4.12) were both present.[171] The addition

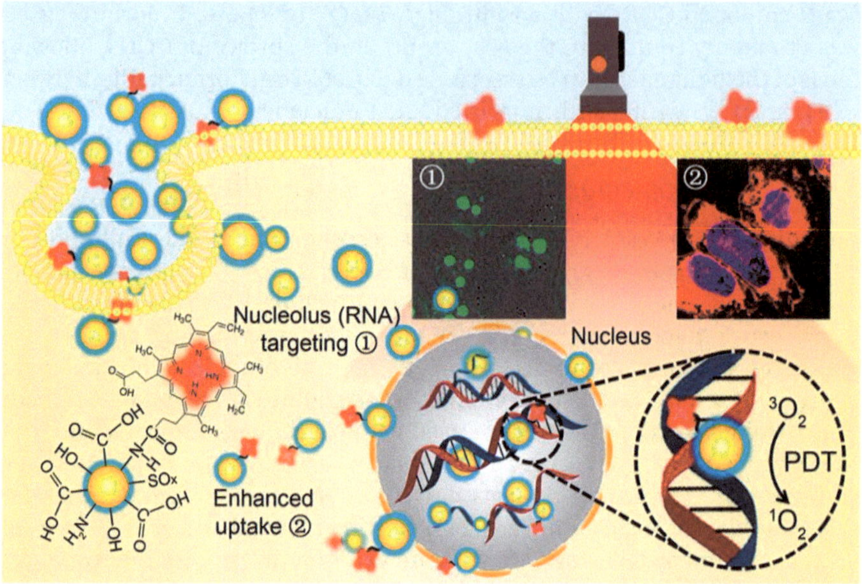

Figure 4.11 Schematic illustration of the synthesis of CQDs, the selective nucleolus imaging of CQDs, and the corresponding enhanced nucleus-targeted photodynamic therapy application from work by Wu *et al.*[130] Reproduced from ref. 130 with permission from American Chemical Society, Copyright 2018.

of Mn to the hypoxic tumour microenvironment increased the photodynamic impact by catalysing H_2O_2 to generate oxygen. The new CQD-based photosensitiser targets hypoxic conditions in solid tumours with a dual imaging mode and oxygenation capability.

Zhao *et al.* recently reported CQDs that could generate singlet oxygen ($^1O_2$), hydroxyl radical ($^{\bullet}OH$), and heat simultaneously under 635 nm laser irradiation, with a $^1O_2$ generation quantum yield of 5.7% and a photothermal conversion efficiency of 73.5% (the highest so far for CQDs).[172] This work showed a possibility of PDT combined with PTT under red-light irradiation.

As a result, some researchers are attempting to incorporate these transition metals into photosensitisers as active centres to improve the materials' photodynamic therapeutic impact. Bi *et al.* prepared a Cu^{2+}-polyacrylate complex and then copper-doped CQDs by *in situ* polymerisation using acrylic acid and copper nitrate as raw materials, ammonium persulfate as the initiator, and hydrazine hydrate as the reducing agent.[173]

4.5.4 Imaging-guided Cancer Cell Therapy Using Carbon Dots

Image analysis technology is now widely used in cancer diagnosis and treatment, and it has three advantages: (1) it can provide a direct and effective method for an early cancer diagnosis; (2) the behaviour of therapeutic

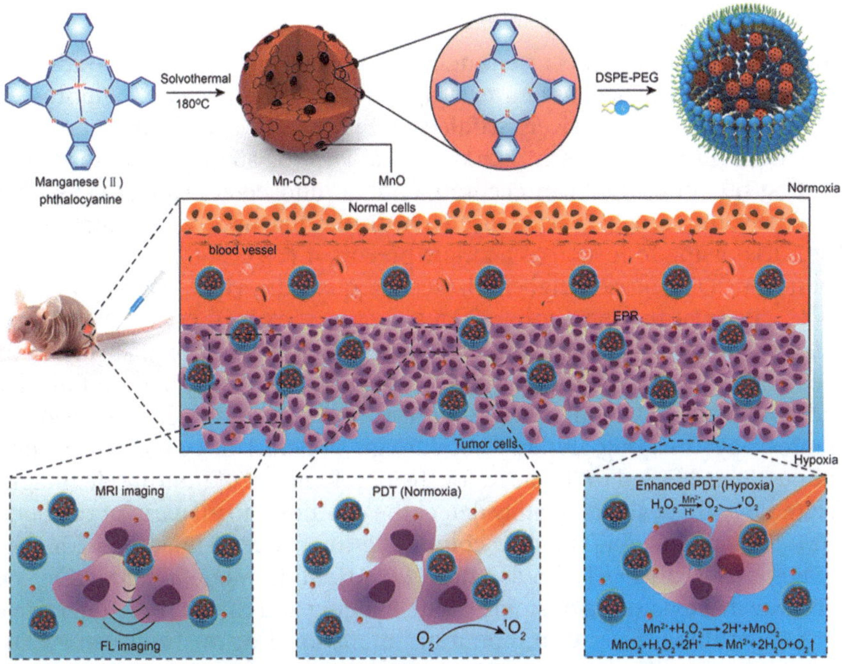

Figure 4.12 Schematic illustration of the assembly of Mn–CQDs as acidic H_2O_2-driven oxygenerators to enhance the anticancer efficiency of PDT in a solid tumour. Reproduced from ref. 171 with permission from John Wiley & Sons, Copyright 2018.

nanoparticles in metabolic pathways can be directly observed; and (3) the therapeutic nanoparticles' response to external stimuli can be controlled.[174] As already noted, carbon dots are a luminous material with low toxicity, high drug loading, facile release, strong biocompatibility, and broad imaging features. Carbon quantum dots also have good optical characteristics, ranging from ultraviolet to near-infrared emission wavelengths.[175] Bi *et al.* have created a method for imaging stress-granule (SG) dynamics and bulk RNA labelling in living cells by employing CDs prepared using microwave-heating (named M-CDs). This method relies on the delicate design of the precursors, which gives M-CDs enhanced RNA selectivity and vivid red PL (22.83%; λ_{em} = 642 nm). Additionally, this fluorescence shows a 2.1-fold increase in response to RNA binding, which enhances the capacity to label RNA. M-CDs are well suited for revealing RNA-related biology, such as SG assembly, due to their high stability, small size, and outstanding biocompatibility.[176] Moreover, the M-CDs with superior RNA selectivity have a higher $^1O_2$ generation capability in response to RNA binding. Therefore, a new type of CDs-based photosensitiser with both high $^1O_2$ generation and RNA targetability has been developed for achieving pyroptotic cancer cell death.

4.5.4.1 Single-mode Imaging

Major challenges in cancer treatment include how to target and identify cancer cells as well as how to deliver drugs with targeted release.[177] Fluorescent CQDs have been extensively studied as fluorescent agents or intracellular imaging agents, used both *in vitro* and *in vivo* for single-mode imaging.[178] Carbon quantum dots can accurately target and detect cancer cells while also delivering drugs. Qu *et al.* used a hydrothermal approach to make CQDs (N-CQDs) using folic acid as a single precursor (Figure 4.13).[179] In aqueous solution, CQDs have a high fluorescence quantum yield, which can reach 94.5%. Because they have folic acid residues on the surface, the CQDs have an intrinsic targeting property and do not require any additional surface modification, leading to precise targeting of cancer cells and effective folic acid receptor-mediated cellular absorption. However, it has been observed that fluorescence imaging alone is not enough to detect the condition in deep tissues. This has led to the need for the development of multimode imaging technology that can adequately detect the condition in all parts of the body.

Multimode imaging involves combining fluorescence imaging with other imaging technologies (such as MRI and PAI). Multi-mode imaging does not have the disadvantages of single-mode imaging and can be more effective in targeting tumour cells. Common multi-mode imaging methods using CQDs are dual-mode fluorescence/MR imaging and dual-mode fluorescence/PA imaging and.[165] These are discussed in the following sections.

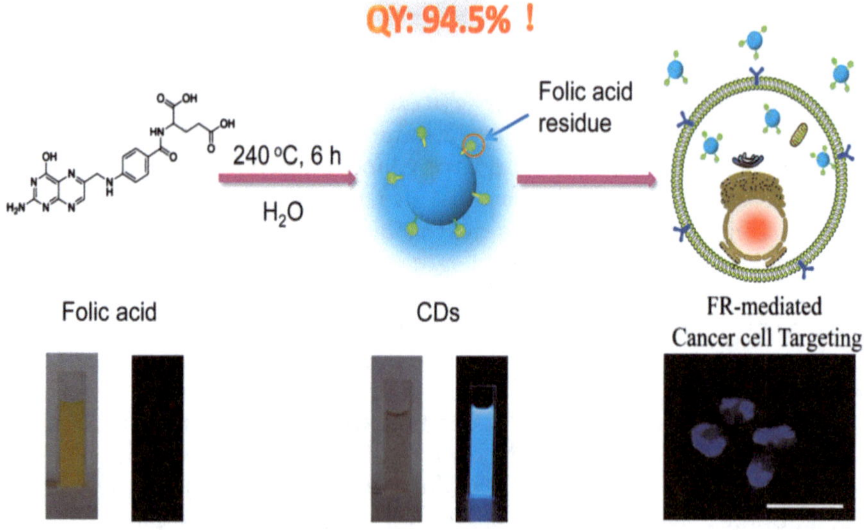

Figure 4.13 Schematic illustration of the synthesis of folic acid-derived CQDs and their application for folate receptor (FR)-mediated cancer cell targeting. Reproduced from ref. 179 with permission from Springer Nature, Copyright 2018.

4.5.4.2 Fluorescence/Magnetic Resonance Dual-mode Imaging

In cancer detection and treatment, fluorescence (FL) and magnetic resonance (MR) imaging can be combined to provide excellent contrast in soft tissues, high spatial resolution, high sensitivity, and rapid data acquisition. Currently, FL/MR dual-mode imaging is most commonly achieved by using Fe^{2+} and Cd^{3+} doping. Zhao's group used GSH, EDTA, and $FeSO_4{\cdot}7H_2O$ to make iron-doped CQDs (Fe-CQDs) in a one-pot hydrothermal synthesis (Figure 4.14). With dual-mode imaging nanoscale probes, Fe-CQDs offer outstanding magnetic resonance imaging and fluorescence imaging capabilities and can be employed in the diagnosis of a variety of disorders. Unpaired electrons in the valence shells of Fe^{3+} and Fe^{2+} give them a high electronic spin. Because iron is a common element in the human body, it can generate a considerable local magnetic field in response to external fields. Because of their biocompatibility, Fe-CQDs can be employed in MRI.[180]

The doping of Mn^{2+} ions in CQDs also allows them to be employed for MRI. Ji *et al.* used a one-pot microwave-assisted heating approach to make Mn-doped CQDs for use as an MR/FL dual-mode imaging nanoscale probe.[181] *In vivo* MR and optical imaging of a mouse brain with a minor glioma shows that Mn-doped CQDs can improve MR and optical contrast in the tumour region, allowing for the detection and surgical localisation of small brain gliomas.

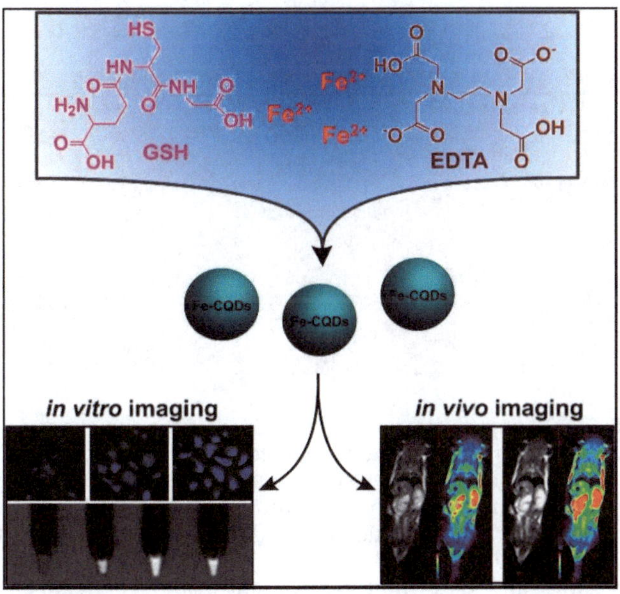

Figure 4.14 Schematic presentation of the synthesis of Fe-CQDs and applications to magnetic resonance and fluorescent dual-mode imaging. Reproduced from ref. 180 with permission from Springer Nature, Copyright 2019.

4.5.4.3 Fluorescence/Photoacoustic Dual-mode Imaging

As noted in Section 4.5.2.4, photoacoustic imaging is a new method for deeper tissue imaging that combines photoexcitation and ultrasound detection to transcend traditional imaging limitations and provide higher spatial resolution. Wu *et al.*[96] have developed a synthetic strategy to prepare porphyrin-implanted carbon nanodots (PNDs) using partial and selective pyrolysis of tetrakis(4-aminophenyl)porphyrin and citric acid (Figure 4.15). These PNDs are said to be stable and biocompatible and have unique properties

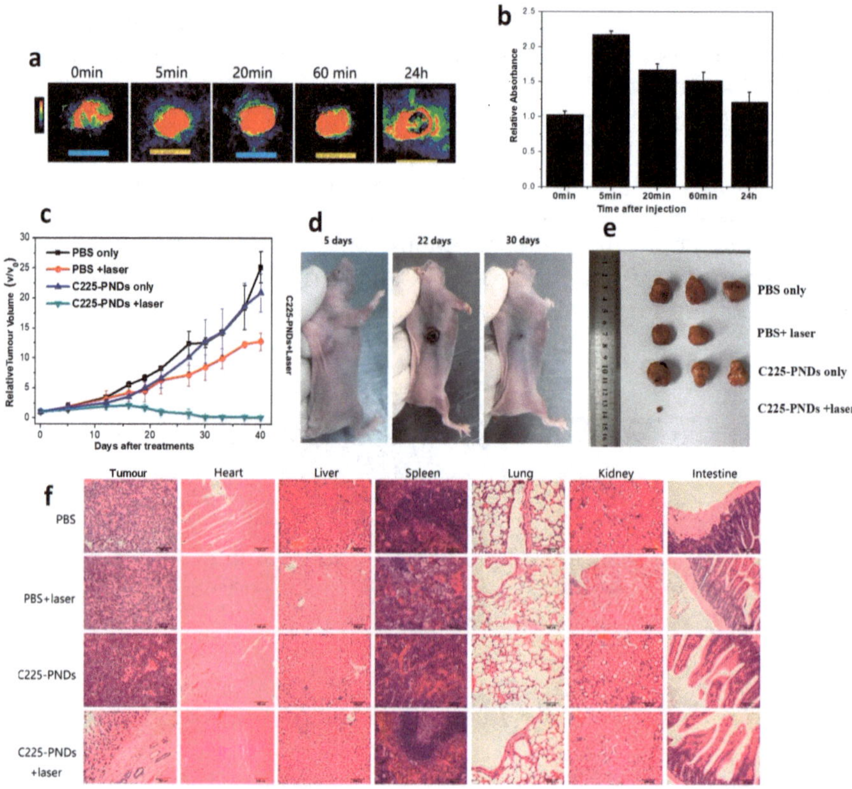

Figure 4.15 (a) Photoacoustic imaging of C225-PNDs in tumour at 0, 5, 20, 60 min, and 24 h after intratumoral injection (7 mg mL^{-1}, 100 μL). (b) Relative PA intensity of C225-PNDs in tumour at 0, 5, 20, 60 min, and 24 h after intratumoral injection (7 mg mL^{-1}, 100 μL). (c) Relative tumour volume of the tumour-bearing mice of the different groups after treatment ($n = 3$, $P < 0.05$ for each group). (d) Photographs of the tumour-bearing mice after different days (5, 22, and 30 days) of treatments with C225-PNDs and laser ($\lambda_{ex} = 808$ nm). (e) Tumour size of the different groups after treatment for 30 days. (f) Hematoxylin and eosin (H&E)-stained slices of the tumour, heart, liver, spleen, lung, kidney, and intestines in mice after treatments. Scale bar: 100 μm. Reproduced from ref. 96 with permission from American Chemical Society, Copyright 2018.

like strong UV-visible and near-infrared absorption. As a result, they are ideal for photodynamic therapy and can even act as a contrast agent for photoacoustic molecular imaging with deep tissue penetration and fine spatial resolution. Interestingly, the Cetuximab-conjugated PNDs have been used to target cancer cells with overexpression of epidermal growth factor receptor (EGFR), resulting in efficient photodynamic therapy.

In other work, Gao's group employed melanin as a carbon source and used hydrothermal techniques to make melanin CQDs (MCQDs) as an excellent fluorescent probe for the diagnosis of triple-negative breast cancer.[182] Melanin is a natural pigment found in most organisms. Not only is it less hazardous to use melanin as a carbon source, compared with small organic dyes such as porphyrin, cyanine and rhodamine, but it also has good biocompatibility and bioimaging capabilities. Moreover, MCQDs have a lengthy excitation peak at 615 nm and a short emission peak at 650 nm, indicating that they can be used for fluorescence imaging, and their absorbance peak at 633 nm suggests that they can be used for PAI. Melanin CQDs outperform regular CQDs in terms of fluorescence and photoacoustic properties, allowing for high-resolution imaging and deep tissue imaging.

4.5.4.4 Imaging-guided Photodynamic Therapy

Photodynamic therapy kills tumour cells by accumulating photosensitisers in a tumour and then exposing it to a specific wavelength of light, which produces reactive oxygen species that kill tumour cells.[183,184] *In vitro* and *in vivo*, CQDs embedded in polymer microsphere or nanosphere (CQDs@polymer) exhibit excellent targeted delivery, achieving simultaneous imaging and targeted PDT after irradiation. The photodynamic action is achieved by creating $^1O_2$ by light irradiation photosensitisers delivered by FA-mediated CQDs@polymer *via* endocytosis.

It is also possible to change the surfaces of CDs to make them behave as a new photosensitiser. Shen *et al.* used a one-step hydrothermal approach to make hyaluronic acid-derived CQDs (HA-CQDs).[185] Under 650 nm excitation, HA-CQDs can generate $^1O_2$ and emit a bright blue photoluminescence. Furthermore, HA-CQDs retain a portion of the HA structure, providing them with a significant advantage in cancer cell targeting. As a result, imaging-guided photodynamic treatment with HA-CQDs has a great deal of promise.

4.5.4.5 Imaging-guided Photothermal Therapy

Photothermal therapy (PTT) converts light energy to heat by photothermal agents, and kills tumour cells. It may play an important role in the treatment of hypoxic tumours. The advantage of image-guided PTT lies not only in its ability to identify the location and size of tumours but also in its ability to monitor the effect of treatment in real-time.[186] Carbon quantum dots can be used as nanocarriers for photothermal agents. Hao *et al.* used watermelon as a carbon source for the first time to synthesise CQDs with NIR-II emission.[187] The CQDs with 900–1200 nm luminescence are under 808 nm

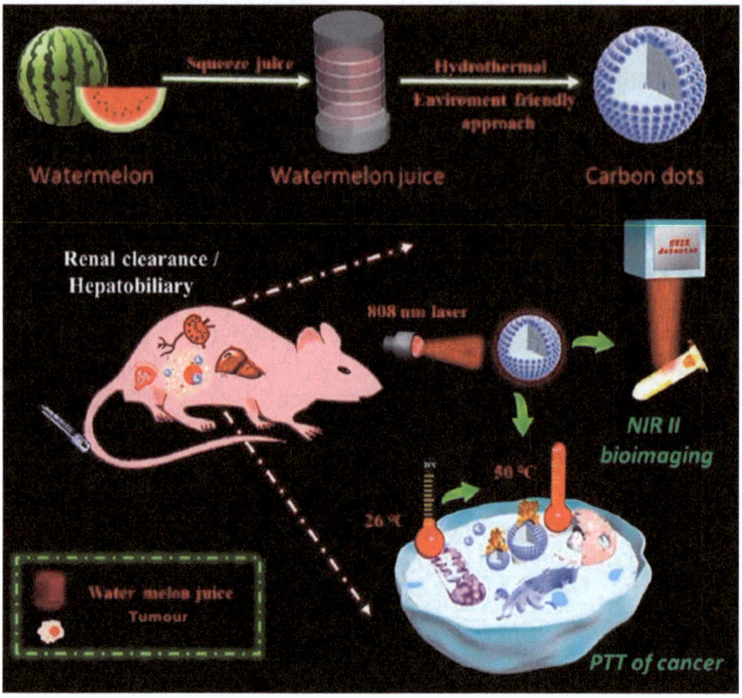

Figure 4.16 Schematic illustration of the synthesis of NIR-II-emitting CQDs for rapid renal clearance, NIR-II bioimaging, and PTT of cancer. Reproduced from ref. 187 with permission from American Chemical Society, Copyright 2019.

laser irradiation (Figure 4.16). In addition, the CQDs had an excellent photothermal conversion efficiency (= 30.6%), ultra-small particle size, and good biocompatibility. It has proven to be an effective probe for *in vivo* NIR-II bioimaging and an ideal material for thermal ablation of cancer. In addition, the designed CQDs have the property of being rapidly cleared *via* the kidneys.

Tumour cells have a different pH from normal cells, which offers a way to improve the treatment effect by designing pH-responsive carbon dots. Park *et al.* developed pH-responsive fluorescent carbon nanoparticles that are composed of a carbonised zwitterionic polymer loaded with a photothermal dye (IR825).[188] When the nanoprobe reaches the tumour cell, its fluorescent switch is turned on and IR825 is released at the same time to convert the light energy into heat energy. It also shows good cellular uptake and biocompatibility *in vitro* and *in vivo*. The combination of imaging and PTT significantly improves the therapeutic effect on tumours and offers a new strategy for tumour therapy.

4.5.4.6 *Synergistic Therapy*

Phototherapy is a novel approach to cancer treatment, as discussed in Section 4.5.3. Photodynamic therapy involves the production of reactive oxygen species by photosensitisers in the presence of oxygen, which brings about the death of cancer cells. Dual-phototherapy (photodynamic therapy/photothermal therapy) functional nanotherapeutic agents have recently received a lot of attention in the field of cancer treatment because they have a greater synergistic therapeutic impact than single phototherapy.[189]

Magnetofluorescent carbonaceous quantum dots (MCQDs) have attracted considerable attention in biomedical studies due to their important role in photothermal cancer therapy. Zhang *et al.* synthesised FeN@CQDs with intrinsic photoluminescent and magnetic properties using a green hydrothermal method. These FeN@CQDs were conjugated with folic acid and riboflavin (Rf-FA-FeN@CQDs) as light-triggered theranostics for simultaneous PTT and PDT. Doxorubicin, an anticancer drug, was further incorporated into the GP-Rf-FA-FeN@CQDs-DOX to enable targeted drug delivery. Results of both *in vitro* and *in vivo* experiments show that the developed anticancer drugs can deliver them to target cells, release them intracellularly under NIR irradiation, and effectively eliminate tumours through chemo-photosynergistic therapeutic action.[190]

Jia *et al.* successfully fabricated CQDs using *Hypocrella bambusae* (HB) as a precursor by a solvothermal method (Figure 4.17). The *Hypocrella bambusae* CQDs had good water solubility, broad absorbance, red emission, and low cytotoxicity. They can produce large amounts of $^1O_2$ and photothermal-conversion efficiency (27.6%) under 635 nm laser irradiation. With the combination of PDT and PTT, this work offers a new oncology drug treatment,

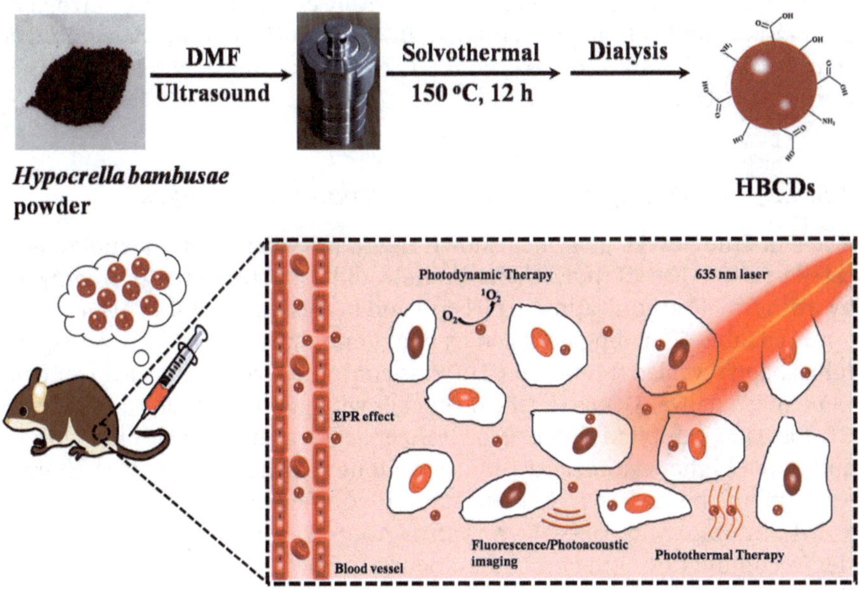

Figure 4.17 Schematic illustration of CQDs derived from *Hypocrella bambusae* for bimodel FL/PA imaging and synergistic PDT/PTT of cancer. Reproduced from ref. 191 with permission from Elsevier, Copyright 2018.

and provides a new method for the synthesis of other carbon-based nanomaterials using natural materials for cancer treatment.[191]

4.5.5 Drug Delivery Systems Based on Carbon Dots

Despite the fact that typical chemotherapy medications like DOX and paclitaxel have great anticancer properties, they also have drawbacks. They can produce systemic toxicity because of their low targeting. Furthermore, due to incorrect medicinal use, the emergence of multi-drug resistance (MDR) can cause low absorption of cancer cells and nuclear infiltration, drastically reducing the therapeutic efficacy. Therefore, it is critical to develop a practical nanomedicine delivery technology. Carbon quantum dots have the potential to overcome MDR and improve the effectiveness of cancer treatments while avoiding the negative effects associated with anticancer medications.[192]

4.5.5.1 Drugs Bound to Carbon Dots

Carbon quantum dots are rich in surface groups and have a large specific surface area and are often combined with drugs through non-covalent bonding, such as accumulation, van der Waals interaction forces, and other physical adsorption phenomena. The combination is simple and effective for drug delivery. Tan and his group found that beer carbon dots (BCD) had high solubility and excellent fluorescence properties, with a quantum yield of about 7.39% and an average particle size of 2.5 nm. Positively-charged doxorubicin (DOX) is charged onto the surface of negatively charged BCD by electrostatic interaction, and adriamycin-coupled BCD (BCD-DOX) induces prolonged cytotoxicity compared to DOX, which is caused by the slow release of DOX from BCD-DOX.[193]

4.5.5.2 Targeting of Drug-loaded Carbon Dot Particles

A recent study by Yu *et al.* has shown that a non-toxic, water-soluble nanocarrier called CD-PEI (polyethylenimine) can be created through microwave hydrothermal carbonisation of glycerol and branched PEI-25k. This nanocarrier is capable of loading both cancer drugs and fluorescence for bioimaging. It has been revealed that when CD-PEI is combined with the anticancer drug DOX, it can deliver DOX to liver cancer cells with good selectivity and efficacy. CD-PEI-DOX exhibited superior anticancer activity, larger levels of apoptosis, and much higher cytotoxicity than DOX alone. Furthermore, *in vivo*, therapy experiments have shown that this combination exhibits a stronger tumour growth inhibitory impact than a control group. The use of CDs as a drug carrier in conjunction with chemotherapeutic medications produces a complex that has a high level of selectivity for tumour cells and is ecologically safe.[192]

Wu *et al.* used *m*-phenylenediamine and L-cysteine to synthesise CDs by a one-pot hydrothermal synthesis, and then linked these with the commonly used drug photosensitiser protoporphyrin IX (PPIX), to create a new class of multifunctional fluorescent CQDs that can successfully target the nucleus.[130] The final CQD-PPIX nanodrug system showed a significant increase in cellular uptake and nucleus targeted features, greatly improving the efficiency of phototherapy. *In vivo* experiments have shown that the negatively-charged CQD-PPIX nanomissile can quickly and specifically target tumour sites after intravenous injection and induce effective tumour ablation after laser irradiation without toxic or side effects.

4.5.5.3 *Releasing the Drug From Loaded Carbon Dot Particles*

Omidi *et al.* synthesised 5-fluorouracil-chitosan-carbon quantum dot-aptamer (5-Fu-CS-CQD-APT) nanoparticles to deliver 5-fluorouracil anticancer drugs. *In vitro*, rapid release of 5-Fu-CS-CQD-APT occurred between 10 min and 6 h below pH 5.4, followed by slow release with a constant release rate of between 6 h and 48 h with 77% of the drug remaining in the nanoparticles at pH 7.4. This indicates better drug release in the tumour microenvironment, reducing drug side effects. Protonation of amines and carboxyl groups alters the electrostatic interactions between chitosan and carbon quantum dots, accelerating drug release.[194] Zhao *et al.* developed a novel drug nanocarrier based on cisplatin(IV) prodrug-loaded charge-convertible CDs (CDs–Pt(IV)@PEG-(PAH/DMMA)) that responds to the tumour extracellular microenvironment. The CDs are equipped with an anionic polymer with dimethylmaleic acid (PEG-(PAH/DMMA)), which can undergo charge conversion to a cationic polymer in a mildly acidic tumour extracellular microenvironment (pH ~ 6.8). This leads to strong electrostatic repulsion and the release of positive CDs–Pt(IV). The positively-charged nanocarrier has a high affinity for the negatively-charged cancer cell membrane, which results in enhanced internalisation and effective activation of cisplatin(IV) prodrug in the reductive cytosol. This innovative technology has the potential to revolutionise imaging-guided drug delivery.[195]

4.5.5.4 *Crosslinking-assembly Delivery System*

Neurodegenerative diseases in an organism occur due to various causes, such as head injuries, defective proteins, toxins, *etc.* The main obstacle in the treatment of these disorders is the difficulty of permeating therapeutics across the blood–brain barrier. Carriers are used to transport drugs to the desired area, and the synthesis of efficient carriers to achieve this goal is an emerging area of research.[196]

Carbon quantum dots are promising drug carriers because of their small size. However, they have similar difficulties crossing the blood–brain barrier as many small molecules. Leblanc *et al.* have successfully synthesised tryptophan CQDs with the aim of using tryptophan as an amino acid to cross the blood–brain barrier *via* LAT1 transporter-mediated endocytosis. The CQDs prepared from these precursors showed excitation wavelength-dependent emission and low toxicity and were observed to appear within the central nervous system of zebrafish (*Danio rerio*). The proposed mechanism for these CQDs to cross the blood–brain barrier involves residual tryptophan molecules attaching to the surface of the CQDs, allowing them to be recognised by the LAT1 transporter. The role of CQDs in transport offers promising pathways for drug delivery and brain health.[197]

Multifunctional zwitterionic nanogels hold great potential for targeted drug delivery and simultaneous imaging in cancer therapy, due to their great stability, bioimaging capability, excellent biocompatibility, controlled drug release, and selective cell targeting.[198] Wang *et al.* prepared a smart nanocarrier (MSNs-SS-CDPAA) with redox-responsive controlled drug delivery and *in vivo* bioimaging properties, binding fluorescent CQDs to the surface of mesoporous silica nanoparticles (MSNs) *via* disulfide bonds.[199] Zeng *et al.* showed that green-emitting CQDs with a carboxyl-rich surface can be used as a trackable drug delivery agent for localised cancer treatment. They are conjugated with the cancer drug DOX *via* non-covalent bonding, using the pH difference between cancer and normal cells as the triggering mechanism for DOX release.[200]

Carbon quantum dots were prepared using citric acid and ethylenediamine *via* a hydrothermal process by Zhu *et al.* The CQDs thus produced showed good biocompatibility and potential proliferative function in L929 cells. The CQDs-DOX drug delivery system showed a relatively high loading capacity and a good drug release profile. In addition, the *in vitro* results showed that the CQDs-DOX complex had higher cellular uptake and a more potent inhibitory capacity for cancer cells than that of free DOX.[201] The drug delivery systems mentioned above have already displayed highly promising properties. It should be assumed that these CQD-based nanomedicines could have great potential as clinical cancer theranostics.

4.5.6 Other Biological Applications of Carbon Dots

4.5.6.1 Tumour Markers

Early detection of cancer cells or associated indicators is critical for the success of cancer treatment. Chemical substances produced by other cells in the body in response to tumours can act as tumour markers and can thus be detected in many body fluids, tissues, and cell lines. They include tissue markers (hormone receptors, cathepsin), genetic markers, serum markers (carcinoembryonic antigen, cytokeratin), and tumour-associated antigens used as biomarkers for cancer diagnosis.[202,203] Carbon quantum dots have significant advantages over traditional organic dyes and semiconductor QDs, and therefore can be used as suitable molecular probes for the low-concentration sensitive detection of tumour markers and cancer cells.[204,205]

Zhang *et al.*[206] developed an immunosensor array for sequential detection of tumour markers using SiO_2@CQDs as signal enhancement markers. The proposed immunoassay showed excellent precision and sensitivity, with CA15-3 being the most commonly used serum marker in breast cancer. Mohammadi *et al.*[207] developed a new method for FRET immune-sensing for detecting CA15-3 tumour indicators (Figure 4.18). The method relies on bio-specific interactions between the antigen, matching antibody, and aptamer. Carbon dots with CA15-3 antibody functionality and Au nanoparticles labelled with polyamidoamine (PAMAM)-dendrimer/aptamer

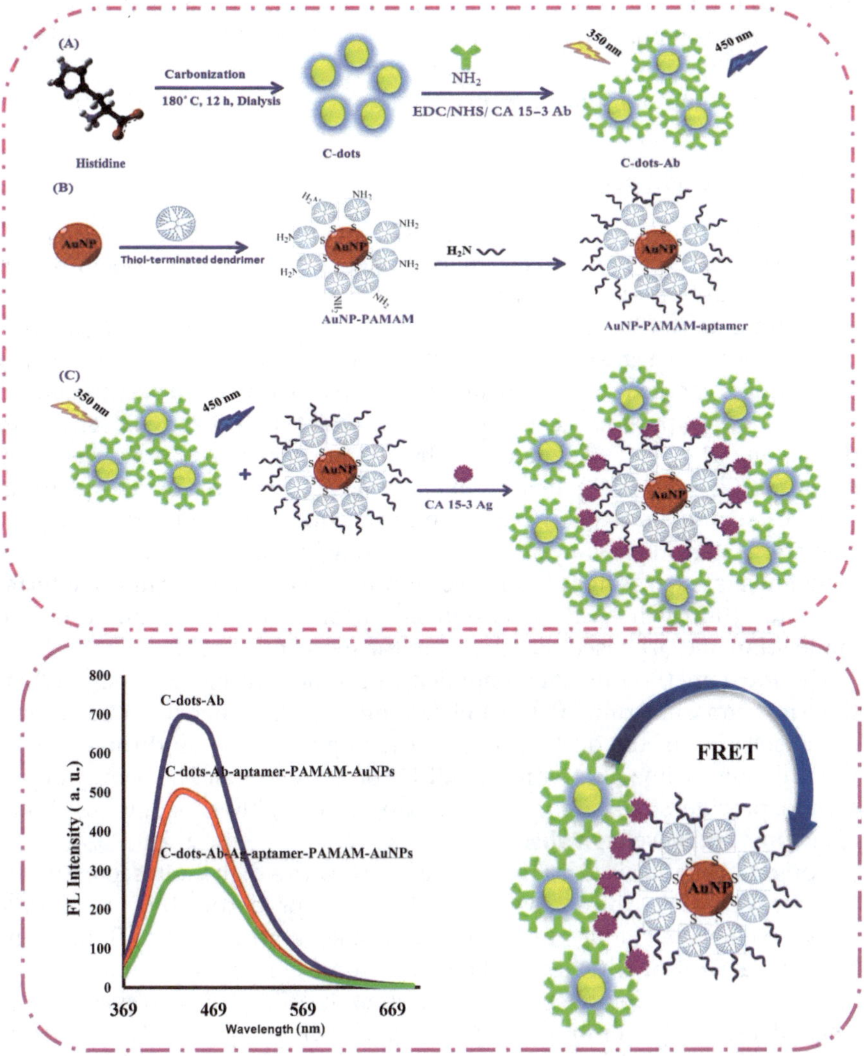

Figure 4.18 (A) Synthesis of Carbon Dots (C-dots) and their conjugation with antibodies (Ab). Histidine is carbonized at 180 °C for 12 hours, followed by dialysis to produce C-dots. (B) Synthesis of AuNP-PAMAM-aptamer conjugates. AuNPs are conjugated with a thiol-terminated dendrimer, followed by PAMAM dendrimer modification. (C) Formation of C-dots-Ab-Aptamer-PAMAM-AuNP complex for FRET analysis. Schematic illustration of the FRET immunoassay for CA15-3 detection. Reproduced from ref. 207 with permission from Elsevier, Copyright 2018.

are used as donors and acceptors in the sandwich-type immunoassay. The strong interaction between the antigen and aptamer brings AuNPs and CDs closer together, reducing the fluorescence signal. This immunoassay has a detection limit of 0.9 U mL^{-1}, good sensitivity and specificity for MDA-MB-231 breast cancer cells, with a correlation value of 0.9955 and a detection limit of 300 cells per mL. It can also be used to dilute human serum, with recovery values for CA15-3 Ag in spiked serum samples ranging from 95.86 to 96.97% and a relative standard deviation (RSD) of less than 7.3%. This method provides a reliable template for developing other immunoassays to identify different cancer tumour indicators using relevant antigens and antibodies.

4.5.6.2 *Biosensing*

Fluorescent materials have gained much attention in the field of sensing. Biosensors have become increasingly popular as diagnostic tools in recent years. Because their surfaces are rich in carboxylic acid groups, carbon quantum dots can be combined with a wide range of polymeric, inorganic, organic, or biological species. The surface chemistry and large surface area of the CQDs can be modified to aid in the construction of a specific capture probe or target species.[208] These advantages enable CQD-based biosensors to detect ions (Hg^{2+}, Cu^{2+}, Ag^{+}, and Fe^{3+}) by monitoring changes in fluorescence signals of CQDs.[209,210] Fluorescent biosensors convert information quantitatively or semi-quantitatively based on interactions between target recognition elements. Changes in fluorescence intensity can be caused by various mechanisms, such as light-induced electron transfer, FRET, or internal filtration effect (IFE), which can be used in biosensors.[211] Here, we summarise the application of CQD-based biosensors in the biomedical field.

The FRET method has been applied in the field of biosensors.[211] When the emission spectrum of the fluorophore (energy donor) overlaps with the absorption of the quenching agent (energy receptor) and the distance is within approximately 10 nm, the FRET process occurs. The energy donor should provide stable fluorescence output, be highly resistant to photobleaching, and have high absorbance.

Gong *et al.* synthesised bright green N-CQDs with a high fluorescence quantum yield of 45%, which have excellent optical properties (Figure 4.19).[212] They developed a simple and reliable IFE based on the internal fluorescent sensor platform of N-CQDs for beta-glucuronic acid glycoside enzyme (G) activity. They also demonstrated the first application of N-CQDs as a novel visual detection sensor for monitoring DOX loading. The N-CQDs can detect DOX loading *via* the fluorescence FRET method, but it is time-consuming and requires complex procedures.

Li *et al.* developed a fluorescence-based detection method for alkaline phosphatase (ALP) using IFE.[213] The fluorophores used in this sensing process are N-doped CQDs, *p*-nitrophenylphosphate (PNPP), and *p*-nitrophenol (PNP), which are enzyme-catalysed derivatives of PNPP. In IFE, ALP substrates and strong absorbents influence fluorophore excitation. Carbon quantum dots are quenched when ALP is present because of the complementary overlap

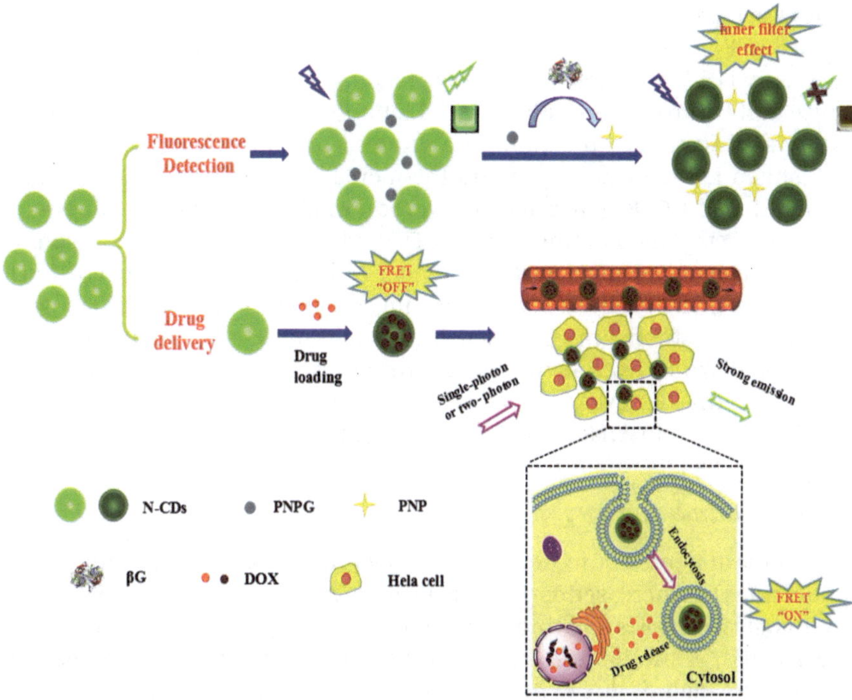

Figure 4.19 Illustration of the application of N-CQDs for ultrasensitive detection of β-glucuronidase based on IFE, visually monitoring anticancer drug loading by FRET and monitoring drug delivery by two-photon fluorescence imaging. Reproduced from ref. 212 with permission from Elsevier, Copyright 2019.

between PNP absorption and CQD excitation. The IFE-based sensing technique eliminates the necessity for a covalent bond between the receptor and the fluorophore, as well as surface modification of the fluorophore, allowing for considerably easier and more flexible probe fabrication.

Carbon quantum dot-based fluorescence sensors can measure ion concentrations in biological samples. Cu^{2+} is essential for enzymatic and biological reactions.[214] Based on the bi-urea reaction concept, Cu^{2+} ions can be detected using CQDs containing hydroxyamino alcohol and surface groups. The fluorescence quenching properties of the dual-emission biological imaging nanoprobe, which was produced using CQDs coated on a silica microsphere core, are favourable for the efficient detection of Cu^{2+} ions. With increasing Cu^{2+} concentration, the response is linear.[215] Polyethylene glycol diamine, ascorbic acid, and rhodamine isothiocyanate B were hydrothermally coupled to make the Fe^{3+} sensor nanoprobe. Fluorescence intensity is increased in the presence of Fe^{3+}, resulting in a more sensitive response than quenched probes. Citric acid and dicyandiamide were dehydrated, polymerised, and carbonised to yield N-doped CQDs that detect Fe^{3+} and F^- ions simultaneously. Carbon quantum dots made without a solvent fluoresce at longer wavelengths, quench when exposed to Fe^{3+}, and directly recover when exposed to F^- ions.[28]

Mercury is a highly poisonous heavy metal. As a result, it is critical to recognise and control it carefully and CQDs fluorescence properties have been investigated in this connection. The fluorescence quenching and recovery mechanism is the principle behind the mercury and iodide detection ability of N-doped CQDs generated by hydrothermal treatment of ammonium citrate and ethylenediamine. These CQDs can detect mercury and iodide ions and have a wide pH stability range. Although there has been much research on CQDs for bioanalytical sensing, there is still a great deal of work to be done on establishing CQD sensing systems on reliable, high-sensitivity platforms. Further work is needed in order to construct sensor systems on stable platforms. Fast verification capabilities across a range of systems indicate CQD-based biosensors have a promising future.

4.5.6.3 *Immunoassay*

In recent years, interest in developing cancer immunoassays for biomedical applications has increased as they can measure very low levels of biomarkers in physiological samples and thus contribute to the early detection of disease.[216] In particular, immunofluorescence has been widely used in clinical diagnosis and nanoanalysis due to its high specificity, sensitivity, and rapidity.[217–221] However, most immunofluorescence techniques require the manipulation of complex fluorescent labels and can increase background interference from primers and affect the specificity of the immune response. The development of label-based fluorescence techniques is of great interest and currently the combination of CQDs with highly specific antibodies to monitor important disease biomarkers is attractive. The immunoassays proposed in recent years can be effective models for establishing other immunoassays to detect various biomarkers with relevant antigens and antibodies. The drawbacks of many labelled antibodies can be eliminated by CQDs-based labelling antibody technology. These include the avoiding of conjugated polymer-induced unlabelled antibody and antigen competition, photobleaching by organic dyes, as well as the size increase of labelled antibody molecules and the ensuing easier and faster cell membranes and tissues penetration.[222]

Wu *et al.* synthesised CQDs by hydrothermal reaction using citric acid and ethylenediamine.[223] A novel proportional immunoassay based on CQDs-doped silica nanoparticles (CQDs-SNP) and fluorescein isothiocyanate (FITC) was developed to target alpha-fetoprotein (AFP). CQDs-SNP capture antibody AB1 and AFP antigen formation and react with AB2-FITC to form an immune complex. The fluorescence emission intensity of FITC and CQDs is measured and the calculated ratio can be used as a signal for quantitative AFP analysis. Huang *et al.*[224] have created new vaccine reagents that show promising results in increasing immune response. By developing quaternary ammonium salt derivatives of carbon quantum dots (CQDs), they were able to generate nanoscale compounds that contain both adsorption and surface antigen-rich ovalbumin (OVA). The new reagents, called BQAS-CQDs, were found to have a better immune response than the original compounds. This study suggests that BQAS-CQDs could be a valuable addition to the toolkit of vaccine researchers. Quaternised cationic functional groups can improve antigen absorption by dendritic cells and boost the formation of antigen-specific antibodies while posing a low risk of immunological irritation and contamination.

Luo *et al.* created uniform-sized carbon dots that were coated with PEG for surface passivation and functionalised with surface groups. These carbon dots were then combined with tumour-antigen OVA. The researchers found that these nanocomposites were able to stimulate the maturation of dendritic cells and the secretion of relevant cytokines. Mature dendritic cells were then able to present tumour antigen messages to T cells, with splenocyte proliferation and interferon gamma (IFN-γ) production happening *in vitro*. Additionally, isolated single spleen cells from C57BL/6 mice showed a high proliferation index and secreted large amounts of interferon gamma IFN-γ cytokines upon re-stimulation with tumour antigen proteins. Carbon dots not only prevented antigens from inactivation in complex physiological environments but also helped in the endocytosis of antigens by antigen-presenting cells (APCs). As a result, the CDs-OVA nanocomposites were able to significantly inhibit the growth of established OVA-expressing B16 melanoma.[225]

4.5.6.4 In Vitro *Diagnosis Agents*

In vitro diagnostics can be applied to blood, urine, perspiration, saliva, and tissue fluid. Because of its particular benefits in the diagnosis of cancer, infectious disorders, and other physiological diseases, the fluorescence detection method is gaining popularity. In recent years, the inherent properties of CQDs have aided researchers in developing more precise detection methods for critical analytes. The luminous emission of carbon quantum dots, for example, gives a distinct detection signal that is sensitive to reactions and interactions in the analyte solution, as we summarise here. It is worth noting that numerous chemical functional groups, such as OH, COOH, C=O, and NH_2, have gradually begun to be employed as 'green' and cost-effective fluorescent nanoprobes in immunoassays, making CQDs a more direct and simple platform than many other alternatives.[226,227]

Xu *et al.* synthesised silylated-CQDs and assembled them with porous silica templates to form a fluorescent CQD-based silica (FCS) colloid. A dendritic silica sphere with a central radial hole was employed to immobilise silanised CQDs. The optimal loading level of CQDs on the inner surface of the spheres promotes the maximum fluorescence intensity. These FCS colloids, combined with the lateral flow band platform, provide an immunoassay with a visual detection limit of 10 pg ml^{-1} for the Zika NS1 protein.[228]

Ag^+ has strong coordination with nitrogen-containing functional groups, binds specifically to N-CQDs, and quenches the fluorescence of N-CQDs by a static mechanism.[229] Fluorescent nanoprobe arrays based on indicator displacement analysis were constructed to take advantage of the different fluorescent responses generated by different binding affinities. The N-CQDs@Ag^+ system can identify bovine serum albumin with 100% accuracy in simulated human serum samples, and a cerebral ischemia disease model has been constructed.

For the selective detection of clenbuterol hydrochloride as a fluorescent probe, Yao *et al.* prepared S/N co-doped CQDs using fructose and L-cysteine hydrochloride monohydrate as precursors. Through an amine–amine coupling procedure, amino groups on CQDs are attached to clenbuterol antibody. The fluorescent system has high specificity and sensitivity, with recoveries ranging from 92.5–108.9%. Doping S and N elements in CQDs significantly increases the fluorescence intensity.[230]

Dong *et al.* developed a fluorescent immunoassay based on CQDs-coated $CaCO_3$ nanospheres and magnetic nanoparticles to detect chloramphenicol in chicken with high sensitivity. Magnetic nanoparticles with chitosan function was synthesised and modified with CAP haptens to serve as a nanocarrier and interface for immunoassay.[231]

4.5.6.5 Nanozyme Applications

In recent years, nanozymes, nanomaterials with enzyme-like properties, have received a great deal of attention because of their great stability, simplicity of synthesis, and tuneable catalytic properties. A revolution in the design of nanozyme systems has taken place since the inherent peroxidase activity of Fe_3O_4 nanoparticles (NPs) was initially revealed in 2007, along with the rapid advancement and ever-deepening understanding of nanoscience and nanotechnology in general. A common characteristic of carbon-based nanomaterials is that their electronic and geometric structures are clearly defined, allowing them to mimic enzyme-like activity. Numerous biomedical implications of nanoparticles with built-in catalytic properties have been documented, including oxygen-dependent tumour therapy, radiation, chemodynamic therapy, bacterial infection disorders, and ROS-related diseases.[232,233] Different microenvironments can affect nanozyme catalytic activity by varying pH, H_2O_2, and glutathione concentrations. Furthermore, by using various stimuli like a magnetic field, light, ultrasound, and heat, nanozymes can be remotely controlled. It is possible to modify these variables together to improve diagnostic and therapeutic efficacy in biomedical applications.[234] For easier understanding, nanozymes are divided into two categories: the hydrolase family, which includes nuclease, esterase, phosphatase, protease, and silicatein; and the oxidoreductase family, which includes enzymes like oxidase, peroxidase, catalase, superoxide dismutase, and nitrate reductase (Figure 4.20).[232,235]

Nanozymes with CDs possess advantages over natural enzymes and other organometallic nanozymes, including affordability, stability, environmental friendliness, ease of functionalisation, exceptional photophysical properties,

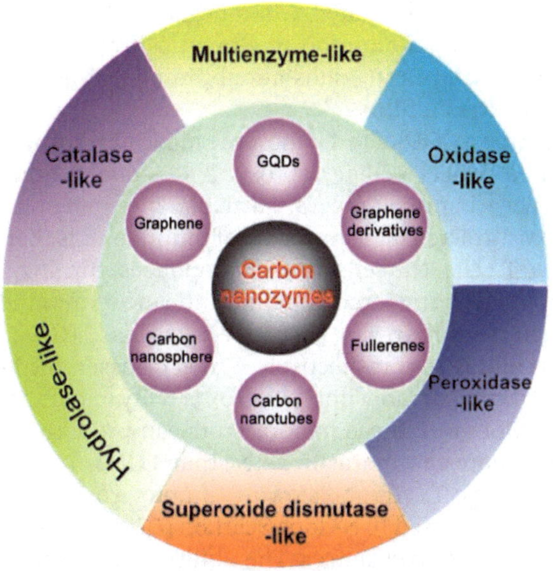

Figure 4.20 Carbon nanomaterials with enzyme-like activity. Reproduced from ref. 232 with permission from Springer Nature, Copyright 2021.

and catalytic activity. Some of their biomedical applications will now be discussed.

The surface modification, heteroatom doping, and combination of CDs with nanoparticles make them promising candidates for nanomaterial composites with nanozymes. In the past few years, many heteroatom-doped CDs have successfully simulated natural enzymes like oxidase, catalase, and superoxide.[236]

Zhang *et al.* synthesised several nitrogen-doped CDs and measured their photo-oxidation activity and phosphorescence quantum yields. These CDs have the maximum activity and can complete oxidation reactions in a matter of seconds, making them a promising oxidase-mimicking nanozyme for photodynamic antimicrobial chemotherapy and other applications.[237]

Liu *et al.* created iron-doped carbon dot (Fe-CD) nanozymes with superior photothermal conversion and photo-enhanced enzyme-like characteristics for synergistic, effective antibacterial therapy and wound healing. Fe doping gives CDs photo-enhanced peroxidase (POD)-like activity, which causes the production of heat and reactive oxygen species. This study shows Fe-CDs have significant wound healing efficacy by avoiding infection and encouraging fibroblast proliferation, angiogenesis, and collagen deposition. Their ultra-small size and superior biocompatibility support clinical translation.[238] To effectively attack inflexible bacterial biofilms, Pan *et al.* created CS@Fe/CD, which selectively activates peroxidase-like catalytic activity and synergistic antibacterial activity, resulting in the successful elimination of exposed bacteria and bacterial biofilm fragments. Interestingly, nanozymes based on

CS@Fe/CD completely eliminated Gram-negative *Pseudomonas aeruginosa* (*P. aeruginosa*) biofilms in addition to Gram-positive *Staphylococcus aureus* (*S. aureus*) biofilms, indicating substantial promise as a viable anti-biofilm agent against bacteria biofilms.[239]

Yao *et al.* reported a CQD-based nanozyme made of chlorogenic acid (ChA), which stimulated cancer cell ferroptosis and reduced tumour growth in HepG2-tumour-bearing mice *in vivo*. Particularly in hepatoma H22-bearing mice, ChA-CQDs attracted large numbers of immune cells that infiltrated the tumour, such as T cells, NK cells, and macrophages, turning 'cold' tumours into 'hot' tumours that triggered systemic anticancer immune responses.[240] To create the chiral carbon dot (CD)-glucose oxidase (GOx) nanoreactors, known as LGOx and DGOx, respectively, Gao *et al.* assembled GOx with L/D-CDs. L/DGOx can considerably improve GOx activity and transport efficiency in cancer cells. Moreover, these nanoreactors can produce hydrogen peroxide, and DGOx is more enzymatically active than LGOx, making it more effective at destroying cancer cells and slowing tumour growth.[241]

Li *et al.* created CDs fused to N-doped carbon (CDs@NC) to detect D-amino acids linked to early gastric cancer. The ability of CDs@NC to function as a peroxidase-like nanozyme for the colorimetric detection of D-amino acids linked to early gastric cancer was demonstrated. Further research suggests that the superoxide radical ($O_2^{\bullet}$) may be the crucial ROS involved in the catalytic events that resemble peroxidases. Excellent saliva test results suggest that patients with early stomach cancer and healthy individuals could be readily separated using the colorimetric approach.[242] Glioblastoma (GBM), which accounts for 50% of all diagnosed malignancies of the central nervous system (CNS), is regarded as the most deadly and common malignant tumour of the brain. Poor clinical prognosis, a high incidence of recurrences, and high fatality rates are characteristics of GBM. The median survival period for individuals with GBM is less than 15 months following diagnosis, and the five-year survival rate is below 5% despite recent progress.[243,244] Muhammad *et al.* studied single-atom nanozyme-mediated catalytic treatment to target drug-resistant GBM with six naturally occurring enzymes. Fe-CDs function as a drug-free nanomedicine that modifies the tumour microenvironment by controlling ROSs and lysosomal-mediated autophagy.[245]

4.5.7 Optoelectronic Devices

4.5.7.1 Light-emitting Diodes

Carbon dots have a great deal of potential in the field of LEDs because of their tuneable photoluminescence, efficient luminescence, and environmental stability. There are two types of photoluminescence devices: photoluminescent and electroluminescent.

Photoluminescent LEDs usually require the excitation of ultraviolet or blue light chips, and photoluminescent multicolour LEDs can be generated by embedding luminescent CDs in solid substrates such as polyvinyl alcohol.

White LEDs (WLEDs) can also be obtained by mixing CDs with complementary colours. The colour rendering index is extremely high, so the light emitted is very close to pure white light. Sun *et al.* reported a series of monochromatic LEDs based on monodisperse CDs, whose emissive colour varies from blue to red depending on the thickness of the down conversion layer and the doping concentration of carbon dots in the polymer matrix.[246] Lu *et al.* discussed the unique structure of full-colour-emission CDs. These CDs are made up of tiny domains of partly sp^2-hybridised carbon atoms around sp^3-hybridised carbon cores. The red-shifting of CDs' emission caused by the change from sp^3 to sp^2 hybridisation was predicted by first-principle calculations. The theoretical predictions served as a guide for the development of a straightforward, practical, and controllable method for hydrothermally producing CDs in a single reaction system. The produced CDs also exhibit extraordinary two-photon emission over the whole visible colour spectrum and full-spectrum emission. Interestingly, minor adjustments to the reaction conditions, such as temperature and pH, can further cultivate these full-colour-emission CDs and change the emission colour from blue to red. This approach has greater industrial potential than conventional fluorophores, as demonstrated by the creation of a flexible LED with full-colour emission utilising epoxy-CD films.[247] To achieve this, citric acid and *o*-PD were used as raw materials, and white-emitting CDs were synthesised in ethanol.[248] Three luminescent CDs of blue, cyan, and yellow were obtained by column chromatography. WLEDs with CIE colour coordinates of 0.37 and 0.39 were successfully prepared with a colour rendering index of 87.8. Zheng *et al.* successfully prepared blue, green, orange, and red carbon dots by adjusting the ratio of two reactants, asparagine, and *p*-phenylenediamine.[249] The as-prepared multicolour CDs exhibited excellent stability and high fluorescence quantum yield (up to 74%). High-quality WLEDs with Commission International de l'Eclairage (CIE) colour coordinates (0.33, 0.33) were successfully fabricated using colour CDs-epoxy composites, with a colour temperature of 5612 K and a colour rendering index of 94.

Ma *et al.* used 1-hexadecylamine passivated CDs to make an electroluminescent device.[250] The hole transport layer material poly(3,4-ethylenedioxythiophene):poly(styrene sulfonate) (PEDOT:PSS) was spin-casted on ITO glass to increase the anode work function and reduce its surface roughness. 1,3,5-Tris(*n*-phenylbenzimidazol-2-yl)benzene electron transport layer was deposited on the emissive layer, followed by thermal evaporation of the LiF and Al electrodes. Yuan *et al.* addressed this limitation by using narrow-bandwidth emitting triangular CQDs prepared from phloroglucinol.[53] These CQDs can achieve high-purity polychromatic emissions from blue to red.

4.5.7.2 Anti-counterfeiting

Carbon dots are ideal for use in security inks due to their superior fluorescence properties and outstanding stability. Jiang *et al.*[251] made a three-mode luminescent CD-based composite material into an anti-counterfeiting ink and loaded it into a gel pen to write Chinese character 'Heng' and English 'A' on banknotes and invoices. After drying, the two characters are almost invisible in sunlight, but under 365 nm UV light and 800 nm femtosecond laser irradiation, the blue 'Heng' and 'A' characters are clearly visible. The CD-based composites have broad application prospects in the security field. Tao *et al.*[84] used two different RTP-emitting CDs and commercial highlighters to paint a butterfly, which exhibited colourful fluorescence under UV light. When the UV light was turned off, the fluorescence of the outer red and orange parts disappeared, and the intermediate blue fluorescence transformed into green and cyan phosphorescence. Along with anti-counterfeiting labels, CDs can also be used to encrypt data. Jiang *et al.*[82] printed a QR code on ordinary A4 paper because the CD ink was almost colourless and the background fluorescence of the A4 paper was strong. Only a small amount of blue fluorescence could be seen under the 365 nm UV lamp, and it was impossible to identify any information. However, the correct QR code appeared when the UV light was turned off. It is also possible to mix the above-mentioned RTP-CD ink with another blue fluorescent ink in order to crack and decode the information. RTP-phosphorescence-CDs sensitive to water molecules also have good practical value. Jiang *et al.*[252] wrote the digital code '13579' on non-fluorescent filter paper with an aggregation-inducing RTP-CD ink, and then used a commercial fluorescent ink to write another set of digital code '02468.' RTP-CD-encrypted information is often invisible in sunlight and UV light. However, when the filter paper is sprayed with water, the code generated by RTP-CDs can be seen, while the code generated by commercial fluorescent ink is invisible. Most crucially, after the code decryption, the non-fluorescent filter paper can be infiltrated with tetrahydrofuran and then re-encrypted, allowing for numerous times of encryption.

4.5.7.3 Fabrication Using Machine Learning

Machine learning has played an important role in material fabrication and the prediction of experimental results, including in the preparation of white LEDs. Kang *et al.* synthesised green and orange CDs using PTCDA (perylene-3,4,9,10-tetracarboxydianhydride).[253] The PL centre of green CDs is located at 493 nm and the photoluminescence quantum yield (PLQY) value is 63.5%, while the PL centre of orange CDs is located at 553 nm, and the PLQY value is 83.6%. White PL emission with correlated colour temperature (CCT) tuneable at 3093–11 018 K was fabricated using machine learning guidance. The model was successful in producing warm white CD LEDs with CCTs of 3107, 4071, and 4548 K and cold white CD LEDs with CCTs of 5632 (CIE coordinates of (0.33, 0.33), external quantum efficiency (EQE) of 1.18%, and luminescence of 598 cd m^{-2}) and 6034 K cool white CD LEDs. This work shows the potential of machine learning models in LED design.

4.5.8 Energy Storage and Conversion

4.5.8.1 Supercapacitors

Supercapacitors (SCs), which are also known as ultracapacitors or electrochemical capacitors, are devices used for storing and releasing energy through the electrostatic separation of charges. These devices are widely recognised for their high-power density, quick charge/discharge rates, and extended cycle life in comparison to conventional batteries. Carbon dots can be anchored on different substrates as electrode materials, which can improve the performance of SCs by providing more active sites for capacitance. Doping CDs can greatly enhance the capacitance. The combination of carbon substrates, carbon dots, and transition metal oxides in composite materials offers a synergistic approach to designing electrodes with improved properties for supercapacitors. This approach is promising for energy storage applications due to the ability of these materials to undergo redox reactions and exhibit pseudocapacitance. Carbon dots with a defect and graphene-layered structure and high conductivity can act as a passageway for ion diffusion and charge transportation ability of the substrate, providing optimisation for the properties of the original substrate. Additionally, CDs can induce morphology changes on metal compound substrates.

Qu *et al.* used 3D graphene (3DG) as a substrate to create GQDs-3DG, which shows a high specific capacitance of 268 F g^{-1}, almost twice as much as the supercapacitor made from pure 3DG electrodes (136 F g^{-1}).[254] Li *et al.* provided a new method for the preparation of GQDs/graphene oxide and N-GQDs/graphene oxide. In this work graphene oxide nanosheets were combined with GQDs and doped with N using NH_3 as the source.[255] The nitrogen doping process modifies the electronic state of graphene to enhance electrical conductivity and accelerate ion adsorption/desorption from the electrolyte, further increasing specific capacitance. Plus, the pseudo-capacitance effect can be improved by nitrogen-containing functional groups attached to the surface of the substrate.[256]

In work by Xiong *et al.*,[257] CDs with different concentrations were added to a $NiCo_2O_4$ precursor to obtain a $NiCo_2O_4$ hybrid with various morphologies. This process was modified by changing the mass ratio of CQDs to the substrate. Carbon dots restrained the anisotropic growth of $NiCo_2O_4$, forming the best structure for capacitance performance. Huang and his group prepared GQDs on the surface of pre-synthesised $NiCo_2O_4$.[258] Due to the strong electrostatic self-assembly effect, the GQDs were adsorbed on the surface of the tremella-like $NiCo_2O_4$. Wei *et al.*[259] synthesised 3D hierarchical CQDs/Co_3O_4 with a pore size distribution of 2–15 nm for supercapacitors. They discovered that micro-nanostructures induced by CQDs could endow stable ion diffusion pathways and short ion transport distances. Carbon dots anchored on a substrate can modify surface properties, and CD-containing composites can be synthesised with high performance. Ghosh *et al.* introduced a controllable anchoring process that formed a network structure with polyaniline (PANI) chains surrounding carbon nanodots.[260] The electrostatic effect enhanced the degree of order of the composite, resulting in great capacitance in cyclic voltammetry and galvanostatic charge–discharge plots. Covalently grafted CND/PANI hybrid formation allowed more access to ions in the electrolyte, further enhancing the capacitance.

4.5.8.2 Lithium and Sodium Batteries

Lithium-ion batteries (LIBs) are secondary rechargeable batteries. They are composed of an anode, a cathode, an ion-conducting electrolyte, and a separator (see also Section 1.10.1). The lithium ions move to the cathode through the electrolyte and are intercalated into the cathode materials. Sodium-ion batteries have been studied as a promising alternative to lithium-ion batteries due to the ready availability of Na, their lower costs, and the similar working mechanism to LIBs. Ji *et al.* examined the effect of coating a Mn_3O_4 composite with CQDs on the discharge capacity and cycling performance of lithium-ion batteries.[261] Through an electrochemical approach, the structure of the Mn_3O_4 matrix changed from nanocrystalline to octahedral. The cyclic performance test revealed the inherent advantages of these CQDs, such as conductivity.

The Mn_3O_4/CQDs electrode exhibited a much greater retention rate, initial capacity, and a higher conductivity compared to the pure Mn_3O_4 electrode.[261] Yu *et al.* prepared C(ZIF-8)@GQDs by adding GQDs to the synthesis process of ZIF-8 derived carbon.[262] The negatively-charged GQDs combined with the positively-charged ZIF-8 precursors by electrostatic adsorption. The specific capacity of C(ZIF-8)@GQDs was much higher than C(ZIF-8) at 100 mA g^{-1}, with a high retention rate after 200 cycles and a specific capacity plateau with different current densities. Anchoring GQDs played a key role in improving specific capacity.

Shen and his team synthesised a composite material consisting of a GQD-coated VO_2 array deposited on graphite foam.[263] This composite material combined the benefits of three components: graphite foam, as an efficient current collector; VO_2 nanobelts with low thickness and dimensions; and VO_2 preventing GQDs from aggregation and dissolution. The capacity and retention of GVG (the whole electrode structure denoted as GVG) was much higher than GV (graphitic foam and VO_2), further indicating the importance of GQDs.

4.6 Outlook and Perspectives

In summary, this chapter has presented a comprehensive introduction to the synthesis, structure, and properties of carbon dots, as well as recent developments in their applications ranging from biomedicine to energy. Despite the great progress achieved, research into accurate synthesis and assembly is still in its infancy.[34,58,264] Moreover, the exact structure and photoluminescence mechanism of carbon dots are not clear. Some of the current challenges and future directions are summarised in the following sections.

4.6.1 Need for a Standard Purification Protocol

Purification has played a crucial role in the field of CDs since their discovery in 2004. In a typical 'bottom-up' synthesis process, the solvothermal treatment of small organic molecules/precursors results in the formation of CDs together with various side-products and unreacted precursors. Inadequate purification, and thus the presence of these species in the final product, can lead to errors in interpreting the observed properties of CDs. The effective removal of these species from the crude product mixture can be challenging. Prato *et al.* have proposed a standard purification protocol for the crude product mixture as follows: (1) filtration; (2) dialysis against milli-Q water for 48 to 72 h; (3) freeze-drying to obtain a solid sample; (4) a further but necessary purification step, using column separation or high-performance liquid chromatography to fully remove the unreacted species from the final products.[61] Unfortunately, such a time-consuming purification process usually leads to a very low product yield and cost efficiency, hampering the large-scale production and practical application of CDs. There is therefore a need for a simple standard purification protocol.

4.6.2 Essential Characterisation/Analysis Techniques

The formation mechanism and exact structure of CDs are as yet unknown. As far as the 'bottom-up' approach is concerned, although it is widely acknowledged that the polymerisation, crosslinking, and carbonisation processes are critical in the formation of CDs, the precise reaction path remains elusive. Furthermore, the origin of photoluminescence and the position of emission centres (core, shell, or both) are still controversial, owing to a lack of understanding of the chemical structure. This has obstructed the rational design of CDs for specific applications. The development of novel characterisation and analysis methods may be of assistance in this regard. For example, *in situ* characterisation techniques could provide more information about the synthetic process by analysis of intermediates and by-products in terms of their size, content, and photoluminescence, which is advantageous for a better correlation between structure and emission properties.

4.6.3 Theoretical Calculations of Carbon Dots

It is essential to create a more reliable method of modelling CDs to systematise experimental findings and facilitate the continued synthesis of CDs for future applications. Boukhvalov *et al.* suggested that oxidation and reduction of synthesised nanoparticles are the most straightforward techniques for adjusting the optical properties of these systems, based on the comparison of available experimental data and simulations of nitrogen-free CQDs.[265] When graphene nanosheets undergo oxidation or reduction, their carbon atoms go through a process called rehybridisation. This process involves a change in the hybridisation state from sp^2 to sp^3, which is crucial to achieving the desired photoluminescence properties. The models presented by Boukhvalov *et al.* will help guide the treatment necessary to achieve these properties. The omission of the contribution from functionalised regions in the middle parts of the graphene sheets, a phenomenon observed in all CQDs is another recurring flaw of the cited theoretical research in terms of its application to CQDs. Boukhvalov *et al.* therefore believe that modelling partially functionalised large graphene sheets may offer a better description of actual CDs based on a brief observation of these new theoretical results.[265]

4.6.4 Precise Control over Carbon Dot Assemblies

Self-assembly of CDs has sparked much interest in recent years because it not only helps us understand some interesting phenomena that occur when CDs go from dispersed to aggregated states (like emission enhancement/quenching, spectral shift/broadening, and fluorescence to phosphorescence transformation), but also offers a promising strategy for imparting or amplifying the specific functions. Nevertheless, many current studies attribute the driving forces of CD assembly simply to electrostatic interactions or hydrogen bonding, lacking the precise control over the binding strength and dynamics, as well as the size, topology and properties of assemblies.[266] We believe that for in-depth investigation of the assembly mechanism, as well as fine-tuning of the structure and properties, a careful selection or judicious design of supramolecular building blocks with well-studied binding modes will be required. In addition, the dynamic and stimuli-responsive (*e.g.* to light or force) behaviours of supramolecular assemblies would also be of current interest as research topics.

4.6.5 Large-scale Production of Carbon Dots

The main incentive for large-scale production of CDs is their potential to improve the mechanical properties of composites, specifically nanofibers, flame retardancy, and energy storage. Carbon dots have many active sites and surface functional groups, which can enhance the electrochemical performance and stability of the electrode materials in energy storage devices. However, the purity of CDs and their aggregation during use are the main drawbacks of these promising applications. Major priorities of current research are the development of high-yield and large-scale synthesis methodologies for high-purity CDs and the superior dispersion of CDs in composites by surface functionalisation.[267]

4.6.6 Expanding the Range of Applications

While the applications based on the unique photophysical features have been extensively exploited, other aspects of CDs should not be overlooked. By using CDs as fillers, Kang and coworkers improved the self-healing efficiency of bulk polymer materials.[268] Bi *et al.* recently reported on the production of CDs with exceptional moisture retention capability, as well as the application potential in skin care.[269] The surfaces of these CDs contain abundant oxygen- and amine-containing functional groups, which can help them covalently link with certain small molecules or polymers. This kind of surface engineering makes them interesting and potentially useful for extensive applications in chemical, mechanical, and electronic engineering. For example, it has been suggested that CDs could be used as adhesives and coatings, expanding their scientific and technological significance.

Despite the limitations and challenges outlined above, the authors feel that CD-based materials have a bright future and are deserving of the attention they are receiving due to their appealing properties, such as low toxicity, good solubility, and high photostability. As a result of interdisciplinary collaboration combining chemists, materials scientists, and engineers, current impediments to CD translation from 'lab to market' will be addressed over time.

Acknowledgments

This work was financially supported by the National Natural Science Foundations of China (Grant No. 52172033, 51772001 and 22005280). We gratefully acknowledge financial support from the National Key R&D Program of China (Grant No. 2021YFA1600202). We acknowledge support of the Key Laboratory of Structure and Functional Regulation of Hybrid Materials of Ministry of Education, Anhui University, Hefei, Anhui 230601, China. We also acknowledge support of the Key Laboratory of Environment Friendly Polymer Materials of Anhui Province, Hefei, China.

References

1. Z. Li, Z. Liu, H. Sun and C. Gao, *Chem. Rev.*, 2015, **115**, 7046.
2. A. Hirsch, *Nat. Mater.*, 2010, **9**, 868.
3. V. Georgakilas, J. A. Perman, J. Tucek and R. Zboril, *Chem. Rev.*, 2015, **115**, 4744.
4. Z. Peng, X. Liu, W. Zhang, Z. Zeng, Z. Liu, C. Zhang, Y. Liu, B. Shao, Q. Liang and W. Tang, *Environ. Int.*, 2020, **134**, 105298.
5. X. Xu, R. Ray, Y. Gu, H. J. Ploehn, L. Gearheart, K. Raker and W. A. Scrivens, *J. Am. Chem. Soc.*, 2004, **126**, 12736.
6. Y.-P. Sun, B. Zhou, Y. Lin, W. Wang, K. S. Fernando, P. Pathak, M. J. Meziani, B. A. Harruff, X. Wang and H. Wang, *J. Am. Chem. Soc.*, 2006, **128**, 7756.
7. F. Arcudi, L. Đorđević and M. Prato, *Acc. Chem. Res.*, 2019, **52**, 2070.
8. B. Wang and S. Lu, *Matter*, 2022, **5**, 110.
9. D. Qu and Z. Sun, *Mater. Chem. Front.*, 2020, **4**, 400.
10. J. Liu, R. Li and B. Yang, *ACS Cent. Sci.*, 2020, **6**, 2179.
11. W. Liu, C. Li, Y. Ren, X. Sun, W. Pan, Y. Li, J. Wang and W. Wang, *J. Mater. Chem. B*, 2016, **4**, 5772.

12. D. Li, E. V. Ushakova, A. L. Rogach and S. Qu, *Small*, 2021, **17**, 2102325.
13. L. Đorđević, F. Arcudi, M. Cacioppo and M. Prato, *Nat. Nanotechnol.*, 2022, **17**, 112.
14. B. Zhao and Z. Tan, *Adv. Sci.*, 2021, **8**, 2001977.
15. L. Ai, Y. Yang, B. Wang, J. Chang, Z. Tang, B. Yang and S. Lu, *Sci. Bull.*, 2021, **66**, 839.
16. M. L. Liu, B. B. Chen, C. M. Li and C. Z. Huang, *Green Chem.*, 2019, **21**, 449.
17. J. B. Essner, C. H. Laber, S. Ravula, L. Polo-Parada and G. A. Baker, *Green Chem.*, 2016, **18**, 243.
18. Y.-W. Zeng, D.-K. Ma, W. Wang, J.-J. Chen, L. Zhou, Y.-Z. Zheng, K. Yu and S.-M. Huang, *Appl. Surf. Sci.*, 2015, **342**, 136.
19. B. B. Chen, R. S. Li, M. L. Liu, H. Z. Zhang and C. Z. Huang, *Chem. Commun.*, 2017, **53**, 4958.
20. J. Kudr, L. Richtera, K. Xhaxhiu, D. Hynek, Z. Heger, O. Zitka and V. Adam, *Biosens. Bioelectron.*, 2017, **92**, 133.
21. B. De and N. Karak, *RSC Adv.*, 2013, **3**, 8286.
22. D. Pan, L. Guo, J. Zhang, C. Xi, Q. Xue, H. Huang, J. Li, Z. Zhang, W. Yu and Z. Chen, *J. Mater. Chem.*, 2012, **22**, 3314.
23. T.-H. Chen and W.-L. Tseng, *Anal. Chem.*, 2017, **89**, 11348.
24. X. Gao, Y. Lu, R. Zhang, S. He, J. Ju, M. Liu, L. Li and W. Chen, *J. Mater. Chem. C*, 2015, **3**, 2302.
25. Y. Su, M. Zhang, N. Zhou, M. Shao, C. Chi, P. Yuan and C. Zhao, *Microchim. Acta*, 2017, **184**, 699.
26. D. Shi, F. Yan, T. Zheng, Y. Wang, X. Zhou and L. Chen, *RSC Adv.*, 2015, **5**, 98492.
27. J. Cheng, C.-F. Wang, Y. Zhang, S. Yang and S. Chen, *RSC Adv.*, 2016, **6**, 37189.
28. J. Hou, W. Wang, T. Zhou, B. Wang, H. Li and L. Ding, *Nanoscale*, 2016, **8**, 11185.
29. D. Qu, M. Zheng, L. Zhang, H. Zhao, Z. Xie, X. Jing, R. E. Haddad, H. Fan and Z. Sun, *Sci. Rep.*, 2014, **4**, 1.
30. S. Kim, S. W. Hwang, M.-K. Kim, D. Y. Shin, D. H. Shin, C. O. Kim, S. B. Yang, J. H. Park, E. Hwang and S.-H. Choi, *ACS Nano*, 2012, **6**, 8203.
31. F. Wang, M. Kreiter, B. He, S. Pang and C.-y. Liu, *Chem. Commun.*, 2010, **46**, 3309.
32. S. F. Parker, *Chem. Commun.*, 2011, **47**, 1988.
33. W. Shi, X. Li and H. Ma, *Angew. Chem., Int. Ed.*, 2012, **51**, 6432.
34. M. J. Krysmann, A. Kelarakis, P. Dallas and E. P. Giannelis, *J. Am. Chem. Soc.*, 2012, **134**, 747.
35. J. Schneider, C. J. Reckmeier, Y. Xiong, M. von Seckendorff, A. S. Susha, P. Kasák and A. L. Rogach, *J. Phys. Chem. C*, 2017, **121**, 2014.
36. L. Song, Y. Cui, C. Zhang, Z. Hu and X. Liu, *RSC Adv.*, 2016, **6**, 17704.
37. Y. Du and S. Guo, *Nanoscale*, 2016, **8**, 2532.
38. J. Xu, Q. Liang, Z. Li, V. Y. Osipov, Y. Lin, B. Ge, Q. Xu, J. Zhu and H. Bi, *Adv. Mater.*, 2022, **34**, 2200011.

39. W. Yang, T. Shimanouchi and Y. Kimura, *ACS Sustainable Chem. Eng.*, 2015, **3**, 591.
40. R. Bandi, B. R. Gangapuram, R. Dadigala, R. Eslavath, S. S. Singh and V. Guttena, *RSC Adv.*, 2016, **6**, 28633.
41. J. Xu, Y. Zhou, S. Liu, M. Dong and C. Huang, *Anal. Methods*, 2014, **6**, 2086.
42. H. Ding, Y. Ji, J.-S. Wei, Q.-Y. Gao, Z.-Y. Zhou and H.-M. Xiong, *J. Mater. Chem. B*, 2017, **5**, 5272.
43. M. Xue, Z. Zhan, M. Zou, L. Zhang and S. Zhao, *New J. Chem.*, 2016, **40**, 1698.
44. M. Saxena and S. Sarkar, *Diamond Relat. Mater.*, 2012, **24**, 11.
45. X. Teng, C. Ma, C. Ge, M. Yan, J. Yang, Y. Zhang, P. C. Morais and H. Bi, *J. Mater. Chem. B*, 2014, **2**, 4631.
46. L. Zhu, Y. Yin, C.-F. Wang and S. Chen, *J. Mater. Chem. C*, 2013, **1**, 4925.
47. H. Zeng, L. Li, Y. Ding and Q. Zhuang, *Talanta*, 2018, **178**, 879.
48. S. Chaudhary, M. Kumari, P. Chauhan and G. R. Chaudhary, *Waste Manage.*, 2021, **120**, 675.
49. C. Xia, S. Zhu, T. Feng, M. Yang and B. Yang, *Adv. Sci.*, 2019, **6**, 1901316.
50. G. Ge, L. Li, D. Wang, M. Chen, Z. Zeng, W. Xiong, X. Wu and C. Guo, *J. Mater. Chem. B*, 2021, **9**, 6553.
51. R. B. González-González, L. T. González, M. Madou, C. Leyva-Porras, S. O. Martinez-Chapa and A. Mendoza, *Nanomaterials*, 2022, **12**, 298.
52. A. A. Kokorina, A. V. Sapelkin, G. B. Sukhorukov and I. Y. Goryacheva, *Adv. Colloid Interface Sci.*, 2019, **274**, 102043.
53. F. Yuan, T. Yuan, L. Sui, Z. Wang, Z. Xi, Y. Li, X. Li, L. Fan, Z. Tan and A. Chen, *Nat. Commun.*, 2018, **9**, 1.
54. B. Wang, Z. Wei, L. Sui, J. Yu, B. Zhang, X. Wang, S. Feng, H. Song, X. Yong and Y. Tian, *Light: Sci. Appl.*, 2022, **11**, 1.
55. I. Singh, R. Arora, H. Dhiman and R. Pahwa, *Turk. J. Pharm. Sci.*, 2018, **15**, 219.
56. P. Zhu, K. Tan, Q. Chen, J. Xiong and L. Gao, *Chem. Mater.*, 2019, **31**, 4732.
57. Y. Park, M. D. Tran, Y. Kim, S. Won, Y.-H. Kim, T.-W. Lee, T. Gregorkiewicz, Y. H. Lee, J.-H. Kim and W. Kwon, *Carbon*, 2022, **188**, 229.
58. F. Rigodanza, M. Burian, F. Arcudi, L. Đorđević, H. Amenitsch and M. Prato, *Nat. Commun.*, 2021, **12**, 2640.
59. F. He, J. Bai, Y. Cheng, K. Weerasinghe, X. Meng, H. Xu, W. Zhang, X. Fang, H.-B. Li and T. Ding, *J. Phys. Chem. C*, 2021, **125**, 5207.
60. Q. Zhang, R. Wang, B. Feng, X. Zhong and K. K. Ostrikov, *Nat. Commun.*, 2021, **12**, 1.
61. L. Đorđević, F. Arcudi and M. Prato, *Nat. Protoc.*, 2019, **14**, 2931.
62. H. Wan, J. Yue, S. Zhu, T. Uno, X. Zhang, Q. Yang, K. Yu, G. Hong, J. Wang and L. Li, *Nat. Commun.*, 2018, **9**, 1.
63. Y. Dong, H. Pang, H. B. Yang, C. Guo, J. Shao, Y. Chi, C. M. Li and T. Yu, *Angew. Chem., Int. Ed.*, 2013, **52**, 7800.
64. L. Bao, C. Liu, Z. L. Zhang and D. W. Pang, *Adv. Mater.*, 2015, **27**, 1663.

65. S. Hu, A. Trinchi, P. Atkin and I. Cole, *Angew. Chem., Int. Ed.*, 2015, **54**, 2970.
66. J. Zhan, B. Geng, K. Wu, G. Xu, L. Wang, R. Guo, B. Lei, F. Zheng, D. Pan and M. Wu, *Carbon*, 2018, **130**, 153.
67. Y. Zhou, E. M. Zahran, B. A. Quiroga, J. Perez, K. J. Mintz, Z. Peng, P. Y. Liyanage, R. R. Pandey, C. C. Chusuei and R. M. Leblanc, *Appl. Catal., B*, 2019, **248**, 157.
68. H. Nie, M. Li, Q. Li, S. Liang, Y. Tan, L. Sheng, W. Shi and S. X.-A. Zhang, *Chem. Mater.*, 2014, **26**, 3104.
69. A. Sharma, T. Gadly, A. Gupta, A. Ballal, S. K. Ghosh and M. Kumbhakar, *J. Phys. Chem. Lett.*, 2016, **7**, 3695.
70. C. J. Reckmeier, J. Schneider, Y. Xiong, J. Häusler, P. Kasák, W. Schnick and A. L. Rogach, *Chem. Mater.*, 2017, **29**, 10352.
71. Z. Feng, Z. Li, X. Zhang, G. Xu and N. Zhou, *J. Mater. Sci.*, 2018, **53**, 6459.
72. I. Y. Goryacheva, A. V. Sapelkin and G. B. Sukhorukov, *TrAC Trends Anal. Chem.*, 2017, **90**, 27.
73. H. Li, X. He, Z. Kang, H. Huang, Y. Liu, J. Liu, S. Lian, C. H. A. Tsang, X. Yang and S. T. Lee, *Angew. Chem., Int. Ed.*, 2010, **49**, 4430.
74. H. Ding, J. S. Wei, P. Zhang, Z. Y. Zhou, Q. Y. Gao and H. M. Xiong, *Small*, 2018, **14**, 1800612.
75. L. Sciortino, A. Sciortino, R. Popescu, R. Schneider, D. Gerthsen, S. Agnello, M. Cannas and F. Messina, *J. Phys. Chem. C*, 2018, **122**, 19897.
76. L. Tang, R. Ji, X. Li, G. Bai, C. P. Liu, J. Hao, J. Lin, H. Jiang, K. S. Teng and Z. Yang, *ACS Nano*, 2014, **8**, 6312.
77. Y. Zhan, T. Geng, Y. Liu, C. Hu, X. Zhang, B. Lei, J. Zhuang, X. Wu, D. Huang and G. Xiao, *ACS Appl. Mater. Interfaces*, 2018, **10**, 27920.
78. X. Yao, R. E. Lewis and C. L. Haynes, *Acc. Chem. Res.*, 2022, **55**, 3312.
79. X. Dong, L. Wei, Y. Su, Z. Li, H. Geng, C. Yang and Y. Zhang, *J. Mater. Chem. C*, 2015, **3**, 2798.
80. Y. Liu, G. Zhan, Z.-W. Liu, Z.-Q. Bian and C.-H. Huang, *Chin. Chem. Lett.*, 2016, **27**, 1231.
81. Q. Li, M. Zhou, Q. Yang, Q. Wu, J. Shi, A. Gong and M. Yang, *Chem. Mater.*, 2016, **28**, 8221.
82. K. Jiang, Y. Wang, X. Gao, C. Cai and H. Lin, *Angew. Chem., Int. Ed.*, 2018, **57**, 6216.
83. K. Jiang, Y. Wang, C. Cai and H. Lin, *Adv. Mater.*, 2018, **30**, 1800783.
84. S. Tao, S. Lu, Y. Geng, S. Zhu, S. A. Redfern, Y. Song, T. Feng, W. Xu and B. Yang, *Angew. Chem., Int. Ed.*, 2018, **57**, 2393.
85. Q. Li, M. Zhou, M. Yang, Q. Yang, Z. Zhang and J. Shi, *Nat. Commun.*, 2018, **9**, 1.
86. P. Long, Y. Feng, C. Cao, Y. Li, J. Han, S. Li, C. Peng, Z. Li and W. Feng, *Adv. Funct. Mater.*, 2018, **28**, 1870263.
87. X. Ma, C. Xu, J. Wang and H. Tian, *Angew. Chem., Int. Ed.*, 2018, **57**, 10854.
88. Z. Zhou, J. Song, L. Nie and X. Chen, *Chem. Soc. Rev.*, 2016, **45**, 6597.
89. J. Du, N. Xu, J. Fan, W. Sun and X. Peng, *Small*, 2019, **15**, 1805087.

90. K. J. Mintz, Y. Zhou and R. M. Leblanc, *Nanoscale*, 2019, **11**, 4634.
91. Y. Su, S. Liu, Y. Guan, Z. Xie, M. Zheng and X. Jing, *Biomaterials*, 2020, **255**, 120110.
92. F. Yuan, S. Li, Z. Fan, X. Meng, L. Fan and S. Yang, *Nano Today*, 2016, **11**, 565.
93. X. Bao, Y. Yuan, J. Chen, B. Zhang, D. Li, D. Zhou, P. Jing, G. Xu, Y. Wang and K. Holá, *Light: Sci. Appl.*, 2018, **7**, 1.
94. Y.-W. Bao, X.-W. Hua, Y.-H. Li, H.-R. Jia and F.-G. Wu, *ACS Appl. Mater. Interfaces*, 2018, **10**, 1544.
95. B. Tian, S. Liu, L. Feng, S. Liu, S. Gai, Y. Dai, L. Xie, B. Liu, P. Yang and Y. Zhao, *Adv. Funct. Mater.*, 2021, **31**, 2100549.
96. F. Wu, H. Su, Y. Cai, W.-K. Wong, W. Jiang and X. Zhu, *ACS Appl. Bio Mater.*, 2018, **1**, 110.
97. D. Li, D. Han, S.-N. Qu, L. Liu, P.-T. Jing, D. Zhou, W.-Y. Ji, X.-Y. Wang, T.-F. Zhang and D.-Z. Shen, *Light: Sci. Appl.*, 2016, **5**, e16120.
98. L. Pan, S. Sun, L. Zhang, K. Jiang and H. Lin, *Nanoscale*, 2016, **8**, 17350.
99. M. Lan, S. Zhao, Z. Zhang, L. Yan, L. Guo, G. Niu, J. Zhang, J. Zhao, H. Zhang and P. Wang, *Nano Res.*, 2017, **10**, 3113.
100. D. Li, P. Jing, L. Sun, Y. An, X. Shan, X. Lu, D. Zhou, D. Han, D. Shen, Y. Zhai, S. Qu, R. Zbořil and A. L. Rogach, *Adv. Mater.*, 2018, **30**, 1705913.
101. Z. Gan, X. Wu, G. Zhou, J. Shen and P. K. Chu, *Adv. Opt. Mater.*, 2013, **1**, 554.
102. Q. Liu, B. Guo, Z. Rao, B. Zhang and J. R. Gong, *Nano Lett.*, 2013, **13**, 2436.
103. K.-W. Chu, S. L. Lee, C.-J. Chang and L. Liu, *Polymers*, 2019, **11**, 689.
104. M. M. Khan, S. F. Adil and A. Al-Mayouf, *J. Saudi Chem. Soc.*, 2015, **19**, 462.
105. Y. Deng, M. Chen, G. Chen, W. Zou, Y. Zhao, H. Zhang and Q. Zhao, *ACS Omega*, 2021, **6**, 4247.
106. S. K. Batabyal, S. E. Lu and J. J. Vittal, *Cryst. Growth Des.*, 2016, **16**, 2231.
107. Z. Zhang, T. Zheng, X. Li, J. Xu and H. Zeng, *Part. Syst. Charact.*, 2016, **33**, 457.
108. C. Sakdaronnarong, A. Sangjan, S. Boonsith, D. C. Kim and H. S. Shin, *Catalysts*, 2020, **10**, 320.
109. Z. Zhang, G. Yi, P. Li, X. Zhang, H. Fan, Y. Zhang, X. Wang and C. Zhang, *Nanoscale*, 2020, **12**, 13899.
110. H. Li, R. Liu, W. Kong, J. Liu, Y. Liu, L. Zhou, X. Zhang, S.-T. Lee and Z. Kang, *Nanoscale*, 2014, **6**, 867.
111. R. Miao, Z. Luo, W. Zhong, S.-Y. Chen, T. Jiang, B. Dutta, Y. Nasr, Y. Zhang and S. L. Suib, *Appl. Catal., B*, 2016, **189**, 26.
112. W. Shi, F. Guo, M. Li, Y. Shi, M. Wu and Y. Tang, *J. Alloys Compd.*, 2019, **775**, 511.
113. X. Hao, Z. Jin, J. Xu, S. Min and G. Lu, *Superlattices Microstruct.*, 2016, **94**, 237.
114. J. Zhang, M. Si, L. Jiang, X. Yuan, H. Yu, Z. Wu, Y. Li and J. Guo, *J. Chem. Eng.*, 2021, **410**, 128336.

115. Z. W. Seh, J. Kibsgaard, C. F. Dickens, I. Chorkendorff, J. K. Nørskov and T. F. Jaramillo, *Science*, 2017, **355**, eaad4998.
116. M. Zhu, Y. Zhou, Y. Sun, C. Zhu, L. Hu, J. Gao, H. Huang, Y. Liu and Z. Kang, *Dalton Trans.*, 2018, **47**, 5459.
117. Z. Zhu, K. Ge, Z. Li, J. Hu, P. Chen and H. Bi, *Small*, 2022, 2205234.
118. S. T. R. Naqvi, T. Rasheed, S. Majeed, Z. Rani, R. Nawaz and M. N. Ashiq, *J. Mol. Liq.*, 2020, **309**, 113111.
119. X. Kou, X. Xin, Y. Zhang and L.-Y. Meng, *Carbon Lett.*, 2021, **31**, 695.
120. M. Green, *Angew. Chem., Int. Ed.*, 2004, **43**, 4129.
121. C. M. Tyrakowski and P. T. Snee, *Phys. Chem. Chem. Phys.*, 2014, **16**, 837.
122. S. Y. Lim, W. Shen and Z. Gao, *Chem. Soc. Rev.*, 2015, **44**, 362.
123. H. Li, Z. Kang, Y. Liu and S.-T. Lee, *J. Mater. Chem.*, 2012, **22**, 24230.
124. S. Zhu, Q. Meng, L. Wang, J. Zhang, Y. Song, H. Jin, K. Zhang, H. Sun, H. Wang and B. Yang, *Angew. Chem., Int. Ed.*, 2013, **52**, 3953.
125. H. Li, X. Yan, D. Kong, R. Jin, C. Sun, D. Du, Y. Lin and G. Lu, *Nanoscale Horiz.*, 2020, **5**, 218.
126. N. Zhou, Z. Hao, X. Zhao, S. Maharjan, S. Zhu, Y. Song, B. Yang and L. Lu, *Nanoscale*, 2015, **7**, 15635.
127. L. Cao, X. Wang, M. J. Meziani, F. Lu, H. Wang, P. G. Luo, Y. Lin, B. A. Harruff, L. M. Veca, D. Murray, S.-Y. Xie and Y.-P. Sun, *J. Am. Chem. Soc.*, 2007, **129**, 11318.
128. G. Gao, Y.-W. Jiang, J. Yang and F.-G. Wu, *Nanoscale*, 2017, **9**, 18368.
129. X.-W. Hua, Y.-W. Bao, J. Zeng and F.-G. Wu, *ACS Appl. Mater. Interfaces*, 2019, **11**, 32647.
130. X.-W. Hua, Y.-W. Bao and F.-G. Wu, *ACS Appl. Mater. Interfaces*, 2018, **10**, 10664.
131. J. Liu, Y. Geng, D. Li, H. Yao, Z. Huo, Y. Li, K. Zhang, S. Zhu, H. Wei, W. Xu, J. Jiang and B. Yang, *Adv. Mater.*, 2020, **32**, 1906641.
132. L. Jiang, H. Ding, M. Xu, X. Hu, S. Li, M. Zhang, Q. Zhang, Q. Wang, S. Lu, Y. Tian and H. Bi, *Small*, 2020, **16**, 2000680.
133. G. D. Luker, K. E. Luker and J. Nucl, *Med.*, 2008, **49**, 1.
134. K. Y. Zhang, Q. Yu, H. Wei, S. Liu, Q. Zhao and W. Huang, *Chem. Rev.*, 2018, **118**, 1770.
135. Z. Li, Y. Zhang, X. Wu, L. Huang, D. Li, W. Fan and G. Han, *J. Am. Chem. Soc.*, 2015, **137**, 5304.
136. A. Abdukayum, J.-T. Chen, Q. Zhao and X.-P. Yan, *J. Am. Chem. Soc.*, 2013, **135**, 14125.
137. Z. An, C. Zheng, Y. Tao, R. Chen, H. Shi, T. Chen, Z. Wang, H. Li, R. Deng and X. Liu, *Nat. Mater.*, 2015, **14**, 685.
138. O. Bolton, K. Lee, H.-J. Kim, K. Y. Lin and J. Kim, *Nat. Chem.*, 2011, **3**, 205.
139. W. Li, S. Wu, X. Xu, J. Zhuang, H. Zhang, X. Zhang, C. Hu, B. Lei, C. F. Kaminski and Y. Liu, *Chem. Mater.*, 2019, **31**, 9887.
140. X. Huang, F. Zhang, L. Zhu, K. Y. Choi, N. Guo, J. Guo, K. Tackett, P. Anilkumar, G. Liu and Q. Quan, *ACS Nano*, 2013, **7**, 5684.
141. H. Chen, G. D. Wang, W. Tang, T. Todd, Z. Zhen, C. Tsang, K. Hekmatyar, T. Cowger, R. B. Hubbard and W. Zhang, *Adv. Mater.*, 2014, **26**, 6761.

142. X. Ren, L. Liu, Y. Li, Q. Dai, M. Zhang and X. Jing, *J. Mater. Chem. B*, 2014, **2**, 554.
143. N. Gong, H. Wang, S. Li, Y. Deng, X. a. Chen, L. Ye and W. Gu, *Langmuir*, 2014, **30**, 10933.
144. Y. Shi, Y. Pan, J. Zhong, J. Yang, J. Zheng, J. Cheng, R. Song and C. Yi, *Carbon*, 2015, **93**, 742.
145. H. Wang, R. Revia, K. Wang, R. J. Kant, Q. Mu, Z. Gai, K. Hong and M. Zhang, *Adv. Mater.*, 2017, **29**, 1605416.
146. H. Chen, G. D. Wang, X. Sun, T. Todd, F. Zhang, J. Xie and B. Shen, *Adv. Funct. Mater.*, 2016, **26**, 3973.
147. F. Du, L. Zhang, L. Zhang, M. Zhang, A. Gong, Y. Tan, J. Miao, Y. Gong, M. Sun and H. Ju, *Biomaterials*, 2017, **121**, 109.
148. H. Chen, Y. Qiu, D. Ding, H. Lin, W. Sun, G. D. Wang, W. Huang, W. Zhang, D. Lee and G. Liu, *Adv. Mater.*, 2018, **30**, 1802748.
149. C. Han, H. Xu, R. Wang, K. Wang, Y. Dai, Q. Liu, M. Guo, J. Li and K. Xu, *J. Mater. Chem. B*, 2016, **4**, 5798.
150. Q. Jia, Z. Zhao, K. Liang, F. Nan, Y. Li, J. Wang, J. Ge and P. Wang, *Mater. Chem. Front.*, 2020, **4**, 449.
151. S. Chung, R. A. Revia and M. Zhang, *Adv. Mater.*, 2021, **33**, 1904362.
152. Y. Xu, X.-H. Jia, X.-B. Yin, X.-W. He and Y.-K. Zhang, *Anal. Chem.*, 2014, **86**, 12122.
153. F. Wang, K. Bae, Z. Huang and J. Xue, *Nanoscale*, 2018, **10**, 5642.
154. Y. Liu, X. Zhi, W. Hou, F. Xia, J. Zhang, L. Li, Y. Hong, H. Yan, C. Peng and J. M. de la Fuentea, *Nanoscale*, 2018, **10**, 19052.
155. H. Ding, D. Wang, A. Sadat, Z. Li, X. Hu, M. Xu, P. C. de Morais, B. Ge, S. Sun, J. Ge, Y. Chen, Y. Qian, C. Shen, X. Shi, X. Huang, R.-Q. Zhang and H. Bi, *ACS Appl. Bio Mater.*, 2021, **4**, 2798.
156. X. Wang, Y. Pang, G. Ku, X. Xie, G. Stoica and L. V. Wang, *Nat. Biotechnol.*, 2003, **21**, 803.
157. A. De La Zerda, C. Zavaleta, S. Keren, S. Vaithilingam, S. Bodapati, Z. Liu, J. Levi, B. R. Smith, T.-J. Ma and O. Oralkan, *Nat. Nanotechnol.*, 2008, **3**, 557.
158. A. de la Zerda, S. Bodapati, R. Teed, S. Y. May, S. M. Tabakman, Z. Liu, B. T. Khuri-Yakub, X. Chen, H. Dai and S. S. Gambhir, *ACS Nano*, 2012, **6**, 4694.
159. J. Ge, Q. Jia, W. Liu, L. Guo, Q. Liu, M. Lan, H. Zhang, X. Meng and P. Wang, *Adv. Mater.*, 2015, **27**, 4169.
160. G. Xu, X. Bao, J. Chen, B. Zhang, D. Li, D. Zhou, X. Wang, C. Liu, Y. Wang and S. Qu, *Adv. Healthcare Mater.*, 2019, **8**, 1800995.
161. G. Gunaydin, M. E. Gedik and S. Ayan, *Front. Chem.*, 2021, **9**, 686303.
162. X. Wu, M. Xu, S. Wang, K. Abbas, X. Huang, R. Zhang, A. C. Tedesco and H. Bi, *Dalton Trans.*, 2022, **51**, 2296.
163. A. Escudero, C. Carrillo-Carrion, M. C. Castillejos, E. Romero-Ben, C. Rosales-Barrios and N. Khiar, *Mater. Chem. Front.*, 2021, **5**, 3788.
164. X. Li, S. Ai, X. Lu, S. Liu and W. Guan, *RSC Adv.*, 2021, **11**, 35392.

165. F. Wu, L. Yue, H. Su, K. Wang, L. Yang and X. Zhu, *Nanoscale Res. Lett.*, 2018, **13**, 1.
166. Y. Choi, S. Kim, M. H. Choi, S. R. Ryoo, J. Park, D. H. Min and B. S. Kim, *Adv. Funct. Mater.*, 2014, **24**, 5781.
167. K. Yang, F. Li, W. Che, X. Hu, C. Liu and F. Tian, *RSC Adv.*, 2016, **6**, 101447.
168. G. Yuan, C. Lv, J. Liang, X. Zhong, Y. Li, J. He, A. Zhao, L. Li, Y. Shao and X. Zhang, *Adv. Funct. Mater.*, 2021, **31**, 2104026.
169. S. Akter, M. S. Khan, E. N. Smith and E. Flashman, *RSC Chem. Biol.*, 2021, **2**, 1384.
170. X. Nie, C. Jiang, S. Wu, W. Chen, P. Lv, Q. Wang, J. Liu, C. Narh, X. Cao, R. A. Ghiladi and Q. Wei, *J. Photochem. Photobiol., B*, 2020, **206**, 111864.
171. Q. Jia, J. Ge, W. Liu, X. Zheng, S. Chen, Y. Wen, H. Zhang and P. Wang, *Adv. Mater.*, 2018, **30**, 1706090.
172. S. Zhao, S. Wu, Q. Jia, L. Huang, M. Lan, P. Wang and W. Zhang, *J. Chem. Eng.*, 2020, **388**, 124212.
173. J. Wang, M. Xu, D. Wang, Z. Li, F. L. Primo, A. C. Tedesco and H. Bi, *Inorg. Chem.*, 2019, **58**, 13394.
174. C. Jin, K. Wang, A. Oppong-Gyebi and J. Hu, *Int. J. Med. Sci.*, 2020, **17**, 2964.
175. M. Behi, L. Gholami, S. Naficy, S. Palomba and F. Dehghani, *Nanoscale Adv.*, 2022, **4**, 353.
176. L. Jiang, H. Cai, W. Zhou, Z. Li, L. Zhang and H. Bi, *Adv. Mater.*, 2023, 2210776.
177. C. Pucci, C. Martinelli and G. Ciofani, *Ecancermedicalscience*, 2019, **13**, 961.
178. K. D. Wegner and N. Hildebrandt, *Chem. Soc. Rev.*, 2015, **44**, 4792.
179. H. Liu, Z. Li, Y. Sun, X. Geng, Y. Hu, H. Meng, J. Ge and L. Qu, *Sci. Rep.*, 2018, **8**, 1.
180. Q. Huang, Y. Liu, L. Zheng, L. Wu, Z. Zhou, J. Chen, W. Chen and H. Zhao, *Microchim. Acta*, 2019, **186**, 1.
181. Z. Ji, P. Ai, C. Shao, T. Wang, C. Yan, L. Ye and W. Gu, *ACS Biomater. Sci. Eng.*, 2018, **4**, 2089.
182. W. Xiao, Y. Li, C. Hu, Y. Huang, Q. He and H. Gao, *J. Colloid Interface Sci.*, 2017, **497**, 226.
183. D. L. Sai, J. Lee, D. L. Nguyen and Y.-P. Kim, *Exp. Mol. Med.*, 2021, **53**, 495.
184. G. Gunaydin, M. E. Gedik and S. Ayan, *Front. Chem.*, 2021, **9**, 400.
185. L. Zhang, Z. Lin, Y.-X. Yu, B.-P. Jiang and X.-C. Shen, *J. Mater. Chem. B*, 2018, **6**, 6534.
186. L. Zhao, X. Zhang, X. Wang, X. Guan, W. Zhang and J. Ma, *J. Nanobiotechnol.*, 2021, **19**, 1.
187. Y. Li, G. Bai, S. Zeng and J. Hao, *ACS Appl. Mater. Interfaces*, 2019, **11**, 4737.
188. E. B. Kang, J. E. Lee, Z. A. I. Mazrad, I. In, J. H. Jeong and S. Y. Park, *Nanoscale*, 2018, **10**, 2512.

189. Y. Bai, J. Zhao, S. Wang, T. Lin, F. Ye and S. Zhao, *ACS Appl. Mater. Interfaces*, 2021, **13**, 35365.
190. M. Zhang, W. Wang, N. Zhou, P. Yuan, Y. Su, M. Shao, C. Chi and F. Pan, *Carbon*, 2017, **118**, 752.
191. Q. Jia, X. Zheng, J. Ge, W. Liu, H. Ren, S. Chen, Y. Wen, H. Zhang, J. Wu and P. Wang, *J. Colloid Interface Sci.*, 2018, **526**, 302.
192. H. Yu, X. Lv, L. Wu, B. Li, K. Huang, Y. Huang, Q. Zhang, C. Mei, X. Ren, R. Zhou, H. Luo, P. Pang and H. Shan, *Nanoscale*, 2020, **12**, 17222.
193. Z. Wang, H. Liao, H. Wu, B. Wang, H. Zhao and M. Tan, *Anal. Methods*, 2015, **7**, 8911.
194. H. S. Zavareh, M. Pourmadadi, A. Moradi, F. Yazdian and M. Omidi, *Int. J. Biol. Macromol.*, 2020, **165**, 1422.
195. T. Feng, X. Ai, G. An, P. Yang and Y. Zhao, *ACS Nano*, 2016, **10**, 4410.
196. S. A. Mathew, P. Praveena, S. Dhanavel, R. Manikandan, S. Senthilkumar and A. Stephen, *RSC Adv.*, 2020, **10**, 24386.
197. K. J. Mintz, G. Mercado, Y. Zhou, Y. Ji, S. D. Hettiarachchi, P. Y. Liyanage, R. R. Pandey, C. C. Chusuei, J. Dallman and R. M. Leblanc, *Colloids Surf., B*, 2019, **176**, 488.
198. W. Li, Q. Liu, P. Zhang and L. Liu, *Acta Biomater.*, 2016, **40**, 254.
199. J. Jiao, C. Liu, X. Li, J. Liu, D. Di, Y. Zhang, Q. Zhao and S. Wang, *J. Colloid Interface Sci.*, 2016, **483**, 343.
200. Q. Zeng, D. Shao, X. He, Z. Ren, W. Ji, C. Shan, S. Qu, J. Li, L. Chen and Q. Li, *J. Mater. Chem. B*, 2016, **4**, 5119.
201. T. Kong, L. Hao, Y. Wei, X. Cai and B. Zhu, *Cell Proliferation*, 2018, **51**, e12488.
202. S. Li, R. Hu, C. Yang, X. Zhang, Y. Zeng, S. Wang, X. Guo, Y. Li, X. Cai and S. Li, *Biosens. Bioelectron.*, 2017, **98**, 325.
203. K. Even-Desrumeaux, D. Baty and P. Chames, *Cancers*, 2011, **3**, 2554.
204. S. Bayda, E. Amadio, S. Cailotto, Y. Frión-Herrera, A. Perosa and F. Rizzolio, *Nanoscale Adv.*, 2021, **3**, 5183.
205. Y. Zhang, M. Li, X. Gao, Y. Chen and T. Liu, *J. Hematol. Oncol.*, 2019, **12**, 1.
206. Y. Zhang, W. Liu, S. Ge, M. Yan, S. Wang, J. Yu, N. Li and X. Song, *Biosens. Bioelectron.*, 2013, **41**, 684.
207. S. Mohammadi, A. Salimi, S. Hamd-Ghadareh, F. Fathi and F. Soleimani, *Anal. Biochem.*, 2018, **557**, 18.
208. N. Azam, M. Najabat Ali and T. Javaid Khan, *Front. Mater.*, 2021, **8**, 700403.
209. N. De Acha, C. Elosúa, J. M. Corres and F. J. Arregui, *Sensors*, 2019, **19**, 599.
210. J. Zhu, H. Chu, T. Wang, C. Wang and Y. Wei, *Microchem. J.*, 2020, **158**, 105142.
211. G. Chen, F. Song, X. Xiong and X. Peng, *Ind. Eng. Chem. Res.*, 2013, **52**, 11228.

212. P. Gong, L. Sun, F. Wang, X. Liu, Z. Yan, M. Wang, L. Zhang, Z. Tian, Z. Liu and J. You, *J. Chem. Eng.*, 2019, **356**, 994.
213. G. Li, H. Fu, X. Chen, P. Gong, G. Chen, L. Xia, H. Wang, J. You and Y. Wu, *Anal. Chem.*, 2016, **88**, 2720.
214. X. Wang, X. Shen, B. Li, G. Jiang, X. Zhou and H. Jiang, *RSC Adv.*, 2016, **6**, 18326.
215. C. Zou, M. F. Foda, X. Tan, K. Shao, L. Wu, Z. Lu, H. S. Bahlol and H. Han, *Anal. Chem.*, 2016, **88**, 7395.
216. V. S. A. Jayanthi, A. B. Das and U. Saxena, *Biosens. Bioelectron.*, 2017, **91**, 15.
217. Y. Chi, L. Yao, X. Hu, S. Huang, N. Huang, S. Li, Z. Shao and J. Wu, *Oncotarget*, 2016, **7**, 14951.
218. H. Zhu, P. S. Dale, C. W. Caldwell and X. Fan, *Anal. Chem.*, 2009, **81**, 9858.
219. S. T. Sanjay, M. Dou, J. Sun and X. Li, *Sci. Rep.*, 2016, **6**, 1.
220. X. Zhang, X. Peng and W. Jin, *Anal. Chim. Acta*, 2006, **558**, 110.
221. C. Hong, R. Yuan, Y. Chai and Y. Zhuo, *Anal. Chim. Acta*, 2009, **633**, 244.
222. A. Khoshroo, M. Mazloum-Ardakani and M. Forat-Yazdi, *Sens. Actuators, B*, 2018, **255**, 580.
223. Y. Wu, P. Wei, S. Pengpumkiat, E. A. Schumacher and V. T. Remcho, *Anal. Methods*, 2016, **8**, 5398.
224. S. Huang, B. Li, U. Ashraf, Q. Li, X. Lu, X. Gao, M. Cui, M. Imran, J. Ye and F. Cao, *ACS Appl. Nano Mater.*, 2020, **3**, 9449.
225. L. Luo, C. Liu, T. He, L. Zeng, J. Xing, Y. Xia, Y. Pan, C. Gong and A. Wu, *Nanoscale*, 2018, **10**, 22035.
226. P. Wu, W. Li, Q. Wu, Y. Liu and S. Liu, *RSC Adv.*, 2017, **7**, 44144.
227. G. Yang, X. Wan, Y. Su, X. Zeng and J. Tang, *J. Mater. Chem. A*, 2016, **4**, 12841.
228. L.-D. Xu, F.-L. Du, J. Zhu and S.-N. Ding, *Analyst*, 2021, **146**, 706.
229. Z.-X. Wang and S.-N. Ding, *Anal. Chem.*, 2014, **86**, 7436.
230. D. Yao, A. Liang and Z. Jiang, *Microchim. Acta*, 2019, **186**, 1.
231. B. Dong, H. Li, J. Sun, Y. Li, G. M. Mari, X. Yu, W. Yu, K. Wen, J. Shen and Z. Wang, *J. Hazard. Mater.*, 2021, **402**, 123942.
232. H. Ding, B. Hu, B. Zhang, H. Zhang, X. Yan, G. Nie and M. Liang, *Nano Res.*, 2021, **14**, 570.
233. X. Ren, D. Chen, Y. Wang, H. Li, Y. Zhang, H. Chen, X. Li and M. Huo, *J. Nanobiotechnol.*, 2022, **20**, 1.
234. D. Jiang, D. Ni, Z. T. Rosenkrans, P. Huang, X. Yan and W. Cai, *Chem. Soc. Rev.*, 2019, **48**, 3683.
235. Y. Huang, J. Ren and X. Qu, *Chem. Rev.*, 2019, **119**, 4357.
236. J. Jin, L. Li, L. Zhang, Z. Luan, S. Xin and K. Song, *Front. Chem.*, 2021, **9**, 800.
237. J. Zhang, X. Lu, D. Tang, S. Wu, X. Hou, J. Liu and P. Wu, *ACS Appl. Mater. Interfaces*, 2018, **10**, 40808.
238. Y. Liu, B. Xu, M. Lu, S. Li, J. Guo, F. Chen, X. Xiong, Z. Yin, H. Liu and D. Zhou, *Bioact. Mater.*, 2022, **12**, 246.

239. T. Pan, H. Chen, X. Gao, Z. Wu, Y. Ye and Y. Shen, *J. Hazard. Mater.*, 2022, **435**, 128996.
240. L. Yao, M.-M. Zhao, Q.-W. Luo, Y.-C. Zhang, T.-T. Liu, Z. Yang, M. Liao, P. Tu and K.-W. Zeng, *ACS Nano*, 2022, **16**, 9228.
241. P. Gao, S. Chen, S. Liu, H. Liu, Z. Xie and M. Zheng, *ACS Appl. Mater. Interfaces*, 2021, **13**, 56456.
242. Z. Li, W. Liu, P. Ni, C. Zhang, B. Wang, G. Duan, C. Chen, Y. Jiang and Y. Lu, *J. Chem. Eng.*, 2022, **428**, 131396.
243. S. Hanif, P. Muhammad, Z. Niu, M. Ismail, M. Morsch, X. Zhang, M. Li and B. Shi, *Adv. NanoBiomed Res.*, 2021, **1**, 2100008.
244. S. Hanif, P. Muhammad, R. Chesworth, F. U. Rehman, R.-j. Qian, M. Zheng and B.-y. Shi, *Acta Pharmacol. Sin.*, 2020, **41**, 936.
245. P. Muhammad, S. Hanif, J. Li, A. Guller, F. U. Rehman, M. Ismail, D. Zhang, X. Yan, K. Fan and B. Shi, *Nano Today*, 2022, **45**, 101530.
246. C. Sun, Y. Zhang, S. Kalytchuk, Y. Wang, X. Zhang, W. Gao, J. Zhao, K. Cepe, R. Zboril and W. Y. William, *J. Mater. Chem. C*, 2015, **3**, 6613.
247. B. Wang, J. Yu, L. Sui, S. Zhu, Z. Tang, B. Yang and S. Lu, *Adv. Sci.*, 2021, **8**, 2001453.
248. B. Wang, H. Song, Z. Tang, B. Yang and S. Lu, *Nano Res.*, 2022, **15**, 942.
249. Y. Zheng, K. Arkin, J. Hao, S. Zhang, W. Guan, L. Wang, Y. Guo and Q. Shang, *Adv. Opt. Mater.*, 2021, **9**, 2100688.
250. F. Wang, Y.-H. Chen, C.-Y. Liu and D.-G. Ma, *Chem. Commun.*, 2011, **47**, 3502.
251. K. Jiang, L. Zhang, J. Lu, C. Xu, C. Cai and H. Lin, *Angew. Chem., Int. Ed.*, 2016, **55**, 7231.
252. K. Jiang, X. Gao, X. Feng, Y. Wang, Z. Li and H. Lin, *Angew. Chem., Int. Ed.*, 2020, **59**, 1263.
253. X. Wang, B. Wang, H. Wang, T. Zhang, H. Qi, Z. Wu, Y. Ma, H. Huang, M. Shao, Y. Liu, Y. Li and Z. Kang, *Angew. Chem., Int. Ed.*, 2021, **60**, 12585.
254. Q. Chen, Y. Hu, C. Hu, H. Cheng, Z. Zhang, H. Shao and L. Qu, *Phys. Chem. Chem. Phys.*, 2014, **16**, 19307.
255. Y. Xu, X. Li, G. Hu, T. Wu, Y. Luo, L. Sun, T. Tang, J. Wen, H. Wang and M. Li, *Appl. Surf. Sci.*, 2017, **422**, 847.
256. Z. Ye, F. Wang, C. Jia, K. Mu, M. Yu, Y. Lv and Z. Shao, *J. Chem. Eng.*, 2017, **330**, 1166.
257. J. S. Wei, H. Ding, P. Zhang, Y. F. Song, J. Chen, Y. G. Wang and H. M. Xiong, *Small*, 2016, **12**, 5927.
258. J. Luo, J. Wang, S. Liu, W. Wu, T. Jia, Z. Yang, S. Mu and Y. Huang, *Carbon*, 2019, **146**, 1.
259. G. Wei, X. Zhao, K. Du, Z. Wang, M. Liu, S. Zhang, S. Wang, J. Zhang and C. An, *J. Chem. Eng.*, 2017, **326**, 58.
260. T. Ghosh, R. Ghosh, U. Basak, S. Majumdar, R. Ball, D. Mandal, A. K. Nandi and D. P. Chatterjee, *J. Mater. Chem. A*, 2018, **6**, 6476.
261. M. Jing, J. Wang, H. Hou, Y. Yang, Y. Zhang, C. Pan, J. Chen, Y. Zhu and X. Ji, *J. Mater. Chem. A*, 2015, **3**, 16824.

262. H. Yu, W. Zhu, H. Zhou, J. Liu, Z. Yang, X. Hu and A. Yuan, *RSC Adv.*, 2019, **9**, 9577.
263. D. Chao, C. Zhu, X. Xia, J. Liu, X. Zhang, J. Wang, P. Liang, J. Lin, H. Zhang and Z. X. Shen, *Nano Lett.*, 2015, **15**, 565.
264. F. Ehrat, S. Bhattacharyya, J. Schneider, A. Löf, R. Wyrwich, A. L. Rogach, J. K. Stolarczyk, A. S. Urban and J. Feldmann, *Nano Lett.*, 2017, **17**, 7710.
265. D. W. Boukhvalov, V. Y. Osipov, B. T. Hogan and A. Baldycheva, *Discover Nano*, 2023, **18**, 1.
266. C. Kang, S. Tao, F. Yang and B. Yang, *Aggregate*, 2022, **3**, e169.
267. L. Chen, C. F. Wang, C. Liu and S. Chen, *Small*, 2022, 2206671.
268. C. Zhu, Y. Fu, C. Liu, Y. Liu, L. Hu, J. Liu, I. Bello, H. Li, N. Liu, S. Guo, H. Huang, Y. Lifshitz, S. Lee and Z. Kang, *Adv. Mater.*, 2017, **29**, 1701399.
269. C. Dong, M. Xu, S. Wang, M. Ma, O. U. Akakuru, H. Ding, A. Wu, Z. Zha, X. Wang and H. Bi, *J. Nanobiotechnol.*, 2021, **19**, 1.

CHAPTER 5

Low Pressure Synthesis of Diamond by Chemical Vapour Deposition and Its Technological Applications

NEIL A. FOX[a,b*]

[a]University of Bristol, UK; [b]HH Wills Physics Laboratory, Tyndall Avenue, Bristol, BS81TL, UK
*Email: neil.fox@bristol.ac.uk

5.1 Introduction

It is now well known that diamond is thermodynamically stable with respect to graphite only at high pressures. This fact has been exploited since the 1950s to produce industrial diamond on a commercial scale by the so-called high temperature, high pressure (HPHT) process involving the squeezing of graphite and metal catalysts in a diamond anvil, that was originally developed by General Electric in the US.[1] But as noted by John Angus[2] 'it has been less well appreciated that, although diamond is thermodynamically unstable with respect to graphite at ambient pressures, the magnitude of this instability, as measured by the difference in Gibbs free energies, is very low. In fact, below (around) one bar and at 1400 K, a typical temperature used in diamond chemical vapour deposition (CVD), the difference in free energy between diamond and graphite is only ~8 kJ mol^{-1}. For comparison, at 1400 K the value of kT is ~11 kJ mol^{-1}. In practical terms, this means that both

Inorganic Materials Series No. 15
Elemental Carbon
Edited by Peter J. F. Harris

Published by the Royal Society of Chemistry, www.rsc.org

phases have essentially the same free energy. Therefore, whether diamond or graphite forms during chemical vapor deposition does not depend on their relative thermodynamic stability, but instead on the rates of nucleation and growth of the two phases.'

The issues of nucleation and growth rate have been largely solved in the past twenty years as improvements to the microwave plasma and hot filament methods of CVD diamond growth have been made, which has enabled high quality, single crystal, as well as polycrystalline films, to be produced on non-diamond substrates,[3,4] and these methods are both being used widely in research laboratories and industry. This chapter will review and discuss some recent diamond materials research at Bristol and elsewhere that has been oriented towards the development of heat to electric power generation and radio-voltaic batteries.

5.2 Synthetic Diamond Synthesis

Diamond growth is achieved by flowing methane diluted (to just a few percent) with hydrogen into a growth chamber. Hot filament CVD reactors are very inexpensive to build compared to plasma-assisted CVD systems and can be used to grow diamond films on large area substrates or single crystal wafers at modest growth rates of 0.5–1.5 microns per hour. Plasma-assisted CVD systems usually employ a radiofrequency source, such as a microwave generator or some form of DC plasma power supply. Plasma-assisted CVD systems are capable of producing diamond films at growth rates of hundreds of microns per hour if kW levels of plasma power are used. This can be understood by considering how the presence of a plasma influences the production of growth species from the feedstock gases. Activation of a hydrogen containing gas mixture produces atomic hydrogen, which starts a cascade of chemical processes (Figure 5.1) in the gas or gas-plasma phase that overall are exothermic and which produce hydrocarbon free radicals, such as methyl and acetyl.

At the surface, atomic hydrogen adds to bare sites, providing a hydrogen-terminated surface that inhibits surface reconstruction. Atomic hydrogen can also abstract a hydrogen atom from a hydrogen-terminated surface. During growth, the relative rates of addition and abstraction lead to a steady state concentration of vacant surface sites where hydrocarbon free radicals can attach. Diamond nucleation proceeds in a very different way depending upon whether the underlying substrate is diamond or non-diamond, crystalline or non-crystalline, but in each case, atomic hydrogen would appear to play a key role.[5] Below a critical nucleus size, fully hydrogen-terminated diamond nuclei are thermodynamically more stable than graphite nuclei, the latter being terminated with hydrogen atoms only along the edges of the basal planes.

If one considers, as an example, the C(100)-2 × 1 surface the reaction proceeds through a mechanism such as that depicted in Figure 5.1; a gas phase H atom reacts with a surface terminating H atom, forming H_2 and leaving an

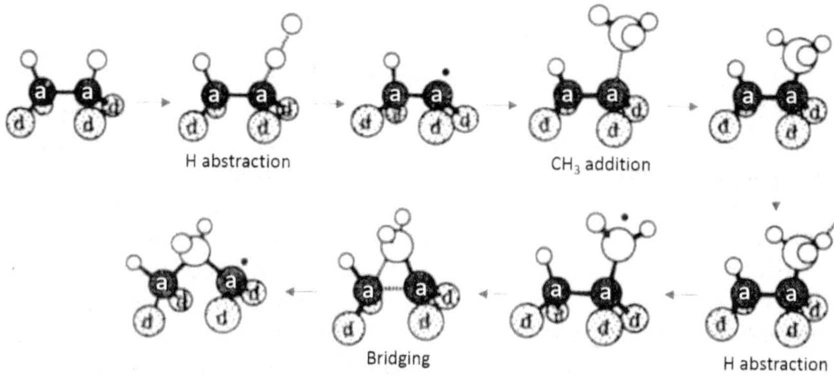

Figure 5.1 An example growth mechanism involving CH_3 addition to the reconstructed (100) surface of diamond. Spotted and dark circles are diamond carbon atoms under the surface (d) and on the surface (a) respectively, small white circles are H atoms and large white circles are adsorbing C atoms. Small dark spots denote unsaturated bonding on a carbon atom. Reproduced and adapted from ref. 37 with permission from A. Croot.

unsaturated carbon atom on the diamond surface (abstraction); this facilitates the attachment of a CH_3 radical to the unsaturated carbon (addition); a second H atom reacts with a hydrogen from the newly attached carbon which it leaves unsaturated (abstraction); the new, unsaturated carbon finds a lower energy position above the centre of the dimer, breaking the bond between the two carbon atoms beneath (bridging); this leaves an unsaturated carbon on the original surface that can facilitate further chemisorption. This mechanism illustrates the importance of atomic hydrogen to the growth process, and suggests that CH_3 and (to some extent) CH_2 will dominate the productive chemisorption reactions that lead to growth. Further mechanisms, such as those that consider more than one CH_x addition and those on other surfaces, have been investigated by numerous researchers but are not outlined here.

For polycrystalline diamond growth onto non-diamond substrates, the choice of material must be one that can support the formation of a stable surface carbide under typical growth conditions, such as tungsten, molybdenum or non-metals, such as silicon. There is a often a requirement to pretreat the substrate surface to accelerate nucleation. This is achieved either by mechanical abrasion, the use of nanodiamond seeds, or the application of a voltage bias.[6–8] For single crystal diamond growth the choice of a non-diamond is complicated by the need to use a single crystal material possessing a compatible lattice match, the desired crystal orientation for the diamond and a coefficient of thermal expansion that is compatible with the low value exhibited by diamond. Practically it has been difficult to achieve all of these requirements and this has led to the use of single crystal substrates on which multilayers of different materials are deposited in order to preserve the crystal orientation while reducing the lattice mismatch between the substrate and the CVD diamond layer. Schreck *et al.*[9] have reported wafer-scale single crystal diamond growth onto a material stack composed of single crystal layers of yttria-stabilised zirconia (YSZ), followed by iridium onto a single crystal silicon substrate. The material layers reduce the lattice mismatch to 7.1% between iridium and the synthesised diamond, while retaining crystal orientation of the underlying substrate, allowing large area crystals up to 90 mm in diameter to be grown. However, the strain control of the diamond layer is much more difficult because the strain induced by the additional YSZ buffer layer between the diamond/Ir and the Si substrate has to be considered and remains an issue. Also, uniform Si removal from the diamond layer remains a barrier to device testing and performance that is still being resolved. Consequently, a number of researchers have investigated Ir/sapphire substrates for diamond heteroepitaxy.[10,11] Sapphire has advantages over alternatives, like MgO and YSZ, such as high crystal quality, low cost, and large wafer size availability of up to 8 inches in diameter.

5.2.1 Recent CVD Development: Static Flow CVD

In conventional CVD diamond growth systems, the feedstock gases of methane and hydrogen are continuously flowed through the reactor. Growth proceeds in an ambient gas pressure of typically between 30 Torr and 200 Torr with the chosen working pressure of a given reactor type being controlled by limiting the pumping rate of the vacuum pump set linked to the CVD system. The typical flow of feedstock gases can be as high as 1 L min^{-1} in a research reactor.

The continual flow of feedstock was understood to be a necessary requirement in order to replenish the supply of carbon in the gas phase due to the short residence time, of the order of milliseconds, for methyl radicals activated by the CVD process.[12] The residence time is defined as the time taken to replenish the entire gas content of the chamber up to the pressure selected for growth. Prior investigations of the effects of gas flow rate on diamond growth rates and morphology have yielded contradictory findings.[13–20]

More recently, studies conducted by the Bristol Diamond Group have suggested that the residence time of methyl radicals is much longer in plasma-assisted CVD using static gas flow conditions: the CVD reactor is pressurised with the feedstock mixture to the working gas pressure and growth proceeds without any further gas addition until the gaseous carbon content is largely depleted.[21]

For the discussion that follows, the spectroscopic notation denoting the intensities of the $C_2(d^3\Pi_g\text{-}a^3\Pi_u)$ $\Delta v = 0$ and $CH(A^2\Delta\text{-}X^2\Pi)$ $\Delta v = 0$ emissions are abbreviated to $I(C_2{}^*)$ and $I(CH^*)$, respectively. These were imaged using a Czerny-Turner spectrograph to establish how the presence of activated carbon species in the gas plasma influenced the deposition of diamond carbon onto a non-diamond substrate of single crystal silicon. In previous work, it has been demonstrated that the C_2(d-a) emission intensity from the hot plasma core region gives a quantitative indication of the local C content in that region,[18,22,23] while the CH(A-X) emission intensity is the better indicator of the concentrations of the CH_x ($0 \le x \le 3$) species responsible for diamond growth.[20]

For this static flow study, spectroscopic measurements of the decay of emission lines associated with activated carbon species in the gas plasma were taken at various points in the gas plasma present above the growth surface (Figure 5.2). It was found that the $I(C_2^*)$ intensity declines exponentially with a time constant, τ, that scales inversely with substrate temperature (T_{sub}). The $I(CH^*)$ intensity drops less steeply and, even at the highest T_{sub} investigated, had only declined to ~40% of its initial value after 90 min.

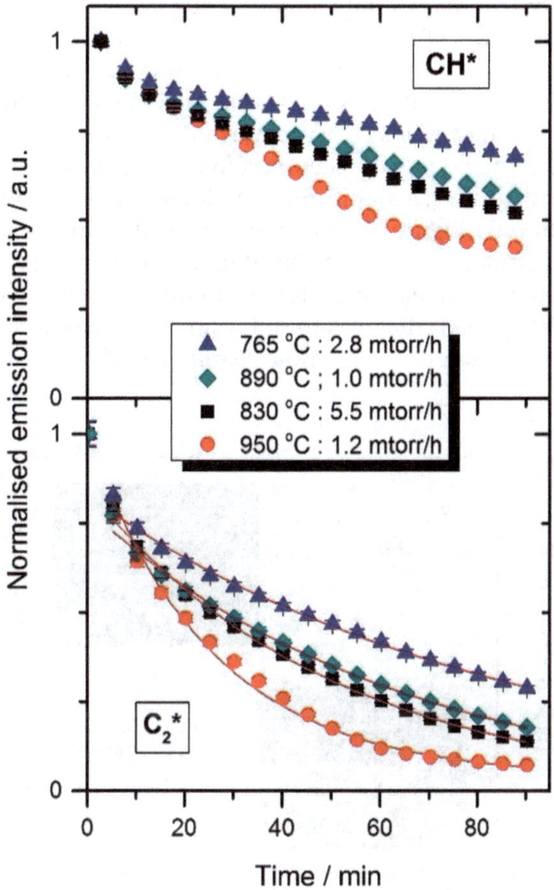

Figure 5.2 $I(CH^*)$ and $I(C_2^*)$ emission intensities measured as a function of time during static mode growth at four different substrate temperatures (shown in the legend, along with the reactor leak rate measured immediately prior to that particular growth run). The solid lines in the lower panel are fitted exponential decays from the second (t = 3 min) data point onwards, by which time the plasma is considered to have reached quasi-steady state. Reproduced from ref. 21 with permission from Elsevier, Copyright 2020.

From this result it was established that $I(C_2^*)$ remains a good proxy for the total C content in the plasma region under static-mode conditions. This would also imply that there might be a correlation between the rate of fall of $I(C_2^*)$ and the material deposition rate.

Figure 5.3 presents the experimental evidence found for the correlation between $I(C_2^*)$ and the deposition rate, where the deposited material thickness after 90 min of growth is determined (by cross-sectional SEM). Using the same measurement point in the plasma, the corresponding $I(C_2^*)$ decay constant for three different growth temperatures has been compared. The SEM images shown as insets to Figure 5.3 show the typical facetted morphology expected for a polycrystalline diamond film.

Although the range of values selected for T_{sub} is small and has very little impact upon the plasma processing and plasma parameters in the gas plasma core, it will influence the gas-surface chemistry required for growth and therefore the rate of loss of carbon from the plasma to the substrate. The measured trends that were recorded in this study suggested that deposition on the substrate is a significant sink for gas-phase carbon under static-mode conditions. Figure 5.4 illustrates how the thickness of diamond samples grown under static-mode conditions at similar T_{sub} values (~880 °C) and with similar reactor leak rates (low, ~1 mTorr h^{-1}) scales with deposition time. A surprising result that was found from this data was a relatively constant growth rate throughout the whole 90 min period of growth, irrespective

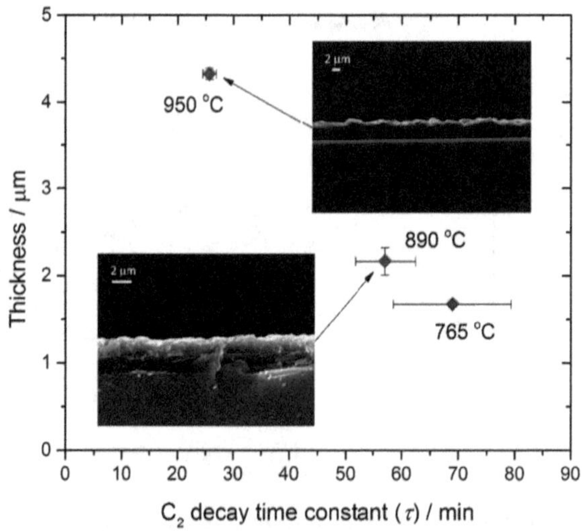

Figure 5.3 Plot of deposited film thickness following static-mode deposition for t = 90 min against the decay constant of the $I(C_2^*)$ emission. Representative cross-section SEM images of the samples grown at T_{sub} = 890 and 950 °C are shown in the insets and the bars on the data points represent the standard error. Reproduced from ref. 21 with permission from Elsevier, Copyright 2020.

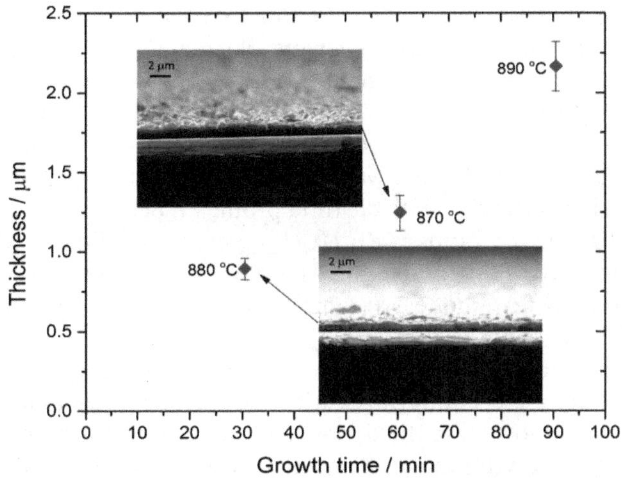

Figure 5.4 Diamond thickness as a function of deposition time for three samples grown at very similar deposition temperatures, T_{sub}. Representative cross-section SEM images of the t = 30 and t = 60 min samples are shown in the insets and the bars on the data points represent the standard error. Reproduced from ref. 21 with permission from Elsevier, Copyright 2020.

of the carbon content of the plasma, which would be expected to become depleted over time.

As Figure 5.2 showed, at T_{sub} ~880 °C, $I(C_2^*)$ in the plasma core has dropped to ~30% of its initial value after 60 min of growth. The small variations in T_{sub} between these three experiments would have had an insignificant effect on the thickness comparison, but the observation made that the growth rate seemed undiminished at later times highlights the challenges trying to unravel the processes that link gas-phase and gas-surface chemistry that control CVD diamond growth. It is understood that the rates of the gas-phase coupling reactions between C_2H_y ($0 \le y \le 6$) and CH_x ($0 \le x \le 4$) species, and the absolute concentrations of these various species, depend on many factors including the local C/H ratio, T_{gas} and, therefore, the location within the reactor volume. The prime factor in the CVD growth is the concentration of CH_x species in the immediate vicinity of the substrate. Referring again to Figure 5.2, it is noted that $I(CH^*)$ declines more slowly than $I(C_2^*)$. Previous plasma chemistry modelling by Bristol collaborators (of flow mode conditions)[22,24,25] has shown that this trend extends to small heights, z above the substrate and that addition reactions involving CH_x species (particularly CH_3 radicals) are important in diamond growth.

Therefore, it is not surprising that growth under static growth conditions can continue for a considerable time but, clearly, growth must cease when the reservoir of gas-phase carbon is fully depleted.

A further perspective on the static-mode deposition may be gained through direct comparisons made with material grown by more conventional standard flow-mode methods. Flow-mode experiments were conducted, at T_{sub} ~890 °C, designed to mimic the time-evolving plasma composition prevalent in the static-mode growth experiments. This was achieved by progressively reducing gas flow rate, $F(CH_4)$ from a set starting point ($F(CH_4)_{init}$) to match the carbon content in the plasma core (as revealed by $I(C_2^*)$) over a 90 min period. The $F(CH_4)$ profile that was used is shown in the inset to Figure 5.5, which also shows the $I(C_2^*)$ *versus* time profiles measured for two different initial CH_4/H_2 mixing ratios, with $F(CH_4)_{init}$ corresponding to 6% (Flow A) and 5.4% (Flow B) of the total input. The $I(C_2^*)$ *versus* time profiles obtained with this tracked flow-mode procedure match well with that measured for static-mode growth at the same T_{sub}.

Optical emission spectroscopy was used to confirm the similarity of the plasma emissions present during static- and tracked flow-mode CVD operation. Figure 5.6 shows that the deposited films have strong similarities. All films exhibit polycrystalline texture, with facetted surfaces and no obvious morphological differences between the samples grown under static- and tracked flow-mode conditions. Notable is the fact that for the film grown under static mode conditions the layer thickness is a little thicker, which

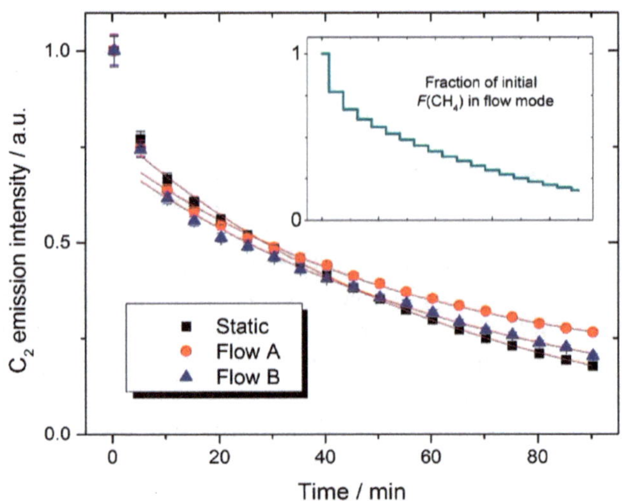

Figure 5.5 $I(C_2^*)$ (z ~ 12 mm) emission intensities measured as a function of time during two tracked flow-mode runs (red circles and blue triangles) compared with that for a static-mode run (black squares) at the same T_{sub} (~890 °C). $F(CH_4)$ was adjusted every 5 minutes in the tracked flow-mode experiments to reduce the carbon content in the plasma core (as revealed by the $I(C_2^*)$ signal) to match that in the static-mode experiment. The initial CH_4 input mole fractions for the tracked flow-mode experiments were, respectively, 6% (Flow A) and 5.4% (Flow B); the $F(CH_4)$ profile employed over the same time interval is shown in the inset. Reproduced from ref. 21 with permission from Elsevier, Copyright 2020.

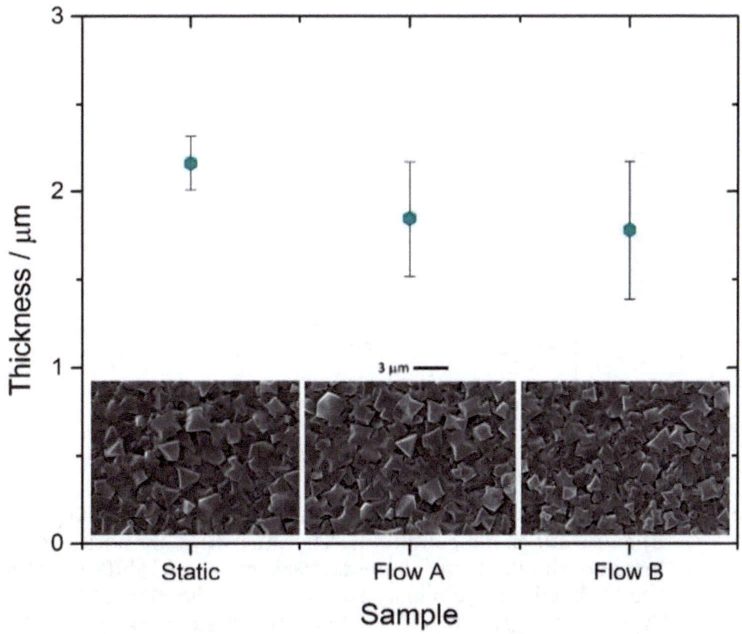

Figure 5.6 Thicknesses of diamond films grown by static- and tracked flow-mode operation along with plan view SEM images of the respective samples (insets). Reproduced from ref. 21 with permission from Elsevier, Copyright 2020.

suggests that it was grown at a slightly higher deposition rate. One plausible explanation for this higher growth rate would be the higher level of nitrogen present in the chamber under static flow conditions due to the small reactor leak. Based on the observations of the similar growth rate, material morphology and also the material quality that is presented in Figure 5.7, static flow CVD does not produce inferior diamond films. Raman spectra show comparable sp^3 and sp^2 carbon ratios and very similar full width at half maximum (FWHM) values for the strong 1332 cm^{-1} feature that is characteristic of sp^3 diamond.

Closer inspection of the SEM images in Figure 5.6 shows that the Flow B sample has a slightly smaller average grain size along with an increase in surface grain boundary density, which may account for the heightened prominence of the sp^2 carbon features in the Flow B spectrum. However, in general as long as the proper measures are taken to limit the air leak rate into the CVD chamber during static flow in particular, there should be little to discriminate between static and continuous flow CVD diamond growth.

The potential benefit of static flow CVD for technological applications of diamond lies in its use for growing isotopically pure diamond material, which requires the use of expensive carbon feedstock gases. Diamond materials synthesised exclusively from pure ^{12}C or pure ^{13}C methane have

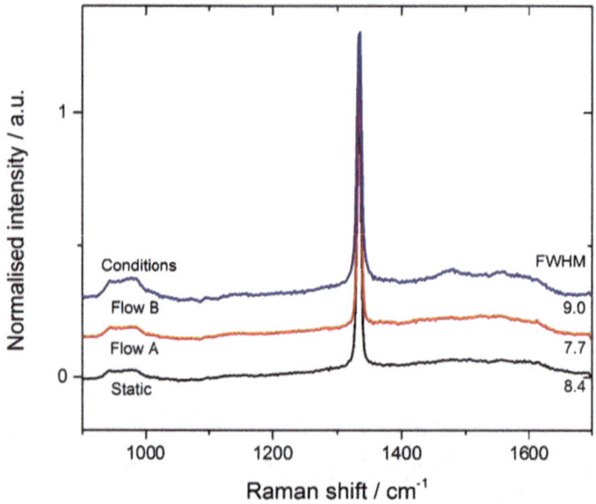

Figure 5.7 Raman spectra of samples grown using tracked flow-mode (Flow A and B) and static-mode conditions. The intensities are normalised to the height of the 1332 cm^{-1} diamond peak and a rigid shift of 0.15 au, has been applied to the intensity axis to separate the spectra. The respective FWHM values (in cm^{-1}) of the 1332 cm^{-1} peak are shown on the right. Reproduced from ref. 21 with permission from Elsevier, Copyright 2020.

potential applications in electronics and photonic devices on account of ^{12}C possessing a smaller band gap and higher thermal conductivity than natural diamond, while ^{13}C exhibits a larger band gap than natural diamond and has been used to confine electronic carriers in isotopically layered, diamond junction devices.[26]

A variation on the static flow CVD technique that involves periodic dosing of the gas plasma with additions of carbon feedstock has been studied to investigate the effect upon growth rate and quality of intrinsic and semiconducting, boron-doped diamond films. Some of the results obtained are presented in Figure 5.8. It was demonstrated that good quality diamond can be grown using static flow with periodic pulses of carbon feedstock injected into the gas-plasma. The recipe for diamond growth using a static flow cycle should be chosen based on which parameters need to be prioritised. For example, if growth rate was prioritised, then flowing methane in every 30 minutes should be the chosen cycle. This research is beneficial to the construction of a diamond beta voltaic battery because it establishes the effects of carbon economy on purity, crystallinity and the growth rate of diamond.

For a diamond battery, the ^{14}C radioisotope is a beta emitter candidate for incorporation into a battery structure but sourcing this from nuclear waste is very expensive. Therefore, the carbon economy parameter should be optimised because the amount of ^{14}C-methane dispensed into the feedstock gas mixture would need to be fully depleted by being incorporated into the battery device structure as diamond. Depending on which other parameters are prioritised a suitable method can be chosen. High crystallinity can improve

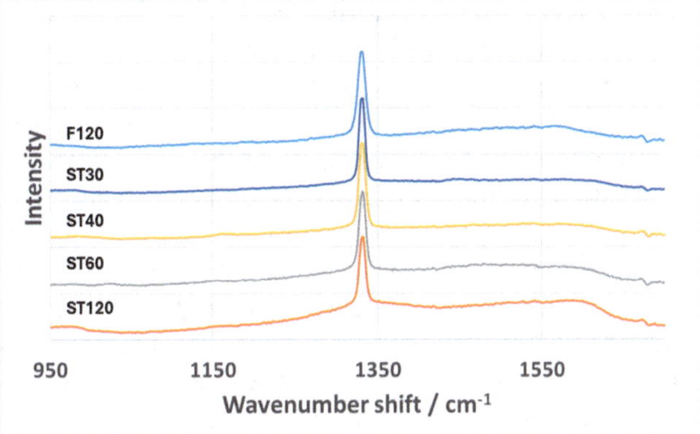

Figure 5.8 Raman spectra illustrating static flow diamond growth performed with periodic dosing of the methane gas to recharge the carbon to hydrogen ratio in the CVD reactor volume to its starting value of 4%. The period measured in minutes between injection of additional pulses of carbon feedstock is colour-coded for growth runs of 120 minutes duration. A continuous flow growth cycle, F120 is shown for comparison. The ST40 growth cycle exhibited the narrowest zero phonon line (1332.2 cm^{-1}) diamond of 9.72 cm^{-1} and highest growth rate of 3.99 μm h^{-1}.

the electronic properties of diamond such as electron mobility. Therefore this would be another parameter that should be prioritised. Figure 5.8 presents the results of static flow growth employing periodic dosing experiments. The ST60 growth run was found to provide the highest crystallinity using the least amount of methane feedstock as only two pulses of gaseous carbon feedstock were injected into the gas-plasma during the growth to recharge the carbon to hydrogen ratio to the initial 4% methane in hydrogen level in the gas-plasma.

5.3 n-Type Semiconducting Diamond

One of the remaining challenges in synthetic diamond synthesis is the identification of an impurity dopant defect that can be incorporated into the diamond lattice and activated to exhibit an efficient n-type property. n-Type diamond is needed to realise diamond-based pn junction devices, which will facilitate applications in diamond micro- and nanoelectronics, optoelectronics, voltaics and thermionics. Nitrogen being the most easily incorporated substitutional dopant produces a relatively deep level of 1.7 eV, which only activates at elevated temperatures above *ca.* 420 °C, to exhibit n-type behaviour due in part to the relatively large electronegativity of nitrogen (3.04) compared with carbon (2.55), which prevents nitrogen releasing its donor electron charge to the surrounding carbon lattice. Therefore, it is not a favourable choice for use as an n-type impurity dopant in diamond devices designed to operate at room temperature.

Over the past 35 years, many candidate elemental dopants including lithium, sodium, oxygen, sulfur, arsenic and phosphorus, and incorporation processes involving ion implantation, transmutation and CVD growth have been studied in an effort to identify an impurity that produces a shallow donor electron level situated close to the conduction band minimum of diamond. To date, none of the candidate dopants have been efficiently incorporated into diamond to yield a high concentration of electron carriers. Although phosphorus can produce a donor level *ca.* 0.6 eV below the diamond conduction band this is too deep to produce a room temperature, n-type diamond material. Furthermore, phosphorus atoms only incorporate substitutionally on the (111) lattice planes of diamond. This severely limits the carrier mobility due to the low concentration of conduction electrons released by activated phosphorus donor atoms.

Previous theoretical work has identified various combinations of elements, in the form of atom clusters that add one electron to the diamond lattice, such as BN_2,[27] Si_4N[28] and LiN_4.[29] The predicted stability, or binding energy, of these clusters is often high, due to electronegativity and size effects, meaning that the co-dopants can couple strongly.[30]

The potential of co-dopants has been revisited in recent years with a study by Croot *et al.*,[31] which explored the use of co-dopant clusters of boron and nitrogen to adjust the position of the donor electron level characteristic of a single substituted nitrogen defect at 1.7 eV below the conduction band minimum (CBM) of diamond. An extensive survey of the possible combinations of dopants was undertaken in this work. The work involved an *ab initio* study of boron–nitrogen clusters in diamond, with both nitrogen-rich and boron-rich clusters, examining the structural differences between substitutional clusters and also their energetics. A summary of the structures studied is presented in Figure 5.9 with their corresponding formation and binding energies shown in Figure 5.10.

In general the boron-rich clusters are less distorted than the nitrogen-rich clusters, resulting in lower formation energies, but have similar binding energies due to the accommodation of single nitrogen defects being very endothermic. For both boron- and nitrogen-rich clusters the number of constituent atoms with the largest binding energy, $E_{b,n}$, is three, provided that they have an accompanying atom of the other element present. When comparing B-BN_2 and N-NB_2 with BN, and N_2 with BN_3 and B-NN_2, it is evident that the addition of a BN unit usually causes only small changes in $E_{b,n}$ when nitrogen is present in the cluster. It was also found that B-N-BN is likely to outnumber its more compact counterparts with the same stoichiometry because its binding energy significantly outweighs the others, with it computed to have the most negative binding energy of any cluster presented in Figure 5.9.

The work also reported that asymmetric N-rich clusters are more stable than the corresponding symmetric clusters for n = 3 and 4, and BN_4 and B-NN_3 are almost energetically equal. This may result in a greater concentration of BN_4 than its smaller counterparts since, for example, an isomeric

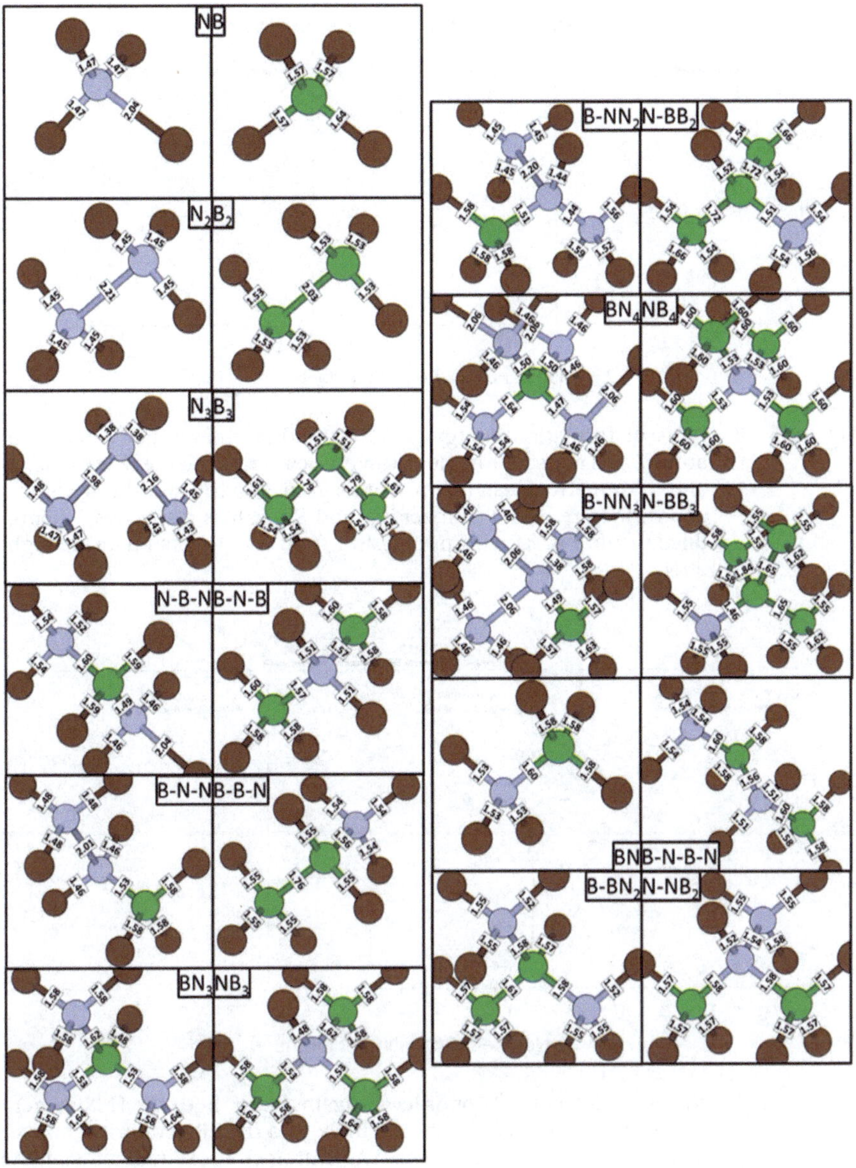

Figure 5.9 Bond lengths (Å) of B- and N-rich clusters with $n \leq 3$. All separations involving dopant atoms (*i.e.* all but C–C) are shown. Reproduced from ref. 37 with permission from the author.

rearrangement of BN_3 favourably results in B-NN_2 formation. The upshot of this observation is that there may not be a clear route to BN_4 formation *via* addition of N atoms to smaller clusters, despite its stability.

Under thermodynamic control and for given appropriate dopant concentrations the clusters with the most negative binding energy will prevail.

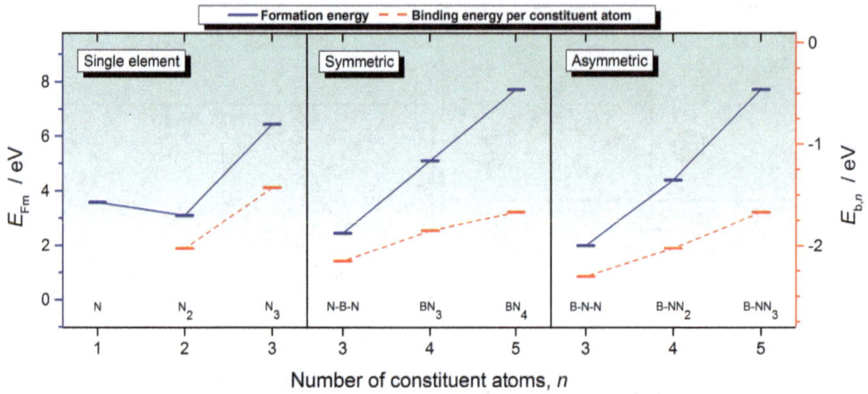

Figure 5.10 Plots of formation energy E_{Fm} and binding energy per constituent atom $E_{b,n}$ against n for the nitrogen-rich single element, symmetric and asymmetric clusters. Present as guides to the eye, the solid blue and dashed red lines connect E_{Fm} and $E_{b,n}$ values respectively. Reproduced from ref. 31 with permission from IOP Publishing, Copyright 2018.

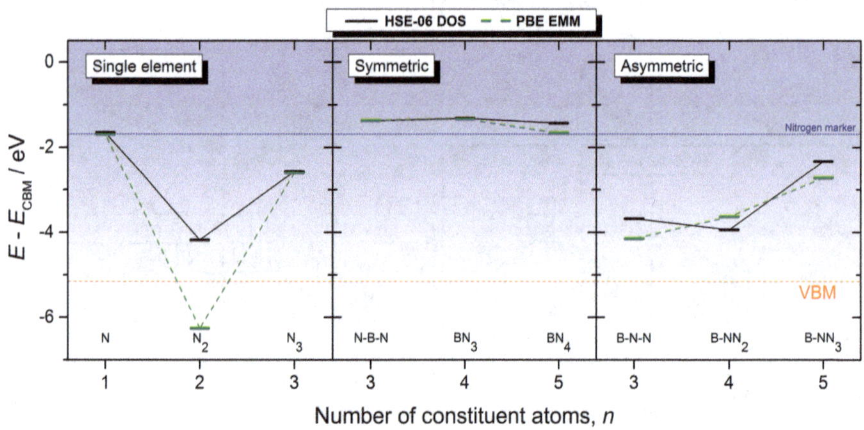

Figure 5.11 Nitrogen-rich cluster donor levels against n for both the HSE-06 density of states method (black solid lines) and the PBE empirical marker method (green dashed lines). The nitrogen EMM marker and the valence band maximum (VBM) are given for reference. Reproduced from ref. 31 with permission from IOP Publishing, Copyright 2018.

Considering the tendency of nitrogen to cluster, a co-doped N-rich sample might contain some combination of (for example) N-N-B, B-NN_2, BN_4 and some purely N-containing clusters, depending on the ratio of nitrogen to boron.

The excitation energies, μ_e, of nitrogen-rich donor clusters are shown in Figure 5.11, where the values are given with respect to the CBM.

A particularly striking difference found between nitrogen-rich and boron-rich clusters was the asymmetric $n = 3$ and 4 clusters. While B-N-N and B-NN_2 have similar μ_e, B-B-N has a shallower level than N-BB_2. The condition seen with B-N-N and B-NN_2 is likely to be associated with the smaller structural distortion in B-N-N relative to B-NN_2, which affects the nitrogen electron density.

A similar effect is not seen with boron-rich clusters. For symmetric nitrogen-rich clusters with $n = 3$ to 5, the addition of N atoms was found not to significantly reduce μ_e as might be expected. However, this effect was not replicated for the asymmetric versions, where each nitrogen addition reduces the donor excitation energy. This difference can be understood by considering N-B-N as $(NB)^0$-N^0 and B-N-N as B^-–N_2^+.[30] The addition of N atoms to the N-B-N form simply causes further N^0 centres in (very) close proximity to an existing deep donor defect, while the addition of N atoms to the asymmetric cluster causes a further concentration of electron density in the already negative region of the donor cluster. This is conceptualised to manifest itself as a stacking of negative charge, which pushes the associated level to higher energies and makes the donation of an electron charge to the surrounding diamond carbon lattice more favourable.

This prediction has been tested experimentally by Liu *et al.*[32] who produced a pin homojunction by a CVD method following the methodology of Croot *et al.*, which predicted that the BN_2 complex could be a donor with a greatly reduced activation energy and an ionisation energy in the range of 0.8–1.2 eV. To realise the BN_2 complex, Liu *et al.* designed a set of CVD growth experiments in which the boron dosage (B_2/C) was fixed at 50 ppm but the nitrogen gas (N_2/C) level was varied between 0 (purely B-doped) and 100 000 ppm to produce a series of N/B ratios in the gas phase during diamond growth up to 2000 : 1.

At a 1000 : 1 (N/B) ratio, a substantial change in morphology, crystalline quality and electrical conduction type was observed in experiments by Croot. Liu *et al.* also recorded a similar change in electrical resistivity and carrier concentration at this N/B ratio, which is presented in Figure 5.12 as a function

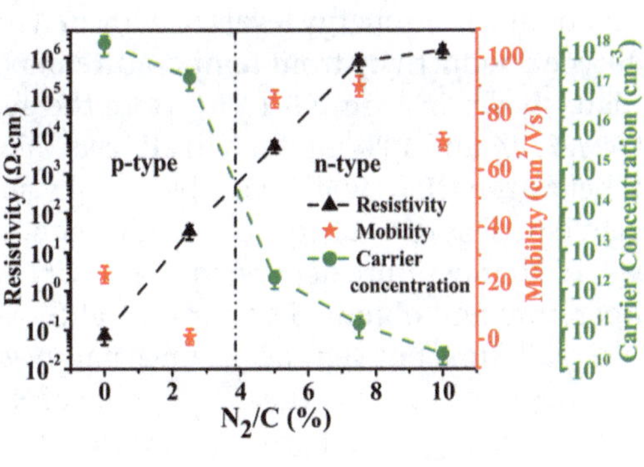

Figure 5.12 Resistivity, mobility and carrier concentration at 773 K for B-doped and B–N co-doped diamond films on single crystal substrates. Reproduced from ref. 32, https://doi.org/10.1063/5.0049151, under the terms of the CC BY 4.0 licence, http://creativecommons.org/licenses/by/4.0/.

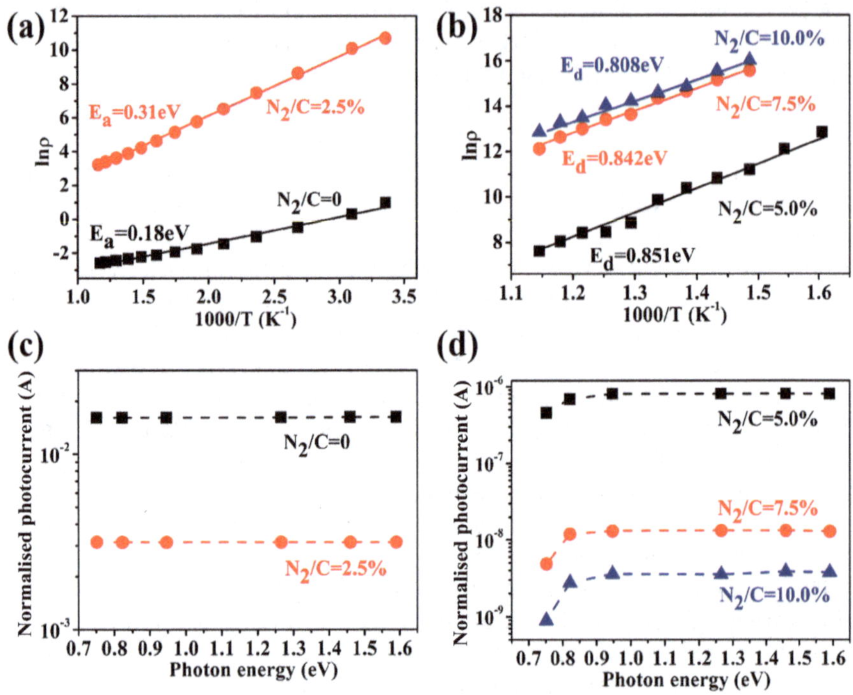

Figure 5.13 (a) The fitting of activation energy by the relation between resistivity and temperature: $N_2/C > 2.5\%$. (b) Photocurrent spectroscopy with illumination wavelength between 780 and 1650 nm on a log scale. The photocurrent spectra are normalised by the light intensity: $N_2/C > 2.5\%$. Reproduced from ref. 32, https://doi.org/10.1063/5.0049151, under the terms of the CC BY 4.0 licence, http://creativecommons.org/licenses/by/4.0/.

of $N_2/C\%$. Liu *et al.* obtained the energy levels of activated donor electrons in films exhibiting n-type conductivity from temperature-dependent resistivity (Schottky–Mott) data shown in Figure 5.13(b). Here the activation energies for N_2/C ratios of 5%, 7.5% and 10% are seen to all have values lying between 0.8 eV and 0.9 eV below the CBM, which is substantially shallower than the 1.7 eV level associated with single substitutional nitrogen.

Croot *et al.* measured temperature dependent resistivity over a larger range of reciprocal temperature and obtained activation values almost a factor of 3 lower for single crystal films but significant uncertainty was placed on the incorporation efficiency of nitrogen at the single value of N_2/C used by Croot of 2.5%; the ratio of B/C being the experimental variable in Croot's study of co-doped films.

Liu *et al.* also measured the photocurrent as a function of the photon energy (Figure 5.13d) for the n-type films and observed abrupt absorption at photon energies of around 0.7–0.8 eV, which is in reasonable agreement with the values that were obtained by fitting the temperature-dependent resistivity and electron concentration.

X-ray photoemission spectroscopy (XPS) may be employed to determine how the excess of nitrogen is incorporated in B–N co-doped films. Liu analysed the C 1s profiles of co-doped films by deconvolving the C 1s peak into components associated with C–B, C–C and C–N bonding.[32] By this method it was possible to make quantitative comparisons of the amounts of B and N impurities present in each co-doped diamond film. When nitrogen is incorporated to levels high enough to produce n-type conductivity, the C–C bond shifts to a lower binding energy (BE) at 284.0 to 284.1 eV. The shift of the BE has previously been associated with combined effects of an upward band bending and a chemical shift due to the charge transfer.[33] Besides the component related to the C–B bond, another peak at around 285.0 eV was included to fit the curve. This peak was assigned by Yang *et al.*, to the various defect-induced C–N bonding states.[34]

The two peaks of the B 1s lines are assigned to the B–C and B–N bonds, and it can be seen in Figure 5.14 that the second peak in the B 1s spectra only appears with nitrogen addition.

The two peaks of N 1s lines are assigned to the N–B and N–C bonds, respectively, based on prior reports published by Korean researchers.[35]

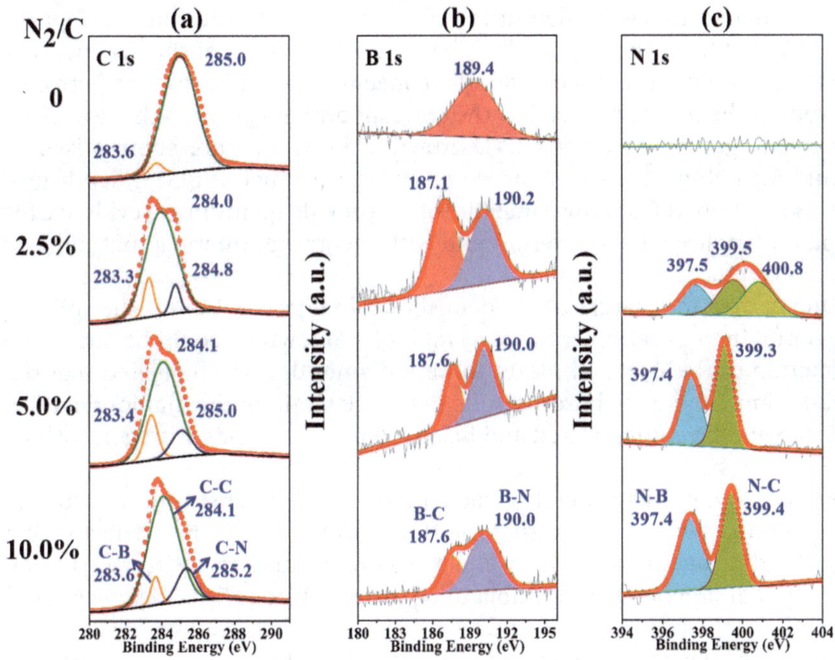

Figure 5.14 Core level spectroscopy of B–N doped diamond compared with B-doped diamond for (a) C 1s, (b) B 1s and (c) N 1s. Reproduced from ref. 32, https://doi.org/10.1063/5.0049151, under the terms of the CC BY 4.0 licence, http://creativecommons.org/licenses/by/4.0/.

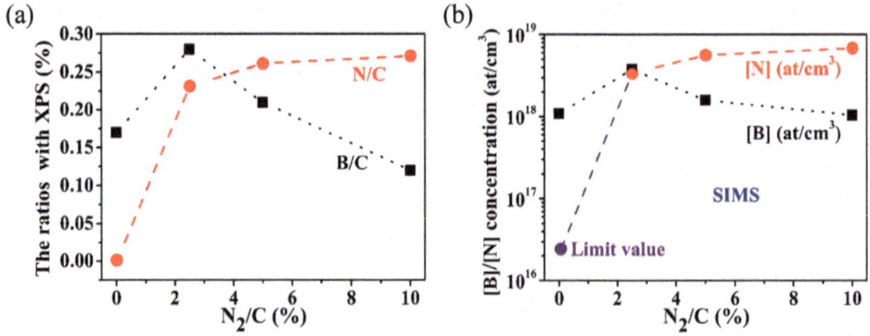

Figure 5.15 (a) Area ratios determined from the deconvolution of core level peaks shown in Figure 5.14 for films grown with different N_2/C gas phase ratios and a fixed amount of boron dopant in the gas phase. (b) The boron and nitrogen concentrations measured by SIMS found in the co-doped diamond films grown at different N_2/C gas ratios. Reproduced from ref. 32, https://doi.org/10.1063/5.0049151, under the terms of the CC BY 4.0 licence, http://creativecommons.org/licenses/by/4.0/.

The area ratio of the boron and nitrogen 1s lines over the carbon 1s lines was calculated by Liu to determine the relative concentrations of dopants in the bulk films. The XPS analysis based on area ratios of the deconvolved peaks, represents the relative doping concentration of boron and nitrogen present in the films produced by the corresponding N_2/C ratio that was present in the gas phase during CVD growth. The results are summarised in Figure 5.15 alongside secondary-ion mass spectrometry (SIMS) depth profile data obtained from the same films that provide quantitative evidence for dopant complexes being present and with incorporation rates that follow a similar trend.

One interesting feature of the data shown in Figure 5.15(a) is the spike in the boron incorporation at the N_2/C ratio of 2.5%, which could be due to the formation of the B–N complexes in bulk diamond. It is speculated that the smaller ionic radius of boron results in tensile strain in the diamond lattice, which could be compensated and balanced by introducing nitrogen with its larger ionic size.

From Croot's first-principles calculations on N–B substitutional clusters in diamond, it was reported that the N–B complex is a simple defect with a large binding energy of −5.52 eV and a negative formation energy of −0.17 eV making it energetically favourable to be formed. Under CVD plasma growth conditions, the presence of B and N in the gas phase as co-dopants would be expected to result in a reduction in the efficiency of the boron incorporation in the deposited diamond layer due to nitrogen suppressing the amount of available boron. But at a certain ratio of B to N in the gas phase there could be enhanced formation of the N–B complex which could explain the spike in boron incorporation. With the addition of larger amounts of nitrogen into the gas phase (Figure 5.15(b)) the B/C ratio reduces while the N/C ratio increases until it reaches a saturation level. Since the n-type conductivity in B–N co-doped diamond was observed by Liu *et al.* at and above 5% N_2/C, it was speculated that N–B complex may form as three or four atom clusters. Croot's calculations on defect formation illustrated in Figure 5.9, suggest that the N-B-N centre may be considered as a 'separate' N adjacent to a valence-satisfied BN. Similarly, in the B-N-N centre, the outer nitrogen atom behaves much like a single N, with its three carbon bonds shorter than in pure undoped diamond by 4% and the remaining bond (to nitrogen) longer by 30%, 2.01 Å. Energetically, both symmetric N-B-N clusters and asymmetric B-N-N clusters have more negative formation energies than nitrogen-only clusters and the difference in binding energy between them is only 0.46 eV, with asymmetric clusters having the lower energies, suggesting that a mixture of these clusters could be the present and represent the dominant cluster type. But Croot *et al.* note that 'Under thermodynamic control and for given appropriate dopant concentrations the clusters with the most negative binding energy will prevail. Considering the tendency of nitrogen to cluster, a co-doped N-rich sample might contain some combination of (for example) B-N-N, B-NN_2, BN_4 and some purely N containing clusters, depending on the ratio of nitrogen to boron'.

Furthermore, as the ratio of boron to nitrogen becomes smaller, there will be a tendency for substitutional nitrogen defects to become more favourable, which appears to be borne out by the XPS measurements of Liu *et al.*, where they noted a N–C peak shift in N 1s XPS at high N_2/C ratios. Unfortunately the consequence of this predominance of the single nitrogen defect over co-dopant clusters at these high N_2/C ratios is to reduce the number of free donor electrons populating the diamond conduction band.

This state of affairs highlights the problem of impurity defect control that researchers are seeking to resolve in order to realise room temperature n-type conductivity in B–N co-doped diamond. Nitrogen-rich B–N atom clusters offer a promising route to efficient n-type conductivity but in order to achieve this goal a number of issues need to be studied in more detail. They include:

1. Identification of the predominant atom cluster responsible for shallow donor level activation.
2. The impact of source material selection upon atom cluster formation in the CVD diamond film.
3. Influence of isotopically pure carbon feedstock on the formation of an efficient n-type material.
4. The choice of diamond substrate material and its preparation.

To date, it has not been possible to identify the B–N atom cluster to target to make efficient n-type diamond. Based on the work conducted by Croot *et al.*[31] and by Liu *et al.*, there is still significant uncertainty as to which B–N complex is responsible. B–N atom clusters containing three and four atoms are all potential candidates, which underlines the need to use higher fidelity, nondestructive spectroscopic methods to determine the signatures of the six different atom clusters that could be present at N_2/C ratios lying between 2.5% and 5%. Candidate analysis tools to obtain this information include high resolution XPS, low temperature photo-luminescent spectroscopy and NanoSIMS (nanoscale SIMS).

The two studies that have been undertaken previously have used gaseous impurity feedstocks to co-dope CVD diamond films, with 5N purity N_2 gas providing one dopant and pure B_2H_6 diluted in hydrogen (5N) providing the other. Both studies also used plasma-enhanced CVD to produce co-doped films. An alternative approach would involve the use of gas–solid combinations, where a solid dopant material, that volatises when brought into contact with the growth plasma containing the carbon–hydrogen feedstock, is employed to accelerate BN cluster formation in the gas phase. Obvious candidate combinations for this approach would be solid BN and a gas containing nitrogen, or solid B and a gaseous form of nitrogen.

The use of isotopic carbon feedstock gases presents an, as yet, untested approach to improve the activation of co-dopant impurities in diamond. The approach would involve the use of static flow CVD to maximise the utilisation of a more expensive carbon feedstock. Due to the larger band gap, higher density and lower thermal conductivity estimated for ^{13}C compared to ^{12}C diamond, the formation and binding energies of dopant clusters would be expected to be altered and potentially in a favourable way. The co-doping of ^{14}C-diamond layers offers a means to study the interaction of beta decay with co-dopant atom clusters with the prospect that free electron populations due to n-type activation and carrier mobility could be markedly improved.

The choice of substrate material for n-type diamond studies would normally be restricted to high quality, single crystal material in which the impurities, such as nitrogen, are controlled to a few parts per billion. In such material the concentration of structural defects can be controlled to be present at a low level. However, it may be advantageous to increase the concentration of vacancy defects in a diamond substrate to promote co-dopant cluster formation either during growth or post-growth.

Salter and collaborators[36] have employed a picosecond IR laser to irradiate micron-sized volumes located below the surface of a diamond substrate to promote the formation of higher densities of charged nitrogen-vacancy (NV) defects. This technique delivers laser energy in the form of picosecond pulses of heat into the diamond carbon crystal lattice. Due to the very high thermal conductivity of diamond, the effect of localised heating also dissipates this energy but on a similar timescale. A train of pulses can therefore be used to disrupt the carbon bonding and where it converts sp^3 carbon bonds to sp^2 bonds a conductivity change occurs. By scanning a focused picosecond laser across a diamond, it is possible to form conductive tracks by this technique. Although similar IR laser techniques has been employed by others to form surface tracks on insulating diamond as reported by Croot *et al.*,[37] the process developed by Salter is able to form such tracks in any plane, anywhere within the volume of a single crystal diamond substrate. Salter's technique has been employed at CERN to produce an array of single crystal diamond radiation detectors in which electrically conductive tracks formed within each diamond are used to improve the charge collection efficiency of the detector. By lowering the laser fluence, it is possible to disrupt carbon–carbon co-ordination without inducing a conversion of the sp^3 bonding to sp^2 and instead promote vacancy formation, particularly in nitrogen-rich diamond.

Since the ratio of $N–V^-$ to $N–V^0$ has been reported by Liu *et al.* to change with increasing N_2/C ratio, it is possible that this effect, coupled with a laser process to promote vacancy formation, could offer a means for the facile production of B–N atom clusters. A practical application of this process would ideally require CVD growth incorporating the use of nitrous oxide gas to enhance nitrogen-vacancy incorporation, and modulation doping techniques to produce diamond layers enriched with co-dopants prior to laser processing. The interaction between B–N clusters and picosecond laser heating would potentially enable nitrogen-rich, B–N atom clusters to form and possibly with a more restricted set of atom cluster types.

5.4 Functionalised Diamond

Diamond exhibits a very high degree of chemical stability even though it is thermodynamically meta-stable at standard temperatures and pressures. This can be evidenced from the fact that at its surface, it is not chemically reactive, preferring to form carbon dimers to minimise surface energy, and will not spontaneously oxidise or chemically interact with the environment under standard conditions. However, the diamond surface may be functionalised through a range of chemical/photochemical, mechanical, electrical and thermal processes to exhibit surfaces that are technologically useful. The simple act of rubbing a diamond surface on an iron block with olive oil will cause the diamond surface not only to be polished by the breaking of surface dimer bonds, but also to take up a surface layer of hydrogen atoms that are covalently bonded to surface carbons.

This gives rise to the formation of a surface dipole that changes the surface electrical properties imparting a p-type conductivity due to the presence of the hydrogen dipole. The bare diamond surface can also be functionalised with other elemental and compound moieties that can impart radically different electronic properties. Diamond is an example of a wide band-gap semiconductor material that has a conduction band minimum with an exceptionally high energy such that it has a value very close to the vacuum energy (~0.3 eV lower). This is in contrast to most semiconductors where the vacuum level lies much higher in energy than the conduction band minimum. This energy difference is a characteristic of a bare (un-terminated) semiconductor surface and is termed the electron affinity (χ). For most semiconductors the electron affinity is positive while for diamond and other wide band-gap semiconductors and insulators, it may be close to zero or even negative. In these cases a dipole subsequently formed on the surface by a surface termination can strongly influence the electron affinity, making it more positive or more negative. The dipole formed by an atom-layer of hydrogen terminating the surface carbons of diamond leads to downward band bending and a large negative electron affinity, where the vacuum level lies below the conduction band minimum, which allows electron carriers to be emitted ballistically from the diamond surface due to the lack of a vacuum energy barrier. In contrast, an oxygen functionalised surface imparts a positive electron affinity and an upward band bending at the diamond surface. Many different functional materials composed of fractional or full monolayers of atoms have been investigated. The interest is particularly focused upon the negative χ materials, as these are technologically useful in electronic devices that require conductive diamond electrodes to generate high levels of electron emission, such as in vacuum microelectronics and thermionics, or to control electron transfer across a diamond/material interface in diamond voltaics or radiation detectors. Activating negative electron affinity (NEA) on a diamond causes a reduction in the effective surface work function. This is because the electron affinity, χ can be related to the surface work function, ϕ by the relation:

$$\chi = \phi + \left(E_{\mathrm{F}} - E_{\mathrm{vbm}}\right) - E_{bandgap} \tag{5.1}$$

where E_{vbm} is the valence band maximum of diamond, E_{F} is the Fermi level, and $E_{\mathrm{band\ gap}}$ is the indirect band gap of diamond.

The earliest reports of the functionalisation of single crystal, p-type oriented diamond by hydrogen were reported by van der Weide *et al.* in a study that revealed that the presence of NEA could be revealed by a large peak in secondary electron emission appearing at low kinetic energies in the energy spectrum produced by Ultraviolet Photoelectron Spectroscopy (UPS).[38] Depending upon the crystalline orientation of the surface, the magnitude and shape of the secondary electron peak has been found to be different due in part to the number of hydrogen atoms co-ordinated with each surface carbon atom being different: two hydrogens for a C(100) surface and one for a C(111) surface, being the two most studied orientations of single crystal material. Representative UPS of NEA diamond formed with hydrogen termination is shown in Figure 5.16.[39]

Hydrogen is not the only element that can be used to terminate p-type diamond to activate NEA. O'Donnell *et al.* conducted computational and experimental evaluations of the dipole formed by functionalising an oxygen-terminated, p-type, C(100) surface with lithium metal and obtained an NEA of −3.9 eV, which is substantially larger than the −1.3 eV measured for a hydrogen-terminated surface.[40] The functionalisation was achieved by first exposing the diamond to ozone to generate an oxygen-terminated surface. Following a vacuum-based anneal at 300 °C, to consolidate the co-ordination of oxygen and promote C–O (ether) bonding, a thick layer of several hundred nanometres of lithium metal was evaporated onto the oxidised diamond. To obtain the desired C–O–Li dipole, the excess lithium was stripped from the

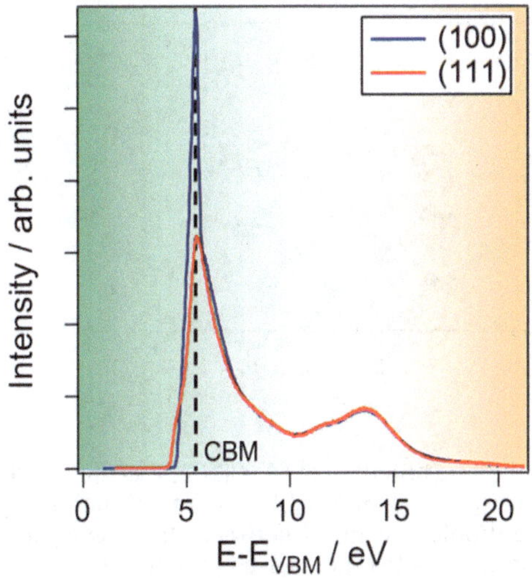

Figure 5.16 Full-wavevector UPS spectra of the hydrogen terminated (100) (blue) and (111) (red) diamond, $h\nu = 21.2$ eV. The x-axis is energy relative to the VBM position (E_{VBM}). The lower and higher energy regions are dominated by the conduction band and valence band structures, respectively. The CBM position is indicated by the black line. Reproduced from ref. 39, https://doi.org10.1002/adfm.202007319, under the terms of the CC BY 4.0 licence, http://creativecommons.org/licenses/by/4.0/.

coated surface by dipping in triply-distilled water followed by a 450 °C vacuum anneal. Subsequent experiments were carried out to refine the method for generating this NEA surface by moving to a UHV-based process involving a thin film evaporation process for forming the lithium metal termination and increasing the annealing temperature to 800 °C.

From previous *ab initio* computational calculations,[41] O'Donnell *et al.* established that the studied C(100)–(1 × 1):LiO surface dipole is not atom-centred and that the Li–O bonds are not sufficiently normal to the surface to induce such a dipole. Instead, there is a delocalisation of surface charge caused by the lowering of electronic states in the presence of the positive Li ion. The result is that the lithium and oxygen sites sit within the positive side of the surface dipole. The negative side of the dipole is located between the oxygen and first layer of carbon sites.

Electronically, the formation of this dipole can be explained through the formation of a delocalised charge enhancement at the dimer-oxygen interface. The presence of the Li atoms lowers the energy of the oxygen lone-pair states, such that they significantly overlap with the higher lying sp^2 bonding states of the dimer. As a result, a further degree of hybridisation and delocalisation of charge occurs associated with the oxygen. This negative side of the dipole is countered by the positive Li atoms giving rise to a very large dipole. The size of the dipole was established through DFT calculations carried out

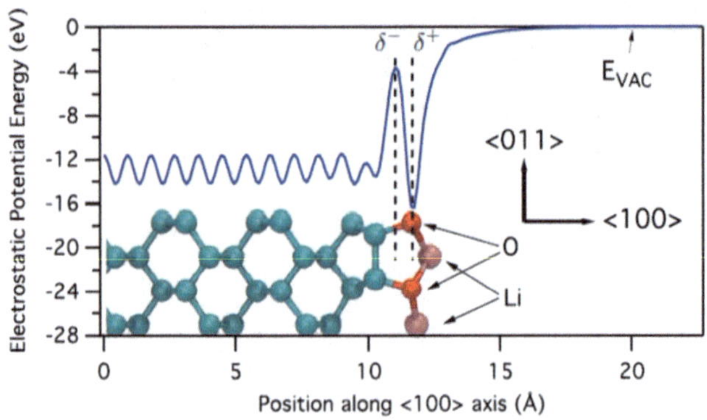

Figure 5.17 Electrostatic potential at the C(100):LiO diamond surface showing the surface dipole generated by the functional group. The negative side of the dipole is centred between the C–O bonds on the surface and not on the O sites themselves causing the dipole to interact with the carbon layers creating a large NEA effect. Reproduced from ref. 42 with permission from John Wiley & Sons, Copyright 2013.

by O'Donnell *et al.* and the result visualised in an electrostatic potential plot of the functionalised surface reproduced in Figure 5.17.

O'Donnell *et al.* also conducted synchrotron-based core level spectroscopy on LiO-terminated diamond prepared under UHV conditions. The results are reproduced in Figure 5.18 as a series of surface-sensitive C 1s (a), O 1s (b), and Li 1s (c) core-level spectra obtained for (from bottom to top) (i) the oxygen-terminated surface after annealing in UHV, (ii) followed by Li deposition of 0.4 Å, (iii) followed by three cycles of annealing at 800 °C for a total time of one hour, and (iv) after further lithium deposition to 2.8 Å and a subsequent 800 °C anneal.

From inspection of the C 1s spectra, the ratio of the areas of the deconvolved peaks representing the surface components, it was estimated that 33% of the surface carbon atoms were oxygenated. Furthermore, the clean oxygenated surface has a simple O 1s core level spectrum dominated by a single peak at 531.6 eV (peak OA) characteristic of a double-bonded oxygen species, such as C=O or C–O–C, but upon 0.4 Å lithium adsorption, the O 1s peak decreases in area by 20%, suggesting that lithium adsorbs over the oxygen-terminated regions. Annealing alters the shape of the core level spectra and induces shifts consistent with a change in the surface dipole. After annealing at 800 °C for a total annealing time of one hour, the NEA was observed to be firmly established on the surface, with both the O 1s and Li 1s spectra shifting by 0.9 eV and 0.7 eV, respectively, to higher binding energies. The large shift to higher binding energy for the O 1s spectrum was considered to be consistent with the surface oxygen and lithium residing within a lowered electrostatic potential, such as the one predicted from the DFT-computed dipole that is shown in Figure 5.17.

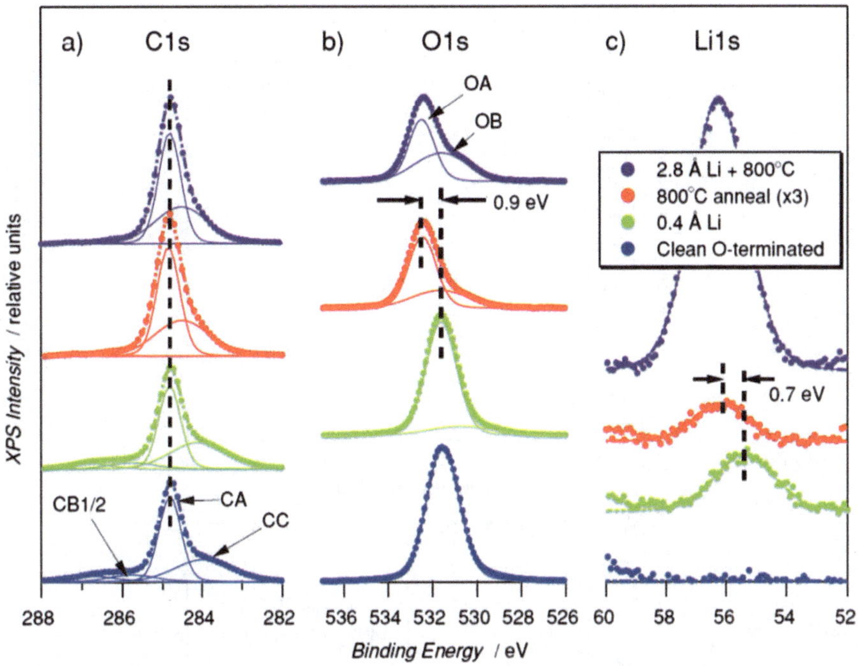

Figure 5.18 XPS core levels obtained from a C(100):O,Li surface prior to and post an annealing step. (a) C 1s core level taken at 330 eV photon energy. (b) O 1s core level taken at 700 eV. (c) Li 1s core level taken at 150 eV. Peak heights are not comparable due to the different synchrotron energies used. Reproduced from ref. 42 with permission from John Wiley & Sons, Copyright 2013.

From these studies, it was established that the NEA surface due to LiO termination was not only temperature stable but also exhibited better chemical stability in air when compared with an NEA surface formed with hydrogen. From a technological point of view, the stability of this NEA surface makes it attractive for use in electron sources, particularly as the secondary electron emission spectrum exhibits a characteristically very narrow energy distribution, useful for applications where quasi-monochromatic electrons are required.

There have also been reports which noted that some fraction of the electrons that thermalised to the conduction band minimum (CBM) actually appear to emerge from the NEA surface at an energy lower than the CBM.[42,43] O'Donnell *et al.*[42] studying photoemission from the true NEA surface produced by LiO on a p-type C(100)-oriented single crystal diamond (Figure 5.19), reported that the large band bending caused broadening and additional structure in the CBM peak that is not seen in the secondary electron emission peak from hydrogen-terminated diamond. It was proposed that the oscillations observed in the photoemission spectrum at energies above the CBM position could be explained using a similar model to the

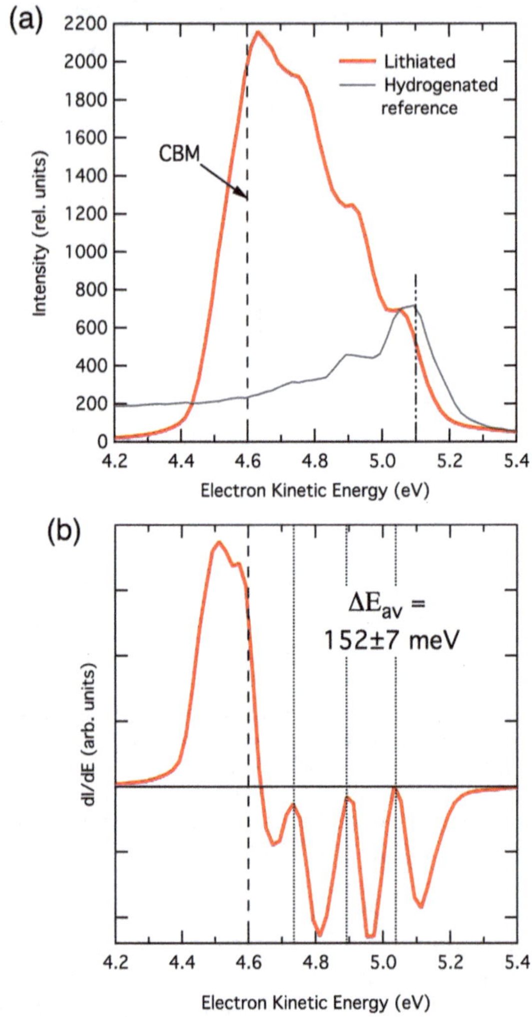

Figure 5.19 (a) Low kinetic energy spectrum for lithiated diamond collected at room temperature. Vertical dashed line is the CBM position as deduced from the C 1s core level. The room temperature hydrogen-terminated spectrum appears for comparison where it is seen that the CBM position is quite different arising from the larger band bending in the LiO surface. (b) The derivative spectrum of (a) with periodic oscillations indicated by dashed lines. Reproduced from ref. 42 with permission from IOP Publishing, Copyright 2014.

previously reported Fan cascade,[44] where sub-CBM emission structure reflects phonon scattering as electrons either thermalise into surface states from which they are emitted into vacuum, or are emitted into vacuum from the CBM with accompanying multi-phonon emission. The process can be understood with the aid of the schematic shown in Figure 5.20. It is well

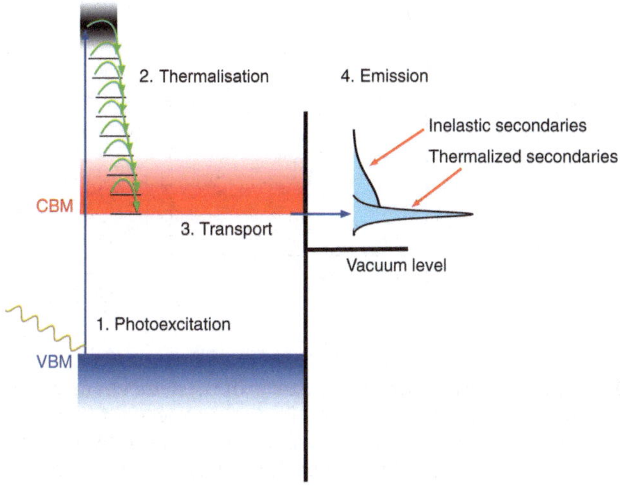

Figure 5.20 Schematic diagram showing the elements of secondary electron emission from a true negative electron affinity diamond (flat bands). Rapid thermalisation and long recombination lifetime at CBM result in most electrons being emitted at the CBM energy. Adapted from ref. 42 with permission from IOP Publishing, Copyright 2014.

known that optical phonon scattering dominates for semiconductors, in general, provided it is allowed energetically. The impact of optical phonon scattering is to provide a much faster pathway to energy loss through the band-bending region. The CBM peak for p-type, C(100):Li,O shown in Figure 5.19 is seen to exhibit an asymmetric distribution due to the periodic features and it was proposed by O'Donnell that the band-bending itself restricts optical phonon scattering.

Electrons starting at the bulk CBM as they enter the band-bending region can only scatter to lower energies when they have traversed a distance such that the CBM has dropped by at least the minimum optical phonon energy. Further scattering processes are subject to the same energy requirement. Since the optical phonon dispersion covers no more than 130–165 meV, with a peak in the phonon density of states at $E_{o,av}$ = 157 meV, the most likely energy distribution of partially ballistic electrons escaping from the band bending region is a series of maxima at multiples of the average optical phonon energy below the bulk CBM. For a triangular band bending approximation, the relative intensities of the maxima are expected to be linear. This is illustrated in Figure 5.21(a) where the intensity of the *j*th phonon peak is given by:

$$I_j = 1 - \frac{x_i}{l} \qquad (5.2)$$

where x_i is the distance that needs to be travelled by an electron of energy E_j to reach the NEA surface, and l is the mean free path of the optical phonon in diamond.

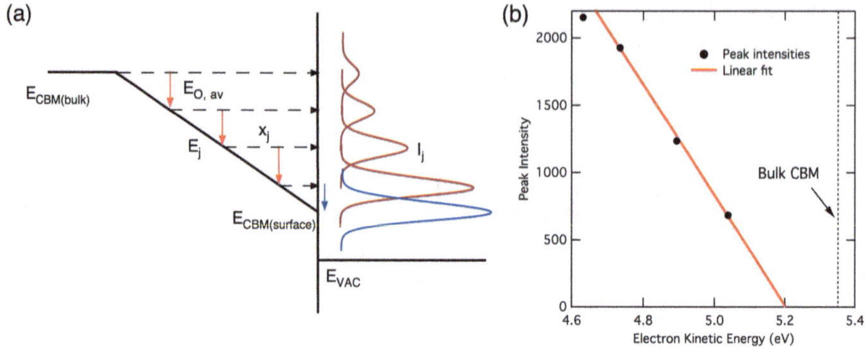

Figure 5.21 (a) Schematic illustrating the proposed mechanism for the oscillations observed in low kinetic energy spectra for lithiated diamond. Electrons are preferentially emitted at multiples of the mean optical phonon energy below the bulk CBM. (b) The fit of the measured peak intensities in (a) to a linear function. Adapted from ref. 42 with permission from IOP Publishing, Copyright 2014.

For the relationship in eqn (5.2) to be correct, $E_j = E_{CBM} - n \cdot E_{o,av}$. Specifically E_j is an integer multiple of the mean optical phonon energy below the bulk CBM, which was precisely the case found when the phonon peaks from the Figure 5.19 data are plotted as a function of energy with a corresponding fit to a linear function (Figure 5.21(b)). The three highest-energy peaks, those spaced by 152 ± 7 meV, all have intensities that scale linearly with energy. The parameters of the fit suggest that the peak at the next highest energy (5.2 eV) would not be observed and nor would the peak corresponding to the bulk CBM itself. In the triangular band approximation, x_j is proportional to E_j, and hence the slope of the peak intensities reflects the ratio of the band gradient to the mean free path for optical phonons.

More recently imaging UPS obtained by NanoESCA (Nanoscale Electron Spectroscopy for Chemical Analysis) of LiO-terminated p-type C(100) diamond has shown that the shape and position of the CBM peak is altered if the NEA condition is activated by changing the surface preparation method of forming the oxygen and lithium co-ordination on the diamond surface. For this study, hydrogen-terminated single crystal material was annealed at 900 °C under UHV (5×10^{-10} mbar) conditions to produce a bare diamond surface. A monolayer of lithium was evaporated onto this surface under UHV prior to the diamond surface being exposed to atomic oxygen source for 10 minutes. The samples were then annealed at 400 °C at a reduced pressure of 2×10^{-5} mbar for thirty minutes.

The activation of NEA on the diamond was confirmed by UPS where asymmetric CBM peaks were observed. Figure 5.22 compares the first derivatives of UPS data obtained from p-type C(100):H-oriented diamond and p-type C(100):LiO-oriented diamond at cryogenic temperatures. There are significant differences between the UPS responses for the two NEA diamond surfaces and a further observation is that the LiO samples show a different

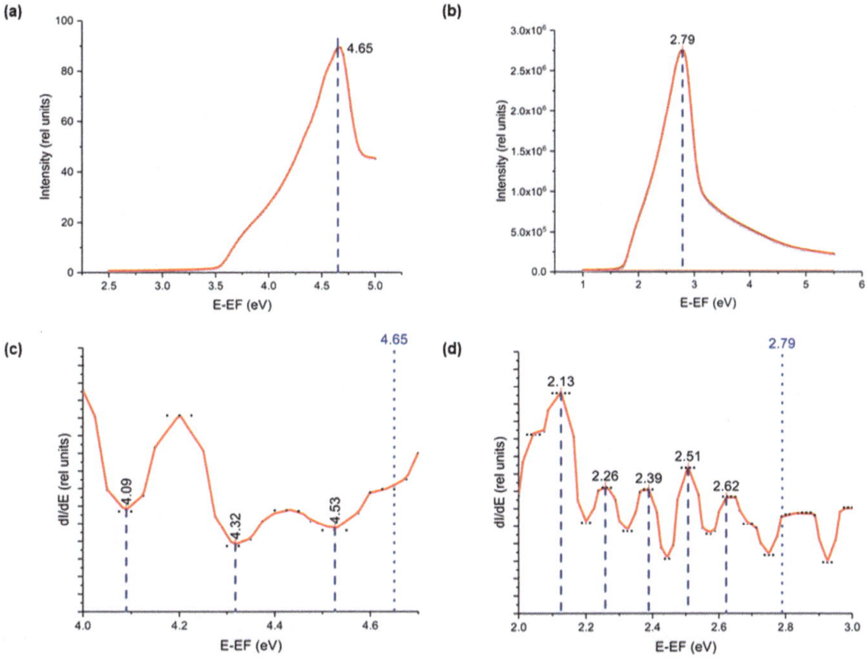

Figure 5.22 (a) UPS low energy spectra for (100) hydrogen-terminated diamond at room temperature. (b) Low energy spectra for (111) hydrogen-terminated diamond at low temperature (34 K). (c) First derivative of the (100) hydrogen-terminated diamond spectra shown in (a). (d) First derivative of the (111) hydrogen-terminated diamond spectra shown in (b). The dashed vertical lines in (c) and (d) indicate where the turning points are for the two spectra.

asymmetry to that reported by O'Donnell and reproduced in Figure 5.19, as well as a phonon peak separation that does not accurately follow the linear relationship for energy separation previously proposed. This suggests that the strength of the dipole activated by the alternative method adopted in this study of forming a LiO-terminated diamond may be much smaller. Also the smaller NEA exhibited by the surface may not be true NEA but instead effective NEA, where the bulk CBM lies above the vacuum level but the CBM at the surface is situated below the vacuum level due to band bending: for true NEA the bulk and surface CBM both lie above the vacuum level. This supposition of effective NEA would be supported by the observation that the oscillations seen in Figure 5.22 for LiO lie to the left of the CBM peak. In contrast, the NEA surface activated on p-type C(111):LiO, featured in Figure 5.23, appears to exhibit true NEA due to the phonon peaks being present at higher energies than the CBM peak in agreement with O'Donnell's findings.

Also, the series of phonons have energies that exceed those that are seen to be produced with the C(100):LiO surface. A potential reason for this

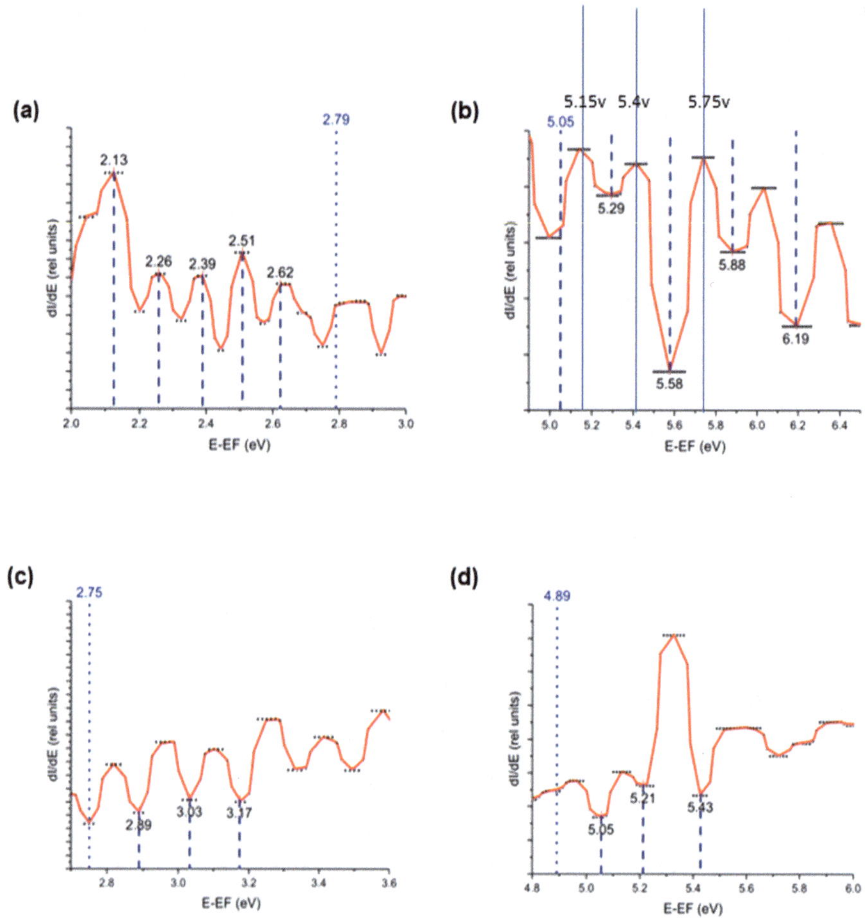

Figure 5.23 UPS low energy spectra for (a) the (100) and (b) the (111) lithiated-diamond samples and the derivative of the spectra all measured at 34 K. The vertical lines in (a) and (b) indicate the maximum of the CBM peak. The dashed lines in (c) and (d) indicate the stationary points present in the derivatives of the UPS spectra.

observation is that it is a consequence of the thermalisation mechanism, whereby the increased band bending of the samples results in the phonons having a greater mean free path within the material and as such, they are unimpeded by external factors for a greater period of time than expected. The consequence of this would be that the phonons resonate for longer and would thus be detected to have a greater energy. Figure 5.23 shows a comparison of the phonon energy gaps recorded for p-type C(100):LiO and p-type C(111):LiO oriented surfaces, which suggests that the mean free path of phonons is smaller in C(100):LiO diamond. Furthermore, it has been reported that phonon mean free paths in (100) and (111) orientations of diamond are anisotropic with the longitudinal phonon modes having a smaller value for the mean free path in the (100) orientation when compared to the (111) orientation.[45]

Such observations would suggest that it would be preferable to utilise an NEA surface formed on a C(111) oriented surface in a device designed to harvest phonons, as well as electron carriers in a diamond-based voltaic junction. Surface terminations that generate a true NEA on oxidised diamond have been extensively studied both computationally and experimentally in recent years, and this has yielded a number of new candidates many of which have been reviewed by James *et al.*, including: Si, Ge, BN, Al, Sc, and Mg.[46]

For some applications, NEA surfaces are required to exhibit high temperature stability. Although a Si termination on bare diamond has been considered, a further candidate that has shown promising high temperature performance is SnO. Ullah *et al.* has conducted computational and experimental work on this 2D material, formed on oriented single crystal diamond, and noted that it is stable to *ca.* 1000 °C.[47,48]

Furthermore, oxide-based 2D materials have been found to be dynamically stable and less susceptible to degradation in air.[49]

One aspect of the study by Ullah *et al.*, which opens up an interesting avenue for future research concerns the combination of SnO and LiO. While a monolayer of LiO is found to be electrically non-conductive, SnO, a 2D material exhibits p-type conductivity. The deposition of Li onto a SnO layer terminating diamond results in a significant change in surface work function measured by monochromated He-I imaging UPS. The work function (WF) reduces from 5.9 eV to 4.1 eV when only SnO was formed on the surface of p-type C(100):O by a half monolayer Sn deposition. This is due to the shift in the charge density between the surface species in a way that imparts NEA to the diamond surface, as shown in Figure 5.24. The work function is further

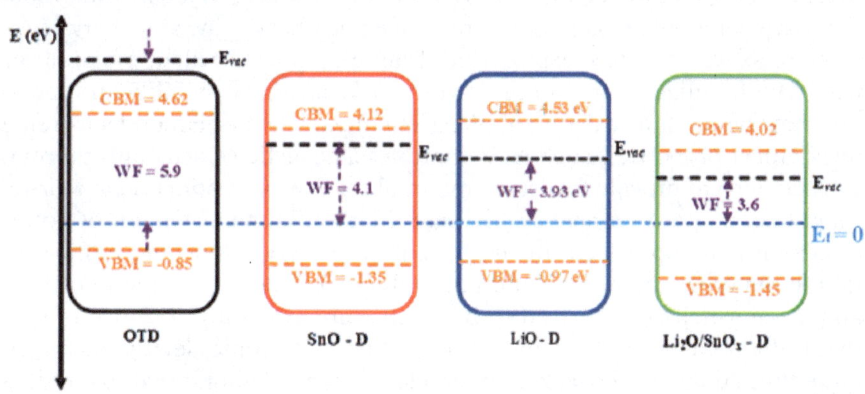

Figure 5.24 Electronic band positions and work function (WF) values deduced from UPS measurements for LiO-terminated and SnO/Li_2O-terminated single crystal C(100) oriented surface after various process steps. The spectra were produced using a monochromatic, focused 21.2 eV He-I discharge light source. WF decreases with the process steps, finally giving a WF reduction of 2.3 eV. Reproduced from ref. 48 with permission from Elsevier, Copyright 2021.

reduced to 3.9 eV when lithium is deposited on the surface. With an increase in annealing temperature to 500 °C, a WF of 3.6 eV was obtained, which is lower than LiO-terminated diamond at that temperature.

This further decrease in WF could be due to the loss of excess lithium during the annealing process, as seen previously in the case of p-C(100):LiO diamond. Furthermore, Ullah *et al.* have calculated the amount of lithium lost with annealing in both samples and it was determined that 47% of the lithium was lost at an annealing temperature of 500 °C for p-C(100):LiO, while only 15% was lost in the case of p-C(100):SnO_x,Li_2O. This would suggest an increased stability of lithium on the surface of diamond with tin monoxide.

As soon as the lithium deposits on the p-C(100):SnO, a reduction of SnO is evidenced by XPS. This is interpreted to occur in the following way, lithium draws the oxygen atoms from tin monoxide and this results in tin atoms being expelled, with the expelled tin atoms migrating to form atomic planes that separate Li_2O layers. The formation of Li_2O and metallic Sn indicates the possibility of the formation of 2D planes between lithium oxide layers in a multilayer structure as suggested by Pedersen.[50] A further observation that was made from analysing the UPS data was that downward-band bending was seen in the case of SnO and Li_2O–SnO_x-terminated diamond, as opposed to upward-band bending in the case of LiO-terminated diamond (seen in a different study), which implies that tin oxide can prevent Li from entering the bulk of diamond (as a dopant) and hence preserve the diamond bulk electronic properties.

5.5 Energy Applications of CVD Diamond

This section presents a brief overview of some diamond device technologies that are being researched intensively at Bristol, as well as in Japan and China, that have potential uses in both terrestrial energy harvesting and energy conversion, as well as space exploration. Diamond is an excellent material for use in high radiation environments due to its chemical stability and radiation hardness. Many existing applications make use of diamond's extreme mechanical properties, such as its hardness, acoustic velocity and thermal conductivity, to provide feedstock material for use as an abrasive, a wear-resistant coating or as a heat spreader in the thermal management of power electronics and optoelectronic devices. Both polycrystalline and monocrystalline materials are produced on an industrial scale for such applications and are normally synthesised to be electrically insulating. The use of synthetic diamond as a semiconductor in active electronic devices has been quite limited due to the lack of an efficient n-type diamond material that is activated and highly conductive at room temperature. Such a material would be required to make pn-junction devices for use in micro/nanoelectronics and photonics.

p-type diamond obtained with the introduction of boron as the dopant impurity into the diamond crystal lattice enables mono-polar junctions, such as Schottky junctions, to be formed, which exhibit high levels of rectification when specific contact metals are chosen.[51] This work has sparked interest in these Schottky diodes being developed for numerous active electronic devices including high voltage power switching, voltaics and electron sources.

In this section, examples of energy-related diamond-materials research are reviewed that illustrate the potential of diamond as an active electronic material.

5.5.1 Thermionic Energy Converters

Nuclear and solar heat to electrical power sources are important elements in planned space missions of explorations to the outer planets and moons, but technology improvements are required to meet the mission requirements. A solid-state technology that can use these heat sources to generate useful and scalable electrical power is the thermionic energy converter (TEC). Caesium vapour based TECs have been in development for decades, the first practical device and innovation from Wilson[52] using caesium vapour-based TECs was followed later by developments at NASA-JPL (Jet Propulsion Laboratory) in the 1980s with the SP-100 projects. However the materials and state of technological development, limited the power to sub-kW levels. Vacuum-based TECs have received less attention in the past 50 years due to the need to identify new materials with the properties that would allow efficient operation across a wide temperature range extending from below red heat up to 850 °C.

Early in the first decade of the 21st century, synthetic diamond material was put forward as a potential electrode material for use in a vacuum TEC. Since that time, research has continued at Bristol University and elsewhere to identify methods for producing diamond materials with the desired thermal and electronic properties for use in a vacuum TEC device. This research has identified a number of novel processes and properties of synthetic diamond materials that when combined would demonstrate the feasibility of diamond-based TEC power sources for use in terrestrial solar thermal power, as well as for use in space exploration.

Heat to electrical energy conversion occurs when electrons thermally emit from a hot surface, traverse a gap, and are efficiently collected by another (cooler) electrode surface. The effect can be observed in a commercial vacuum diode if it is allowed to operate without the usual applied voltage between emitter and collector (Figure 5.25). This process, starting with thermionic emission, produces an electrical current that can subsequently drive an electrical load. Figure 5.26 shows how a vacuum diode that has been modified to operate without an applied voltage or space-charge barrier can achieve significant power generation if the collector work function is very much smaller than the emitter work function. Particularly well suited for high-temperature applications, since it was first proposed by Schlichter,[53] vacuum thermionic emission has been pursued as a power generation method for over a century and yet very few applications have been realised in either space or terrestrial applications.[54,55] However, recent developments in advanced materials and nanotechnology, as well as a deepening understanding of the physics of

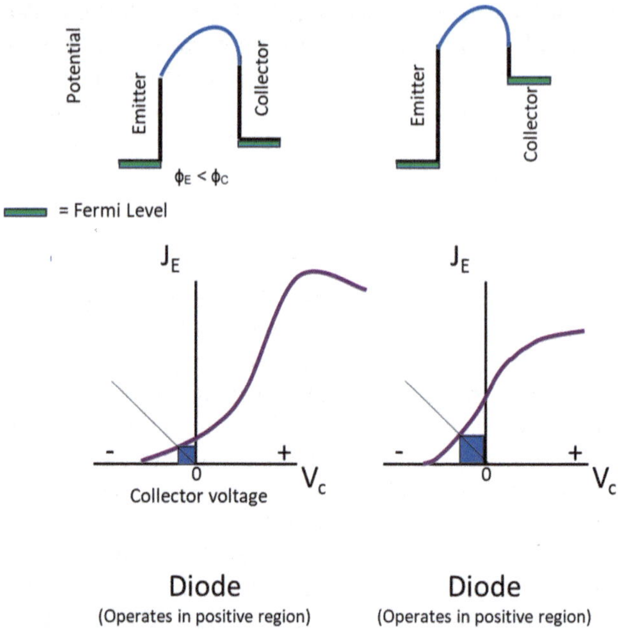

Figure 5.25 Representation of the inter-electrode potential and the corresponding current–voltage characteristic of a vacuum diode with a thermionic cathode emitter, (left) with an emitter work function lower than the collector electrode work function, (right) representation of the inter-electrode potential and the corresponding current–voltage characteristic with a collector work function lower than the emitter work function. The blue shaded area indicates where power would be generated if the externally-applied voltage between the emitter and collector was switched off while the emitter remained thermionically heated.

surfaces and material interfaces, open up new opportunities to develop practical vacuum thermionic convertors, potentially opening up the field to new applications in renewable energy generation.

Thermionic emission is described by the Richardson–Dushman equation,[56] which relates thermionic current density (j) to the temperature of the emitter (T_e) as follows:

$$j = A \bullet A_0 \bullet T_e^2 e^{-(\phi/k_B T_e)} \left(\text{A cm}^{-2}\right), \tag{5.3}$$

where, A_0 = 120 A cm^{-2} is the Richardson constant and k_B is the Boltzmann constant. The parameter A in eqn (5.3) is a correction factor to the Richardson constant for a specific emitter material. If an electric field (E) is applied between the emitter and the collector, the potential barrier is lowered. The resulting emission current is described by eqn (5.4), which is known as the Schottky equation:[57]

$$j_s = j \cdot \exp^{\left(\left[\frac{E \bullet q^3}{4\pi\varepsilon_0}\right]^{1/2} / k_B T\right)} \left(\text{A cm}^{-2}\right) \tag{5.4}$$

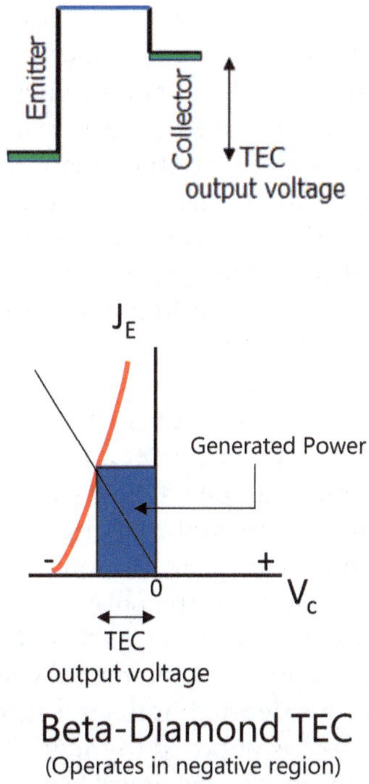

Figure 5.26 Representation of the inter-electrode potential and the corresponding current–voltage characteristic of a vacuum diode in which space charge has been mitigated, and with a collector work function that is significantly smaller than the emitter work function. The blue shaded area shows where power would be generated. This exhibits a much larger current output than a standard diode due to the internal voltage produced by the work function difference between the emitter and collector and the lack of space charge.

where q is the electron charge and ε_0 is the permittivity of free space.

From the electromotive diagram of a vacuum TEC in Figure 5.25, it is seen that the maximum voltage (at maximum power) that can be achieved is essentially the difference between the work functions of the emitter and collector, and the power density, P, is given by eqn (5.5).

$$P = j \cdot V_{\text{out}} = \frac{j(\phi_e - \phi_c)}{e}\left(\text{W cm}^{-2}\right) \tag{5.5}$$

The efficiency (η) is the power density produced by the thermionic convertor divided by the net heat flux supplied to the emitter (Q_e), which includes heat losses due to electron emission, conduction down the leads and other paths, and heat exchange (predominantly radiation) from the hot emitter to the collector.

$$\eta = \frac{j \cdot V_{\text{out}}}{q''_e} = j(\phi_e - \phi_c)/e \cdot Q_e. \tag{5.6}$$

From eqn (5.6) and the preceding equations, it becomes clear that there are a number of factors that directly influence the amount of power produced by a thermionic convertor, as well as its efficiency, that are often in direct competition with each other. From eqn (5.3), the current density is seen to increase when the emitter work function is small. However, to generate high power, large work function differences between the emitter and collector are needed (Figure 5.27), and thus the work function of the collector must be even smaller than that of the emitter, which is a significant materials challenge. Practically, the collector work function should be ~0.5 eV, which is extremely small compared with the work function of most natural materials. In addition, to generate sufficient current for conventional emitter materials, the temperature of the emitter must be very high and typically much greater than 1500 K. However, this leads to greater thermal losses due to both radiation and conduction, decreasing the efficiency of the thermionic convertor. There are also additional fundamental challenges in TEC that do not appear directly in these equations, and prime among these is the space-charge effect. The effect of space charge on the output voltage of a diode TEC with no space-charge mitigation is shown in Figure 5.27. At sufficiently high current density, electrons build up in the interstitial gap between the electrodes and create a space-charge barrier that limits the current to that described by the Child–Langmuir law,[58,59] as opposed to the Richardson–Dushman equation. Hence, managing the electron transport across the electrode gap is also an essential aspect of vacuum TEC design.

For diamond materials to be adopted in thermionic energy conversion applications, the issues of stable operating temperature range and greater than 1 W cm^{-2} power output levels, need to be demonstrated. This also requires some consideration of the methods and materials to harvest heat

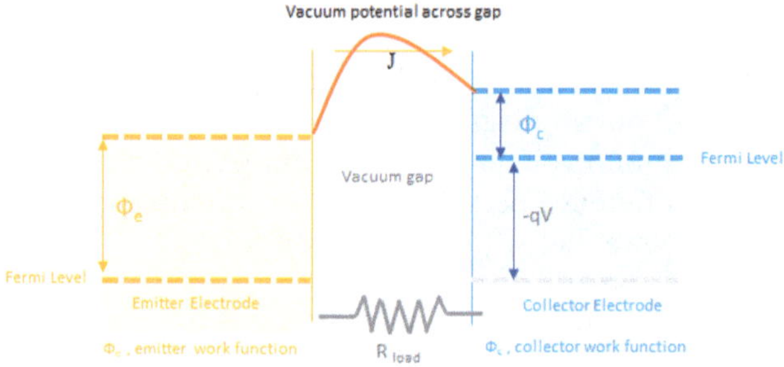

Figure 5.27 Electromotive diagram showing electron energy levels during TEC operation. The yellow energy levels represent the hot thermionic emitter and the blue energy levels represent the cool or cooled collector electrode. The thermionic current, J, generated by the thermionic emitter, has to overcome a space charge barrier at the emitter to reach the collector. The voltage difference between electrodes is $-qV$, which corresponds to the voltage dropped across the electrical load placed between the collector and emitter.

and ensure its efficient transfer into diamond emitters, such as through the use of plasmonic, meta-materials.[60]

To maximise the work function difference between the emitter and collector, previous work at Bristol on the co-doping of diamond with boron and nitrogen is being combined with functional coatings that exhibit a strong surface dipole supporting NEA, such as the Li–O–Sn systems, as a means to realise a tuneable work function. The excellent thermal properties of diamond also allow a collector electrode made from it, to reject heat highly efficiently compared to refractory metals or even copper.

5.5.2 Beta-enhanced TEC (BTEC)

To date, vacuum TEC research has considered the use of photon-enhancement thermionic emission (PETE)[61] and ionisable residual gas ambients,[62] as approaches to increase the efficiency of converter operation. The PETE concept is focused on improving the utilisation of concentrated radiation incident on a GaAs-based TEC by the incorporation of a photovoltaic element to harvest visible and near-infra-red radiation to augment the power generated by the thermionic elements. The gas ionisation concept is focused on mitigating space charge to improve the emissivity of a diamond thermionic electrode. A further concept that has been reported to enhance electrode emissivity of hydrogen-terminated, nitrogen-doped nanocrystalline diamond (HTND), is the use of beta-radiation.[63] This has been evaluated by incorporating a weak source of beta-radiation into the surface of the cooled molybdenum collector of a TEC prototype so that primary electrons from the source irradiate the emitting surface of the thermally heated diamond electrode. Once incident on the emitter they dissipate their energy to the hot electron population promoting measurable increases in thermionic output. The heated electrode was composed of a nitrogen-doped diamond film deposited onto a molybdenum substrate. The diamond film was formed on the molybdenum substrate by microwave-assisted CVD growth and terminated with hydrogen to exhibit a NEA surface. The collector electrode was made from molybdenum foil, which had been perforated to allow beta-radiation to pass through it from an embedded nickel-63 source (Ritverc GmbH, Germany).

Nickel-63 (^{63}Ni) has a 99% beta-radiopurity, with a 17.4 keV average and 70 keV maximum particle energy. A preformed molybdenum cap held a 3 mm × 3 mm × 0.1 mm foil of either ^{63}Ni (2.6 MBq) or ^{59}Ni (zero activity control), as shown in Figure 5.28a. The perforated holes in the cap plate exposed 50.3% of the Ni foil, so that the sample was exposed to a total calculated activity of 1.3 MBq using the Ni beta-source. The TEC diode housing, shown in Figure 5.28b, could be removed to exchange both the embedded Ni foil and the emitter sample. When replaced, the inter-electrode gap was controlled by a linear motor (Z825V, Thorlabs, Germany) to ensure a constant spacing of 200 μm. A 25 V bias between the electrodes ensured that emitted electrons contributed to the measured current, while producing no measurable

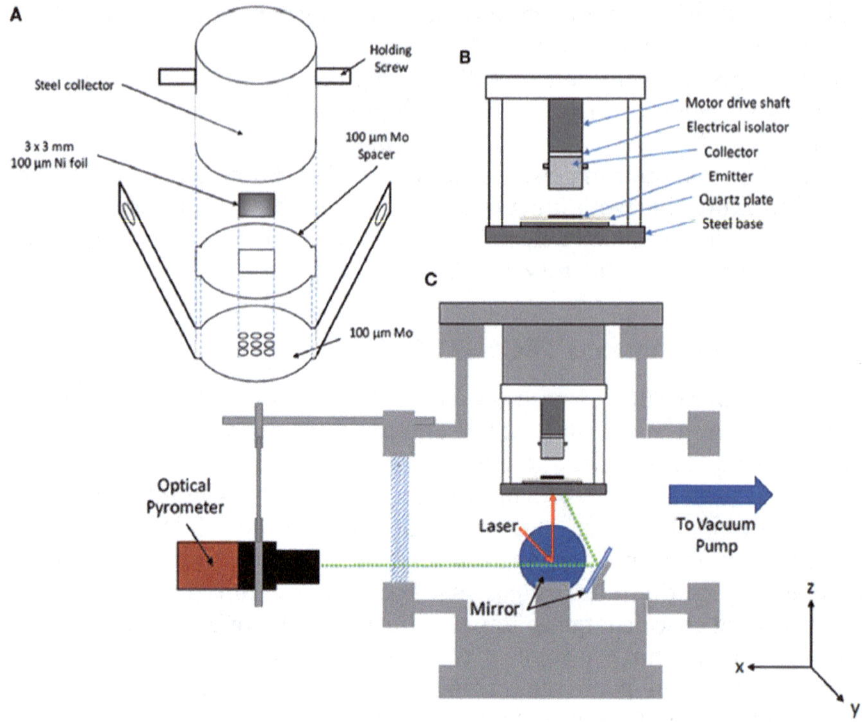

Figure 5.28 Schematic representations of the experimental equipment for measuring beta radiation enhanced thermionic emission from diamond thin films. (A) shows the molybdenum housing for the Ni foil within the collector, (B) shows each of the parts of the thermionic diode structure with electrical connections omitted for clarity, and (C) shows the entire test set-up, showing how both the laser (y-axis) and pyrometer (x-axis) are directed at the sample *via* mirrors. Reproduced from ref. 63, https://doi.org/10.3389/fmech.2017.00013, under the terms of the CC BY 4.0 licence, http://creativecommons.org/licenses/by/4.0/.

field emission. The emitter assembly was encased in a stainless-steel chamber (Figure 5.28c) and held under vacuum with a base pressure of 10^{-6} Torr. The back (non-diamond) surface of the sample was laser micromachined to ensure high radiation absorption heating by a polarised 40 W CO_2 laser (Firestar V series, Synrad, UK) with a wavelength 10.6 μm and spot-size of ~6 mm. A two-colour optical pyrometer (Spotmeter R160, Land-Ametek, UK, <2.5 K uncertainty) monitored the sample temperature through a borosilicate window.

A set of eight HTND samples was used to test thermionic emission under beta-irradiation. Each sample was tested four times, twice under beta-irradiation from ^{63}Ni (referred to as 'beta'), and twice under no irradiation ('non-beta'). The order of beta and non-beta test runs was balanced such that any resulting bias was eliminated. Sample set A was examined by repeated cycling of the cathode temperature and showed the magnitude and stability of emission current both with and without beta-irradiation. Whereas, sample set B was used in a direct comparison of the emission threshold temperature.

Figure 5.29 shows the average peak emission current of all the beta and non-beta tests in sample set A. The 1.3 MBq exposure induced on average a ratio of beta-enhanced to non-beta-enhanced current, or 'enhancement factor,' of 2.7 between the 2nd and 20th cycles. Regarded as unrepresentative, the first measurement cycle was ignored, an omission considered valid due to peak current and peak temperature being off-set far more than the average. This is likely due to the least stable hydrogen surface species being easily desorbed even at the modest temperatures used here. Enhancement is, however, still present in the first cycle current measurement. Sample set B was tested purely as a measure of the enhancement effect of beta-irradiation on the emission threshold temperature T_t. Figure 5.30 shows an example of the data in the region close to the threshold. Of those examined, all samples showed lower T_t values when exposed to beta-radiation than without exposure.

The average threshold temperature $T_{t,av}$ of all the non-beta runs was 566 ± 32 °C, whereas that of the beta runs was 504 ± 22 °C, where the quoted errors are standard errors on the mean of sample set B. When the threshold temperature change of each sample is considered individually, the threshold temperature change average $(\Delta T_t)_{av}$ is a clear reduction of 58 ± 11 °C from non-beta to beta in sample set B.

It may be reasonable to expect the enhancement factor experienced in the emission to be caused by secondary electron emission (SEE). The

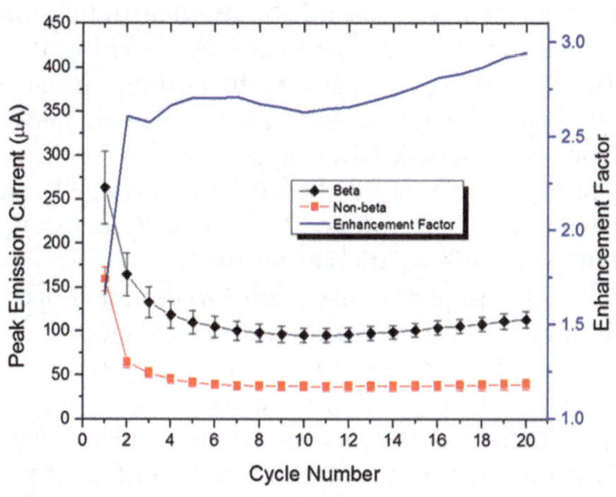

Figure 5.29 Average temporal peak emission current values for sample set A, showing the beta (^{63}Ni) current to be consistently higher than the non-beta (^{59}Ni) current. The enhancement factor (defined in the main text) is also shown on the right-hand axis. The error bars represent the standard error on the mean of all runs in sample set A. Reproduced from ref. 63, https://doi.org/10.3389/fmech.2017.00013, under the terms of the CC BY 4.0 licence, http://creativecommons.org/licenses/by/4.0/.

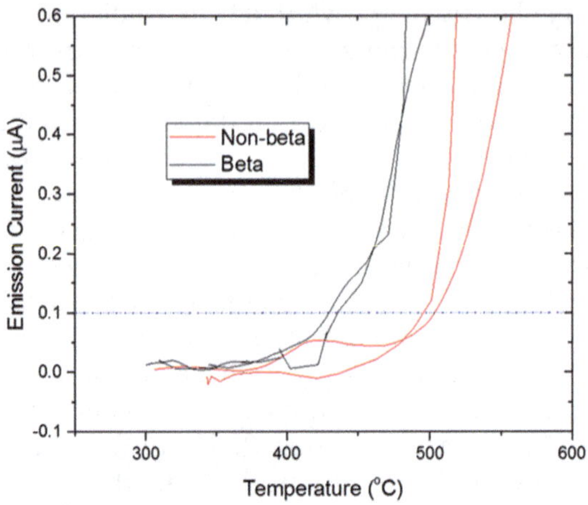

Figure 5.30 Example characteristic emission current as a function of temperature in the threshold region for a sample in set B, with (black line) and without (red line) beta radiation exposure. The blue dashed line shows the current threshold considered to be thermionic emission. Reproduced from ref. 63, https://doi.org/10.3389/fmech.2017.00013, under the terms of the CC BY 4.0 licence, http://creativecommons.org/licenses/by/4.0/.

beta-particles could, upon contact with the surface and subsurface electrons, transfer sufficient energy to allow emission from the surface and capture by the biased collector. While SEE will most likely occur, it is trivial to show that no enhancement of this size is possible by secondary emission alone.

With a nominal 50.3% of the beta-particles reaching the emitting diamond surface and assuming 100% successful entrance into the material and interaction with charge carriers within it, then there are just over 13 million primary electrons per second. Given a generous secondary electron yield (secondary to primary ratio) of 100 based on reports by Dvorkin,[64] the current returned would be of the order of 20 pA, which is a result far lower than that obtained in the beta-irradiation experiment.

With a deep donor level and consequent low carrier concentration, nitrogen-doped diamond has a low room temperature electrical conductivity.[65] As high-energy beta-particles enter and travel through the diamond lattice, they generate electron–hole pairs. These extra charge carriers could increase the electrical conductivity of the diamond, in turn, increasing charge carrier replenishment at the surface. This mechanism could be responsible for the increase of thermionic current obtained herein and would be applicable in the case of nitrogen-doped diamond, as it has a low carrier concentration. In order to ascertain the validity of this hypothesis, a new sample was grown on an insulating diamond substrate under conditions identical to those of sample sets A and B. In the same vacuum apparatus used for thermionic tests, the resistance of a diamond thin film was measured at both 300 and 600 °C, with and without beta-irradiation. This test produced no change in resistance within the precision of the experiment on the sample studied. To support this experimental evidence, one can make a calculation based on the properties of electron–hole pairs found by previous researchers,[66] who reported a room temperature exciton lifetime of 350 ns, and an electron–hole creation energy[67] of 13.3 eV; fewer than 600 electron–hole pairs will be activated at any given time within the diamond material.

The hydrogen termination may be considered a possible factor in the emission changes observed in the data. There are two possibilities regarding this hypothesis: first, chemically bonded H species on the surface could be desorbed by the high energy incoming radiation or in fact by outgoing secondary electrons. And second, any desorbed H atoms could become ionised, by beta primary or a secondary electron produced by it and thus, add an ionic component to the existing electronic current. Both of these conceivable options are effects that would exist in the case of beta-irradiation, but not in the non-beta case. Addressing the first point, it has been stated here, and by others,[68] that the H termination of HTND is necessary for its thermionic emission properties. Thus, the assumption that as H desorbs from the surface the current will decrease, as is apparent at high temperatures, which negates this enhancement mechanism. The second point is improbable, since electron impact ionisation of H atoms is most likely to result in H^+ ions, which would act to lower the current generated in the system. The H^- ion product may be a small component of the electron impact ionisation; however, its effect on the current will be far outweighed by that of the H^+. Production of H^- from secondary electrons would cause no change in the current, since emitted secondary electrons will contribute to current whether bound, or not, to a H atom. It can therefore be concluded that the H-termination on the diamond surface was not playing a role in the thermionic current enhancement that was reported.

Another possibility is the alteration of bulk diamond by radiation damage giving rise to defects. Vacancies and interstitials are produced in diamond by beta-radiation of considerably higher energy than is released during the decay of ^{63}Ni, and although nitrogen-vacancy defects are known to be produced at lower energies, none of these point defects are likely to increase conductivity in nominally n-type material studied in these beta-enhanced experiments reported by Croot *et al.*[63] Since the major surface disruption known for HTND is H desorption, interaction with H is unlikely to produce a stronger dipole. Furthermore, sub-surface defects are predicted, at least for substitutional N, to reduce the H bond strength.[69] Irradiation damage is, therefore, likely to result in a lower emission current in HTND. As beta-particles interact with the pre-existing valence band and conduction band electrons within the diamond, they will undergo inelastic energy transfer, removing the native electrons from equilibrium. The non-equilibrium electrons, also called hot electrons by some authors,[70] will cause a distortion of the Fermi–Dirac distribution of electrons, such that energy states with high energy will be populated. The barrier to emission, due to upward band bending at the surface, results in a proportion of conduction band electrons, which in an NEA material are considered to emit easily, being trapped within the bulk. When given a non-zero energy input, some number of these electrons could be promoted to a level above the barrier height. Skewing the conduction band electron distribution in this way could add a component to the emission-energy electron population and would result in an enhanced thermionic current, without significantly increasing conductivity. This is in contrast to the work of Schwede *et al.*,[71] where photon radiation enhanced current by increasing the total number of conduction band electrons.

At sufficiently high current density, thermionically emitted electrons build up in the interstitial gap between the electrodes and create a space-charge barrier that limits the current to that described by the Child–Langmuir law, as opposed to the Richardson–Dushman equation. Hence, managing the electron transport across the electrode gap is also an essential aspect of TEC design. The issue of space charge may be mitigated to some degree by beta-irradiation but a more conventional approach might involve adopting a scheme employed in high power tetrodes termed electron beaming, a concept tested at Bristol and illustrated in Figure 5.31.[72] By arranging a dual-gate electrode arrangement to be positioned in the vacuum gap between the emitter and collector, it is possible to mitigate space charge at the emitter by modulating the dual-gate voltages to draw pulses or beamlets of electron current away from the space charge region in front of the emitter. This can be achieved using a small fraction of the power output or taking advantage of a beta-emitter material, such as a ^{14}C diamond coating on the grid electrode, to charge it to a specific bias voltage. This concept is being developed to look into another novel aspect of this approach to enhance TEC efficiency, which involves the use of an external resonant circuit tuned to the electron transit time of the TEC. This provides a high Q oscillator mode set by the inter-electrode spacing of the TEC to maximise output power. Such a TEC design approach would use standard thermionic theory to produce the first pulsed voltage TEC, giving enhanced TEC efficiency and with all the associated advantages of AC operation.

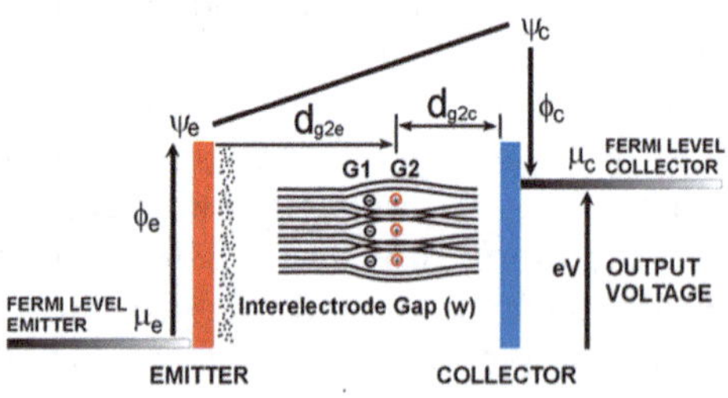

Figure 5.31 Concept diagram of the dual-grid thermionic energy convertor, grid 2 (G2) is the attractor and grid 1 (G1) is the deflector. The grids work in conjunction to attain electron transport and space charge mitigation with minimal power loss to G2, the device geometry and voltages/signals are important to attain efficient operation. Reproduced from ref. 72, https://doi.org/10.3389/fmech.2017.00014, under the terms of the CC BY 4.0 licence, http://creativecommons.org/licenses/by/4.0/.

5.5.3 Beta-voltaic Batteries

Research into the harvesting of beta-radiation dates back to the work of Rappaport[73] at RCA Labs, who in 1953 published the results of a semiconductor junction irradiated by a Sr-90/Y-90 beta-source that produced a measurable output at an efficiency of only 0.2%. This spurred further developments at other organisations, such as Donald W. Douglas Labs, where Olsen's group[74] were able to increase battery efficiency to 4% to allow the device to be used as a power source for a commercial heart pacemaker.

Due to its large indirect bandgap of 5.475 eV, diamond, unlike silicon, is not a candidate for terrestrial photovoltaic applications. However, under electron irradiation diamond semiconducting junctions possessing an active volume of only a few square millimetres have the potential to deliver microwatt levels of power output.

The beta-voltaic energy conversion process is similar to that of solar cells and consists of three main steps. They include: (1) beta-particles emitted from the decay of radioisotope source irradiate the semiconductor to generate electron–hole pairs by impact ionisation; (2) the generated electron–hole pairs diffuse into the depletion region of the diode junction and are separated to form free holes and electrons by the built-in electric field; (3) the electrons and holes drift in opposite directions toward the junction contacts and are collected at the anode and cathode, respectively generating a flow of electrical current as shown in Figure 5.32.

Unlike a solar cell, a single beta-particle can create multiple electron–hole pairs through a series of interactions. These interactions are subject

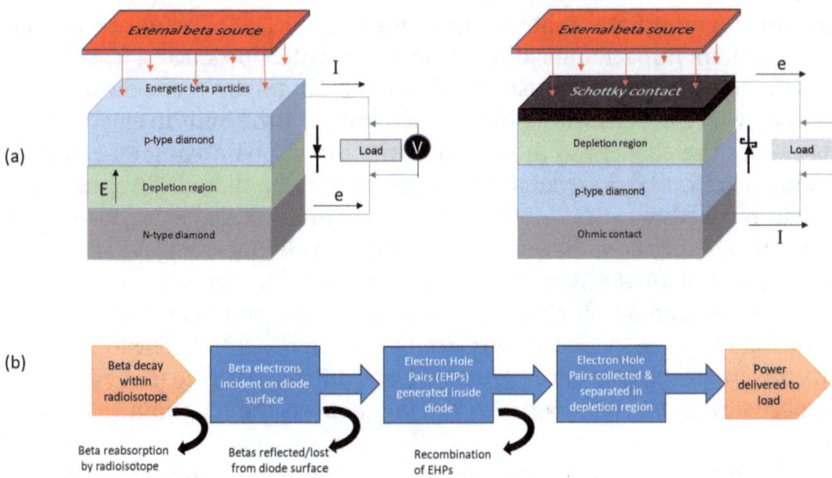

Figure 5.32 Schematic illustration of (a) operation principle of the beta voltaic device based on p–n junction and Schottky barrier diode; (b) various stages of energy conversion in a conventional beta voltaic battery with external beta source.

to energy losses in the form of optical and acoustic phonons. It has been established that only one-third of the energy of the beta-particles, entering a beta-voltaic junction, is used to generate electron–hole pairs, with the average creation energy for a single electron–hole pair being equal to $2.8E_g + 0.5$ eV.[75] This finding implies that the efficiency of energy conversion for beta-voltaic devices should be much higher for diamond materials compared with most other semiconductors due its larger band gap.

Most of the reports on diamond voltaics have been conducted using p-type Schottky junctions synthesised from synthetic single crystal material, reporting conversion efficiencies of up to 9.4%.[76,77] The predominant use of Schottky junctions is largely due to the lack of an efficient n-type diamond material that would enable a pn junction device to be used instead. However, higher conversion efficiency of 28% has been reported for a diamond pn diode irradiated by a monoenergetic electron beam used to simulate beta-irradiation, despite the fact that the use of phosphorus to dope diamond n-type is not fully activated at room temperature.[78] The main attraction of a pn junction device is the fact that it can exhibit a built-in potential of ~4.5 V, more than double the value of the built-in voltage of a typical Schottky device. This allows the maximum power point to appear at an output voltage that is attractive in applications that would require a like-for-like replacement for a chemical primary battery cell that can typically output up to 3.3 V. A representation of the power and current voltage characteristics for a pn beta-voltaic device are shown in Figure 5.33.

Beta-voltaic operation is achieved by exposing these junctions to a source of primary electrons generated by radioactive decay. The source is arranged to direct beta-radiation onto one face of a planar diode structure, where it is absorbed through an electron transparent electrical contact into the depletion region of the Schottky junction (Figure 5.34). Primary electrons are subject to impact ionisation processes, which produce a cascade of electron–hole pairs that diffuse into the depletion region formed within the junction, where the electron–holes are separated by the built-in electric field. This causes the carriers to drift in opposite directions driving an electrical current. Electron beam irradiation (Figure 5.35) of a diamond voltaic junction can be used to estimate the performance of a voltaic device. Using this approach Delfaure *et al.* have reported a single cell efficiency of 9.4% and a power output of 10.48 μW under a simulated beta-flux equating to 32 GBq.[76] The diode utilised an aluminium Schottky contact and generated an open circuit voltage of 1.85 V and a short circuit current of 7.12 μA. The low conversion efficiency of this research prototype was due primarily to the high losses that occur when the beta-particles are incident on the junction, referred to as source loss.

An alternative diamond beta-voltaic design has been proposed by Bristol researchers, that incorporates pure beta-emitting radioisotopes within the diode structure near to the depletion region, to dramatically increase the source efficiency and the electron-carrier generation within the depletion region of the junction. For ^{14}C batteries, the radioisotope is fixed within the

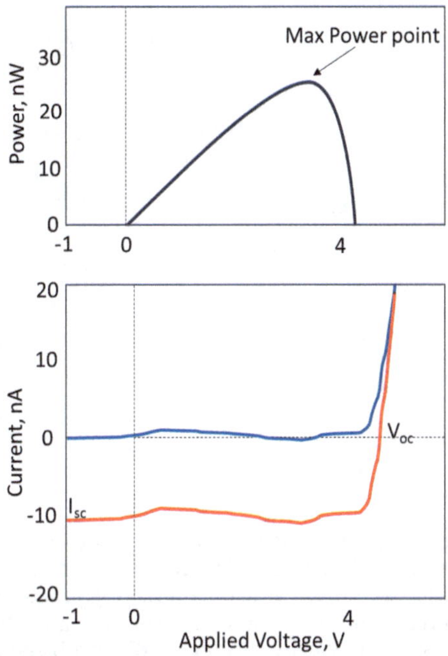

Figure 5.33 Representation of the *P–V* and *I–V* curves of a diamond pn diode at an accelerating voltage of 15 kV. The blue line shows the (dark) current–voltage characteristic obtained in the absence of electron beam irradiation. The red line shows the change in current obtained under electron irradiation. The characteristic open circuit voltage and short circuit current values that can be extracted from an experimental *I–V* are identified with appropriate labels on the diagram.

junction by forming a layer by CVD composed of ^{14}C diamond. In the case of a tritium diamond battery, the tritium is infused into the diamond junction, where it becomes trapped at charged vacancy defects created to localise the radioisotope within a diffusion length of the depletion region. This approach for fabricating a beta-voltaic battery has the advantage that the amount of radioactive material required to produce a specific output can be substantially reduced, assuming that material and interface defects can be minimised. For example, 10 μW of power from a ^{14}C battery can be reduced from 30 GBq to 3 GBq. Furthermore, due to the radioactivity being localised at atomic sites deep within the semiconducting junction structure, beta-decay energy is converted into electron–hole pairs and secondary electron cascades. This also means that such a diamond battery device, particularly one fashioned from ^{14}C, cannot release radioactive contaminants and is inherently safe, even when subjected to extreme operating temperatures and pressures, at which a chemical battery would be subject to a catastrophic failure.

The incorporation of radioisotopes into diamond may be accomplished by a number of methods. For example, ^{14}C may be localised within a

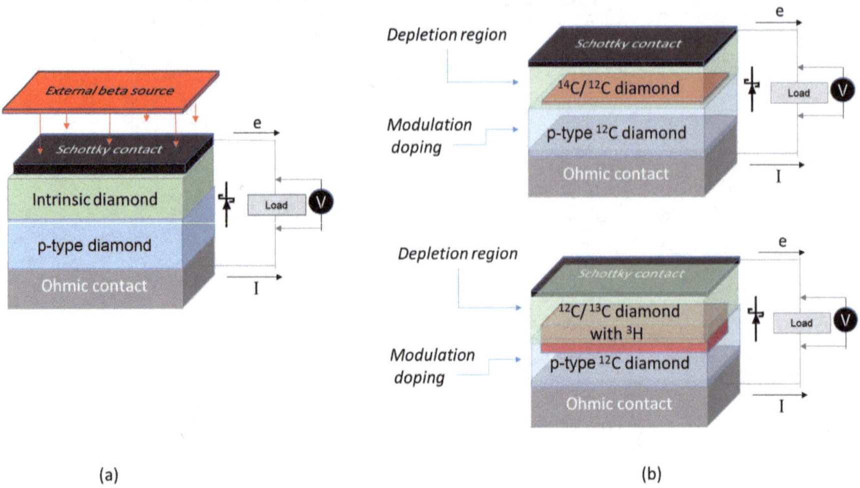

Figure 5.34 Examples of diamond beta-voltaic diodes with (a) an external beta source, (b) with beta-emitting radioisotopes incorporated within the diamond structure. The semiconducting junction is constructed from a p^+ layer with an ohmic metal contact, an intrinsic p^- layer and a Schottky metal contact. Under irradiation by beta radiation, the high energy primary electrons penetrate the junction layers where they generate electron–hole pairs by impact ionisation. The electron–holes diffuse into the depletion region formed between the intrinsic diamond layer and the Schottky contact where they are separated by the built-in electric field and flow in opposite directions generating an electrical current when connected to an electrical load.

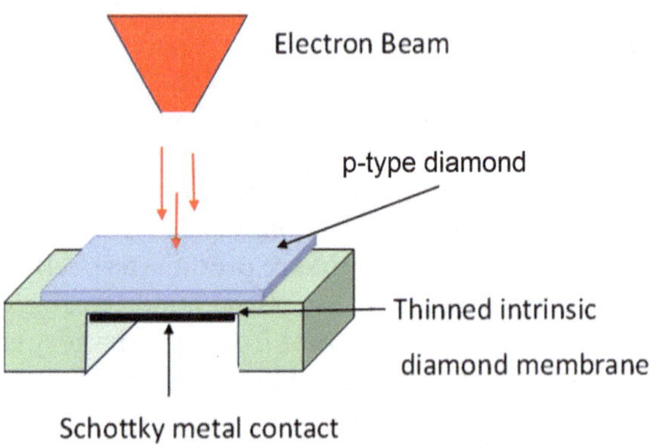

Figure 5.35 Representation of the Schottky diode structure evaluated by Delfaure that used an electronic grade, single crystal diamond selectively thinned to 5 microns to operate as the intrinsic region of a piM diode. The degenerately doped, p-type diamond with a thickness of 200 nm was grown by microwave plasma CVD, and the Schottky contact was formed selectively on the oxygen-terminated surface of the diamond membrane by depositing 200 nm of aluminium metal by e-beam evaporation under high vacuum conditions.

diamond diode either by CVD using a methane feedstock enriched with ^{14}C, or by transmuting a region of the diamond diode that is rich in ^{13}C diamond and nitrogen impurities: two isotopes each with favourable neutron cross sections for thermal neutrons. In principle, tritium can be loaded into diamond by field-assisted, high temperature and pressure infusion. Prelas[79] has reported hydrogen loading into synthetic diamond. In this work, a diamond substrate was placed between two graphite electrodes and heated to 800 °C in a hydrogen ambient of 70 Torr. A DC bias of up to 20 V was applied across the diamond for the duration of the 8-hour process. This resulted in 758 μg of hydrogen being loaded into the diamond substrate equivalent to 18 at%.

More recently, work at the University of Bristol has explored an adaptation of this method to the loading of several hundred microgrammes of tritium gas into diamond at elevated pressures of several atmospheres, as well as at field-assisted, high temperatures to enable the process to be used ultimately to localise tritium into a Schottky diode. In principle, a loading of tritium, *ca.* 300 μg, into diamond would equate to an internal source activity of 90 GBq with a potential electron generation rate of approximately 4×10^{13} carriers per second. The actual power output of a tritium-loaded diamond diode will depend on a number of linked factors, including the efficiency with which the tritium loading may be localised within a specific bulk region of the diode structure, as well as the collection efficiency of the diode's depletion region for the electron–hole pairs generated by the beta-electrons *via* impact ionisation with the carbon atoms.

5.6 Summary

In this chapter, recent research on diamond synthesis by CVD has been reviewed, with a focus on the development of static-flow CVD for the growth of isotopic carbon diamond using expensive carbon feedstocks, and the renewed interest in co-doping to realise efficient semiconducting, n-type diamond, specifically BN cluster studies. The functionalisation of diamond surfaces, using oxygen and metals, is discussed and its technological importance to new developments in energy harvesting and storage are discussed in the later sections of the chapter, which feature some exciting developments in diamond thermionics and diamond batteries.

From the examples presented in this section on the energy-related applications of diamond materials, it is evident that synthetic diamond will continue to attract significant R & D directed towards its use in terrestrial micropower batteries and concentrated solar power generation. Its use in the rapidly expanding space power industry, where there is mounting interest in long life, solid-state power sources, whose operation is not adversely affected by temperature extremes or high-radiation environments, will also continue to gain interest.

Acknowledgements

I would like to acknowledge the contributions made by past members of my research team to the diamond materials research work presented in this chapter including: Dr Kane O'Donnell, Dr Tomas Martin, Dr Hugo Domingues Andrade, Dr Zamir Othman, Dr Alex Croot, and Dr Gary Wan. I also acknowledge the support and advice of academic colleagues in the Bristol Diamond Group, including Professor Mike Ashfold, Professor Paul May and Dr James Smith, who have contributed valuable advice that helped to formulate the material presented in this chapter.

References

1. F. P. Bundy, H. T. Hall, H. M. Strong and R. H. Wentorf, *Nature*, 1955, **176**, 51.
2. J. C. Angus, *Diamond Relat. Mater.*, 2014, **49**, 77.
3. T. Tabakoya, S. Kanada, Y. Wakui, Y. Takamori, T. Yamada, M. Nagai, Y. Kojima, O. Ariyada, S. Yamasaki, C. E. Nebel, T. Matsumoto, T. Inokuma and N. Tokuda, *Phys. Status Solidi A*, 2019, **216**, 1900244.
4. S. Gsell, T. Bauer, J. Goldfuß, M. Schreck and B. Stritzker, *Appl. Phys. Lett.*, 2004, **84**, 4541.
5. D. Huang and M. Frenklach, *J. Phys. Chem.*, 1992, **96**(4), 1868.
6. X. Jiang, K. Schiffmann and C.-P. Klages, *Phys. Rev. B: Condens. Matter Mater. Phys.*, 1994, **50**, 8402.
7. P. Malshe, R. A. Beera, A. A. Khanolkar, W. D. Brown and H. A. Naseem, *Diamond Relat. Mater.*, 1997, **6**, 430.
8. S. Yugo, T. Kanai, T. Kimura and T. Muto, *Appl. Phys. Lett.*, 1991, **58**, 1036.
9. M. Schreck, S. Gsell, R. Brescia and M. Fischer, *Sci. Rep.*, 2017, **7**, 44462.
10. Z. Dai, C. Bednarski-Meinke and B. Golding, *Diamond Relat. Mater.*, 2004, **13**, 552.
11. Y.-H. Tang and B. Golding, *Appl. Phys. Lett.*, 2016, **108**, 052101.
12. K. K. Hirakuri, T. Kobayashi, E. Nakamura, N. Mutsukura, G. Friedbacher and Y. Machi, *Vacuum*, 2001, **63**, 449.
13. H. Yamada, A. Chayahara and Y. Mokuno, *Diamond Relat. Mater.*, 2020, **101**, 107652.
14. V. Ralchenko, I. Sychov, I. Vlasov, A. Vlasov, V. Konov, A. Khomich and S. Voronina, *Diamond Relat. Mater.*, 1999, **8**, 189.
15. F. G. Celii Jr., D. White and A. J. Purdes, *J. Appl. Phys.*, 1991, **70**, 5636.
16. F. G. Celii Jr., D. White and A. J. Purdes, *Thin Solid Films*, 1992, **212**, 140.
17. J. Su, Y. Li, Y. Liu, M. Ding and W. Tang, *Diamond Relat. Mater.*, 2017, **73**, 99.
18. S. V. Nistor, M. Stefan, V. Ralchenko, A. V. Khomich and D. Schoemaker, *J. Appl. Phys.*, 2000, **87**, 8741.
19. L. F. Hei, J. Liu, C. M. Li, J. H. Song, W. Z. Tang and F. X. Lu, *Diamond Relat. Mater.*, 2012, **30**, 77.
20. I. A. Martorell, W. D. Partlow, R. M. Young, J. J. Schreurs and H. E. Saunders, *Diamond Relat. Mater.*, 1991, **8**, 29.

21. A. Croot, E. J. D. Mahoney, H. Dominguez-Andrade, M. N. R. Ashfold and N. A. Fox, *Diamond Relat. Mater.*, 2020, **109**, 108011.
22. E. J. D. Mahoney, B. S. Truscott, M. N. R. Ashfold and Yu. A. Mankelevich, *J. Phys. Chem. A*, 2017, **121**, 2760.
23. A. Croot, G. Wan, A. Rowan, H. D. Andrade, J. A. Smith and N. A. Fox, *Front. Mech. Eng.*, 2017, **3**, 17.
24. B. S. Truscott, M. W. Kelly, K. J. Potter, M. N. R. Ashfold and Yu. A. Mankelevich, *J. Phys. Chem. A*, 2016, **120**, 8537.
25. J. C. Richley, M. W. Kelly and M. N. R. Ashfold, *J. Phys. Chem. A*, 2012, **116**, 9447.
26. H. Watanabe, C. E. Nebel and S. Shikata, *Science*, 2009, **324**, 1425.
27. H. Katayama-Yoshida, T. Nishimatsu, T. Yamamoto and N. Orita, *Phys. Status Solidi B*, 1998, **1998**(210), 429.
28. D. Segev and S. H. Wei, *Phys. Rev. Lett.*, 2003, **91**, 126406.
29. J. E. Moussa, N. Marom, N. Sai and J. R. Chelikowsky, *Phys. Rev. Lett.*, 2012, **108**, 226404.
30. R. J. Eyre, J. P. Goss, P. R. Briddon and M. G. Wardle, *Phys. Status Solidi A*, 2007, **204**(9), 2971.
31. A. Croot, M. Z. Othman, S. Conejeros, N. A. Fox and N. L. Allan, *J. Phys.: Condens. Matter*, 2018, **30**, 425501.
32. D. Y. Liu, L. C. Hao, Y. Teng, F. Qin, Y. Shen, K. Tang, J. D. Ye, S. M. Zhu, R. Zhang, Y. D. Zheng and S. L. Gu, *Appl. Phys. Lett.*, 2021, **9**, 081106.
33. M. Chandran, S. Shasha, S. Michaelson and A. Hoffman, *Appl. Phys. Lett.*, 2015, **107**, 111602.
34. B. J. Matsoso, K. Ranganathan, B. K. Mutuma, T. Lerotholi, G. Jones and N. J. Coville, *New J. Chem.*, 2017, **41**, 9497.
35. J. H. Yang, S. Y. Lee, M. H. Yum, Y. S. Shin, C. H. Yang, N.-K. Park, C.-Y. Park, Y.-R. Jang, M. H. Kwon, C. Jo, J.-I. Lee, C.-W. Yang, J.-B. Yoo, K. Ihm, T.-H. Kang, Y.-W. Jin and J.-M. Kim, *J. Korean Phys. Soc.*, 2007, **51**, 125.
36. Y. C. Chen, B. Griffiths, L. Weng, S. S. Nicley, N. N. Shazeaa Ishmael, Y. Lekhai, S. Johnson, C. J. Stephen, B. L. Green, G. W. Morley, M. E. Newton, M. J. Booth, P. S. Salter and J. M. Smith, *Optica*, 2019, **6**, 662.
37. A. Croot, *PhD Thesis*, University of Bristol, 2018.
38. J. van der Weide, Z. Zhang, P. K. Baumann, M. G. Wensell, J. Bernholc and R. J. Nemanich, *Phys. Rev. B: Condens. Matter Mater. Phys.*, 1994, **50**, 5803.
39. G. Wan, A. Croot, N. A. Fox and M. Cattelan, *Adv. Funct. Mater.*, 2020, **31**, 2007319.
40. K. M. O'Donnell, M. T. Edmonds, J. Ristein, A. Tadich, L. Thomsen, C. I. Q-H.WuPakes and L. Ley, *Adv. Funct. Mater.*, 2013, **23**, 5608.
41. K. M. O'Donnell, T. L. Martin, N. A. Fox and D. Cherns, *Phys. Rev. B: Condens. Matter Mater. Phys.*, 2010, **82**, 115303.
42. K. M. O'Donnell, M. T. Edmonds, J. Ristein, K. J. Rietwyk, A. Tadich, L. Thomsen, C. I. Pakes and L. Ley, *J. Phys.: Condens. Matter*, 2014, **26**, 395008.
43. L. Diederich, A. P. O. Kűttel and L. Schalpabach, *Surf. Sci.*, 1999, **424**, 314.
44. H. Y. Fan, *Radiat. Eff.*, 1970, **4**, 7.

45. I. I. Kuleev, *Phys. Solid State*, 2017, **59**, 682.
46. M. C. James, F. Fogarty, R. Zulkharnay, N. A. Fox and P. W. May, *Carbon*, 2021, **171**, 532.
47. S. Ullah and N. A. Fox, *J. Phys. Chem. C*, 2021, **125**, 25165.
48. S. Ullah, G. Wan, C. Kouzios, C. Woodgate, M. Cattelan and N. A. Fox, *Appl. Surf. Sci.*, 2021, **559**, 149962.
49. K. J. Saji, K. Tian, M. Snure and A. Tiwari, *Adv. Electron. Mater.*, 2016, **2**, 1.
50. A. Pedersen and M. Luisier, *ACS Appl. Mater. Interfaces*, 2014, **6**, 22257.
51. S. Koné, G. Civrac, H. Schneider, K. Isoird, R. Issaoui, J. Achard and A. Gicquel, *Diamond Relat. Mater.*, 2010, **19**, 792.
52. V. C. Wilson, *J. Appl. Phys.*, 1959, **30**, 475.
53. W. Schlichter, *Ann. Phys.*, 1915, **352**, 573.
54. G. N. Hatsopoulos and E. P. Gyftopoulos, *Thermionic Energy Conversion, Volume I: Processes and Devices*, The MIT Press, Cambridge, MA, 1973.
55. G. N. Hatsopoulos and E. P. Gyftopoulos, *Thermionic Energy Conversion, Volume II: Theory, Technology, and Application*, The MIT Press, Cambridge, MA, 1979.
56. O. W. Richardson, *The Emission of Electricity from Hot Bodies*, Green and Company, Longmans, London, 1921.
57. W. Schottky, *Ann. Phys.*, 1914, **44**, 1011.
58. C. D. Child, *Phys. Rev. (Series I)*, 1911, **32**, 492.
59. I. Langmuir, *Phys. Rev.*, 1913, **2**, 450.
60. H. Cen, S. Nunez-Sanchez, A. Sarua, I. Bickerton, N. A. Fox and M. J. Cryan, *Photon. Energy*, 2020, **10**, 024503.
61. J. W. Schwede, I. Bargatin, D. C. Riley, B. E. Hardin, S. J. Rosenthal and Y. Sun, *et al.*, *Nat. Mater.*, 2010, **9**, 762.
62. W. Paxton, S. Ravipati, M. M. Brooks, M. Howell and J. L. Davidson, *Front. Mech. Eng.*, 2017, **3**, 18.
63. A. Croot, G. Wan, A. Rowan, H. D. Andrade, J. A. Smith and N. A. Fox, *Front. Mech. Eng.*, 2017, **3**, 17.
64. V. V. Dvorkin, N. N. Dzbanovsky, N. V. Suetin, E. A. Poltoratsky, G. S. Rychkov and E. A. Il'Ichev, *Diamond Relat. Mater.*, 2003, **12**, 2208.
65. J. Mort, M. A. Machonkin and K. Okumura, *Appl. Phys. Lett.*, 1991, **59**, 3148.
66. H. Morimoto, Y. Hazama, K. Tanaka and N. Naka, *Diamond Relat. Mater.*, 2016, **63**, 47.
67. D. A. Dimitrov, R. Busby, J. R. Cary, I. Ben-Zvi, T. Rao and J. Smedley, *J. Appl. Phys.*, 2010, **108**, 073712.
68. W. F. Paxton, M. Howell, W. P. Kang and J. L. Davidson, *J. Vac. Sci. Technol., B: Nanotechnol. Microelectron.: Mater., Process., Meas., Phenom.*, 2012, **30**, 021202.
69. Z. Yiming, F. Larsson and K. Larsson, *Theor. Chem. Acc.*, 2014, **133**, 1.
70. M. L. Brongersma, N. J. Halas and P. Nordlander, *Nat. Nanotechnol.*, 2015, **10**, 25.
71. J. W. Schwede, I. Bargatin, D. C. Riley, B. E. Hardin, S. J. Rosenthal and Y. Sun, *Nat. Mater.*, 2010, **9**, 762.

72. I. Bickerton and N. A. Fox, *Front. Mech. Eng.*, 2017, **3**, 14.
73. P. Rappaport, J. Loferski and E. Linder, *RCA Rev.*, 1956, **17**, 100.
74. L. Olsen, S. Seeman and B. Griffin, *Beta Voltaic Nuclear Electric Power Sources*, McDonnell Douglas Corp, Richland, Washington, 1969.
75. R. Duggirala, A. Lal and S. Radhakrishnan, *Radioisotope Thin-Film PoweredMicrosystems*,SpringerScience&BusinessMedia,NewYork,2010,p.89.
76. C. Delfaure, M. Pomorski, J. de Sanoit, P. Bergonzo and S. Saada, *Appl. Phys. Lett.*, 2016, **108**, 252105.
77. V. Bormashov, S. Troschiev, A. Volkov, S. Taralkin, E. Korostylev, A. Golovanov, M. Kuznetsov, D. Teteruk, N. Kornilov, S. Terentiev, S. Buga and V. Blank, *Phys. Status Solidi A*, 2015, **212**(11), 2539.
78. T. Shimaoka, H. Umezawa, K. Ichikawa, J. Pernot and S. Koizumi, *Appl. Phys. Lett.*, 2020, **117**, 103902.
79. M. A. Prelas, T. K. Ghosh, S. K. Loyalka and R. V. Tompson, *J. Wide Bandgap Mater.*, 2002, **10**, 99.

Subject Index